Michael M. Widdersheim

Circulation of Power

Current Topics in Library and Information Practice

Michael M. Widdersheim

Circulation of Power

The Development of Public Library Infrastructure in Greater Pittsburgh, 1924–2016

DE GRUYTER
SAUR

ISBN 978-3-11-101225-4
e-ISBN (PDF) 978-3-11-101340-4
e-ISBN (EPUB) 978-3-11-101414-2
ISSN 2191-2742

Library of Congress Control Number: 2023932465

Bibliographic information published by the Deutsche Nationalbibliothek
The Deutsche Nationalbibliothek lists this publication in the Deutsche Nationalbibliografie; detailed bibliographic data are available on the internet at http://dnb.dnb.de.

Printing and binding: CPI books GmbH, Leck

www.degruyter.com

To Eliza and Eve

Acknowledgements

I would like to thank the University of Pittsburgh for supporting my research during the early stages of this project. I am thankful to Masanori Koizumi, Kip Currier, Sheila Corrall, and Leanne Bowler for their helpful guidance and feedback during this time.

This project would not have been possible without the participation of numerous volunteer interviewees involved with the public library infrastructure in greater Pittsburgh. Thank you all for lending your time, memories, and experiences to make this account as complete and multiperspectival as possible. I would especially like to thank Marilyn Jenkins for facilitating access to field observation sites and institutional archives.

The following people offered invaluable research assistance over the course of this project: Andrew Brown, David Frank, Jon Klosinski, and Miriam Meislik, University of Pittsburgh Library System; Greg Priore and Marilyn Holt, Carnegie Library of Pittsburgh; Michael Lear and Jesse Noonan, State Library of Pennsylvania; Andrew McWilliams and Suzanne Stasiulatis, Pennsylvania State Archives; Catherine Loikits, Lehigh University; Lorelei Broskey, Lehigh County Law Library; Audrey Rasmusson, Allegheny County Law Library; Richard Hudic, Allegheny Regional Asset District; and Amanda Baxa, Darian Bishop, Beth Hendrickson, Betty Kemboi, Brady Lund, and Megan Mahoney, Emporia State University.

I wish to thank my colleagues at Emporia State University's School of Library and Information Management for their help and encouragement.

Thank you to peer reviewers for their helpful feedback and suggestions.

Thank you to my editors, Claudia Heyer, Georg Bucher, John Ryan, and Adriana Stroe, for their time and care.

Finally, I am eternally grateful to my family and friends for their unyielding support.

All errors and omissions in this work are entirely my own.

Contents

Preface

I began researching the regional public library infrastructure in greater Pittsburgh in 2014. I was a PhD student then and, at the time, I had no idea such an infrastructure even existed. My mentor and I were "casing" venues in search of a new research project. Our general ambition was to understand public libraries and the public sphere and, as with any research, we needed data to do this. Together, we started asking around to see which libraries might give us access to data. Carnegie Library of Pittsburgh said no, they weren't interested in us hanging around and bothering them. But Allegheny County Library Association was another matter. They were just coming out of a re-envisioning process, they were proud of it, and someone whispered in our ear that they might have something interesting there. Of course, we didn't know what that would be, but it didn't matter, because they weren't put off by my questions. I began attending meetings, interviewing people, and digging into whatever archives I could find. One thing led to another. At some point, my mentor moved on to other projects, but I remained hooked. At first, I had no idea how deep my research would go, but I knew I needed to get to the bottom of it, whatever "it" was. Every time I discovered a new piece of information about the library association, it just led to another, then another, until I realized that this public library "apparatus," or whatever it was, was vast, complicated, and it had a longer history than people realized. I recall saying that it felt like I was an archeologist, using my fingers to feel along the edges of some object buried beneath the soil. I couldn't fathom how big the object was yet because I could only view part of it. I never guessed it at the time, but years later, I would still be working to unearth the same object.

The research that started in 2014 became my dissertation, which I defended in 2017.[1] In several subsequent articles, I elaborated my findings. These included a new research design called historical case study.[2] I also described new communicative structures—tessellations and decision cycles—that advance our understanding of the public sphere.[3] Based on these models, I was able to create a new communicative theory of public library development, one that explains why public libraries develop as they do.[4] And from this theory, I proposed change management strategies that library leaders can use to enact the changes they imagine.[5] Looking back, these were all important findings, and I am glad to have produced them using data from Pittsburgh.

Still, despite these various findings, it did not yet feel that I had accomplished all that I had set out to do. Somehow, the infrastructure I had studied in my research was not yet fully revealed. What was missing? I came to realize that what I had overlooked about the case was the case itself. What was missing was its story.

https://doi.org/10.1515/9783111013404-001

To me, this was a grave omission because it is just this type of narrative knowledge that is essential for grasping how things really are and how they come to be. I decided that this additional knowledge needed to be written and shared. Knowing that a journal article was not conducive for presenting this narrative, I instead undertook a book project.

As it turned out, the process of creating a detailed narrative account of public library infrastructure in greater Pittsburgh, from 1924 to 2016, was a much longer and much more intricate process than I had anticipated. Crafting the narrative, bit by bit, detail by detail, was a painstaking process that took several years. The previous results that I had produced only required that I know the broad strokes of the narrative, not its details, so when completing this new narrative, I had to find the many missing pieces. In addition, it became apparent that contextual information about the case was much more important than it had been. It took considerable time to fully grasp the state and national landscapes—economic, legal, cultural, technological, demographic—that influenced issues in Pittsburgh. To understand the national context, there were secondary sources I could rely on, but in the case of Pennsylvania, no historical account of public library development in Pennsylvania after World War II existed, so it was left to me to research it on my own using primary sources. For this extensive additional research, I visited additional archives and conducted further interviews. Disruptions from the COVID-19 pandemic caused further delays. In the end, the new information I gathered was integrated with the substantial data that I already possessed, and the narrative was revised and revisited several times over to package it into book form.

The result is this story. It is the story of the regional public library infrastructure in greater Pittsburgh, from 1924 to 2016. It is my hope that this story offers a comprehensive view of the object that I have been excavating for the past several years.

Michael M. Widdersheim — Emporia, Kansas

1 Michael M. Widdersheim, "Libraries and the Circulation of Power: A Historical Case Study of Pittsburgh, 1924–2016" (PhD thesis, University of Pittsburgh, 2017), http://d-scholarship.pitt.edu/32925/.

2 Michael M. Widdersheim, "Historical Case Study: A Research Strategy for Diachronic Analysis," *Library & Information Science Research* 40, no. 2 (2018).

3 Michael M. Widdersheim, "From Two-Track to Tessellation: A Revised Circulation of Power Model," *Journal of Documentation* 78, no. 6 (2022).

4 Michael M. Widdersheim, "A Political Theory of Public Library Development," *Libri* 68, no. 4 (2018).

5 Michael M. Widdersheim, Brady D. Lund, and Betty J. Kemboi, "Change Management in Public Libraries: Research-Based Political Strategies," *Journal of Library Administration* 59, no. 7 (2019).

Introduction

This is the story of the development of a regional public library infrastructure in greater Pittsburgh, Pennsylvania, United States. The spatial coverage of this narrative corresponds to Allegheny County, Pennsylvania, the county where the City of Pittsburgh is located. Additional accompanying information is also provided about state and national contexts. This is to situate the county libraries within those larger movements. Pittsburgh is located in the center of Allegheny County, and other cities and suburban areas surround it. The City of Pittsburgh has its own public library system, known as the Carnegie Library of Pittsburgh (CLP), or as it is known locally, "The Carnegie" (pronounced CAR-ne-gee). This story includes CLP libraries, but also extends beyond the city. There are dozens of public libraries outside the city and in its environs. This story is about all of them, the city and county public libraries combined. And while there are other types of libraries found within the county—academic, school, special, religious—this story focuses exclusively on public libraries. This is not a story about any particular library, or some subset of libraries. It is not a story collection. Rather, this is the story about how all of the public libraries in the county, taken as a whole, came together to function as a single regional public library infrastructure. The purpose of this book is to describe and to explain how this infrastructure developed from the time it was first imagined in 1924 to the close of 2016.[1]

The term infrastructure is used in this book to describe the arrangement of public libraries in Allegheny County. This word choice is deliberate. Other terms could have been used instead, such as system, association, or consortium. These terms are more commonly used today in the context of library groups. In fact, the term infrastructure is not one that many people in Pittsburgh typically use or have ever used to describe their public libraries. Yet, infrastructure is the term that best fits this story. There are several reasons for this. Libraries in greater Pittsburgh today are considered a federated system in the eyes of the State Library of Pennsylvania. This is for the purposes of state aid. But this federated arrangement was not always the case. While it might make sense to call the arrangement today a system, it would certainly be a mistake to describe previous arrangements that way. A further difficulty with the term system is that, even today, not everyone in greater Pittsburgh agrees that public libraries there are a system. This is because, despite state recognition as a federated system, the libraries in Allegheny County operate according to a semi-non-centralized governance structure, not a fully centralized one. The term system is problematic, therefore, not only because it describes a contingent arrangement, but also because even the nomenclature sometimes used today is disputed. As for the term association, the county library association is a relatively new phe-

https://doi.org/10.1515/9783111013404-002

nomenon. The formation of the association in 1994 marked a major milestone, but that turning point specifies a particular legal arrangement, one that did not always exist. These same problems hold true for the term consortium, which suggests a limited degree of resource-sharing whose nature might change over time. Due in part to these difficulties with other terms, infrastructure remains the best fit. Unlike the terms system, association, and consortium, the term infrastructure can be used to describe the public libraries consistently at any time during their connected and interwoven development. The term infrastructure, unlike other terms, focuses not on some specific aspect of the arrangement, such as governance structure, legal status, or a reciprocity agreement. These specifics constantly change. Instead, infrastructure suggests only a general cohesion. The term also does not lay claim to some endpoint, and it therefore sidesteps debates about what the preferred goal is for public library development. It is agnostic with respect to the nature of progress. The term infrastructure recognizes some single unit, some whole, while at the same time acknowledging the connected sub-parts that work together within that whole. The term treats these sub-units together as a single unit, and it attributes to them some semblance of order and cohesion, yet it does not isolate a particular aspect of the connection or prescribe what the arrangement should be. Infrastructures in general may be rigidly arranged or loosely fitting, they may be tightly structured or roughly cut. This makes the term appealing for describing a case such as public libraries in Pittsburgh where library configurations have been loose and variable.

Besides its flexibility, there is another reason that makes the term infrastructure the most suitable. What makes the term infrastructure especially fitting is its connotation of hiddenness and invisibility. Infrastructures are essential: they operate in the background to make other things work. While they do this, however, their structures and processes remain unnoticed. The histories of infrastructures, especially, are difficult to fathom because their development is not always observed and tracked. It is only post facto, when things go awry, that infrastructures become visible at all. Even then, they may blend into the background, appearing to be natural fixtures of life, something that was always already there, an ever-present and immutable reality. Of course, infrastructures have histories and tendencies, and it is crucial that they be fully unearthed. As the term implies, an infrastructure exists below the surface. It is largely invisible. Yet, it is precisely those seldom-noticed qualities that deserve attention because they determine what goods and services are provided, from where and to whom. This book is the story not of the surface-level goods and services of public libraries, but of their undergirding, the public library infrastructure that has evolved unseen.

Public libraries in Allegheny County, then, are an infrastructure in two senses. In one sense, public libraries resemble other types of infrastructures, such as roads, bridges, water lines, the electrical grid, and gas pipelines. Like these, public libraries in greater Pittsburgh supply goods and services to people across the region. The utilities supplied by libraries may appear relatively ephemeral by comparison—they include intangibles such as information, education, and entertainment. Nevertheless, these products require a physical substrate that must be maintained through deliberate human interventions. Contrary to how public library services may appear, they do not manifest out of the ether. Instead, their infrastructure must be created and sustained through some process of contestation.

Infrastructures can be hidden in plain sight. The regional public library infrastructure in greater Pittsburgh is especially so. There are several reasons for this. The main reason is that discussions about public libraries in the Pittsburgh region are overshadowed by the legacy of Andrew Carnegie. As is well known, Andrew Carnegie, the steel magnate and philanthropist, donated a portion of his wealth for the construction of public library buildings. Pittsburgh was the epicenter of this library building project that spanned the globe. Carnegie grew up on the North Side of what is today Pittsburgh, using an apprentice library as a boy. Carnegie's steel mills were later located in greater Pittsburgh, and several library buildings were built for use by the workers at these mills. Some of the earliest public library buildings that were funded by him were in Pittsburgh. The result of Carnegie's presence is a density of eponymous buildings in the area. Due to this, it is tempting for observers to remain fixated on the name Carnegie and on those individual library buildings he donated. In contrast to this popular account of public libraries in Pittsburgh, however, a view of the public library infrastructure on a regional scale requires a new perspective, one that brings a different set of features into focus. The story presented in this book includes the history and configuration of all public libraries in the region, including those unrelated to Carnegie. It is necessary to move beyond Carnegie's legacy to view the connected root work that extends throughout the region.

Public libraries in greater Pittsburgh also represent an infrastructure in another, more figurative sense. When applied more metaphorically, the term infrastructure may refer to subterranean social structures, structures that operate behind everyday life. These social structures represent patterns that describe or explain the visible features of sociality. This holds true for the structures underlying public libraries. Analogous to how public libraries circulate books and other media to their clients, libraries also circulate something else: power. Specifically, it is a set of underlying communicative structures that circulate power and are responsible for public library development. These communicative structures determine how public libraries deliver goods and services. The circulation of power, therefore, is the infrastructure of note here. It is an infrastructure of an infrastructure, so to speak. This circulation of power is the process that moves public libraries forward in time, and it is this meta-infrastructure that is the focus of this book.

For as long as public libraries have existed in the U.S., attempts have been made to explain why they develop as they do. The origins and distribution of public libraries in late nineteenth and early twentieth century in America attracted particular scholarly attention. This was the time of America's great public library movement, the period when social libraries and private lending libraries transitioned into the government-sponsored agencies we know today. Traditional accounts of public libraries list several factors to explain public library development.[2] These include population, education, literacy, urbanization, geological features, class, race, and charismatic leadership. Existing accounts are diverse. There is no consensus among them. For this reason, there is no definitive account for why public libraries developed as they did. While a number of factors are listed as possible causes for public library developments, existing accounts do not specify the necessary and sufficient conditions for development. This makes it difficult to determine which factors really matter and which ones are optional. Existing accounts also tend to focus on library genesis only, thus ignoring any subsequent developments. This preoccupation with library origins limits the generalizability of any developmental theory. And the unit of analysis in existing accounts is nearly always the individual brick-and-mortar library. This myopia thus fails to explain how public library development happens generally, beyond an isolated physical existence. As public libraries moved forward through the twentieth century, it is apparent that public libraries must be considered as part of a larger whole, whether it be as members of consortia or networks. It is also essential to account for their increasingly virtual presence. In other words, what is missing from existing developmental accounts is an infrastructural view, one that is flexible and scalable. A developmental theory must imagine public libraries as parts of larger units, at scale, and in various formations, both physical and digital. The term infrastructure best captures this arrangement. Adopting an infrastructural view makes is possible to address the questions of how and why public libraries develop as they do.

Recent evidence suggests that demographic and other variables like those mentioned in previous accounts do not serve as productive starting points for identifying the causes of public library development. Research should instead focus on a more subtle variable: communication. In other words, it is the effects of communicative practices that determine how public libraries develop. Such research that focuses on communicative discourse may be loosely classified as research related to the public sphere. A public sphere approach focuses on the various modes of communication surrounding the library and within it. The public sphere related to public libraries includes three main arenas of discourse: 1) those discursive arenas where public libraries are legitimated and supported by their publics over time; 2) discourse where a library's publics participate in the governance of the library; and 3) discussion where, through the library's provision of physical and virtual spaces, services, and collections, a library facilitates opinion and will-formation by its users about issues of common concern.[3] It is in this third arena that the public library functions as a communicative platform. Through the analysis of all the communicative interactions in all three arenas, public sphere research about public libraries analyzes how libraries promote in individuals the development of personality, identity, and taste, the establishment of norms, and the transmission of culture.

What is distinctive about a public sphere approach is that it attempts to understand public library development through a communicative lens. To this end, several public sphere models have been proposed as frameworks for analyzing public sphere discourse as it applies to public libraries.[4] These frameworks can be used to describe the communicative structures—the infrastructure—of public libraries. Understanding these structures is necessary to understand public library development. A communicative approach is infrastructural in the sense that it describes structures and processes that otherwise remain obscured. Factors like population, philanthropy, technology, and urbanization, while relatively easy to observe and measure, are considered secondary if not tangential causes. Instead, communication is the primary driver. The challenge remains how best to identify and measure those communicative factors that influence public library development. Therefore, it is important to choose an appropriate model to serve as the guiding framework when describing and explaining public library development. A research framework for the public sphere determines which features of communication are observed and recorded. In other words, the framework determines what data are measured and analyzed.

In the area of research related to the public sphere, the leading descriptive model of communication is the circulation of power model. Following its inception, enhancements and additions to this model have allowed it to become scalable, nuanced, and better suited for describing distributed governance arrangements.[5] These revisions are especially important for analyzing political landscapes like those found outside Europe and in the U.S., notably greater Pittsburgh, whose political structure is complex, layered, and polycentric.[6] The updated model was designed using data from public libraries in greater Pittsburgh, and it is therefore equipped to account for a strong third sector and many diverse local governmental actors that vary in size, jurisdiction, and function. The revised circulation of power model, described below, is best suited for analyzing public library communication, and it is this model that is adopted and applied in this book.

There are two closely linked sides to the circulation of power model, one related to the organization of the public sphere and one related to its communicative process. On the organizational side, the basic building block of the circulation of power model is the circuit. A circuit is a social structure that describes the arrangement or configuration of the public sphere. It is a set of positions and relationships that actors inhabit throughout the course of a decision-making process. Actors occupy the positions of a circuit, and, from these positions, actors establish different relationships with one another. The positions in a circuit determine what those relationships are. In other words, the positions in a circuit dictate how actors in the public sphere communicate with one another. Actors in a circuit can be individual people or they may be groups or organizations. The purpose of using the circuit structure in this book is to describe and map out in a consistent and reliable way how communication travels in the public sphere.

A circuit can be somewhat complicated and difficult to visualize. It is therefore helpful to imagine a circuit as analogous to a molecule. Just as molecules are composed of atoms and are held together by ionic or covalent bonds, circuits are composed of positions held together by power relationships. Actors occupy one of four positions in a circuit—core, inner periphery, outer periphery, or diffuse public—just as various atoms occupy the positions in a single molecule. In their positions in a circuit, actors send and receive different types of power. Power transmissions may be sent either to one of the other positions within one's circuit or to one's own position in the circuit. There are four types of power recognized in the model: communicative power, administrative power, social power, and formal decision power. The type and directionality of a power transmission depends in part on what relative positions the actors occupy. This is like how the types of bonds in a molecule vary depending on the nature of its atoms. Just as different atoms in a molecule form different bonds, different positions in a circuit share different types of power. Figure 1 shows an illustration of a circuit with actors and positions. In the figure, which shows a hypothetical and generalized circuit, Actor 2 inhabits the core, Actor 3 inhabits the outer periphery, and Actor 1 is moving to occupy the inner periphery. In these positions, the actors will use various forms of power to communicate with one another to raise, decide on, and implement issues of mutual concern.

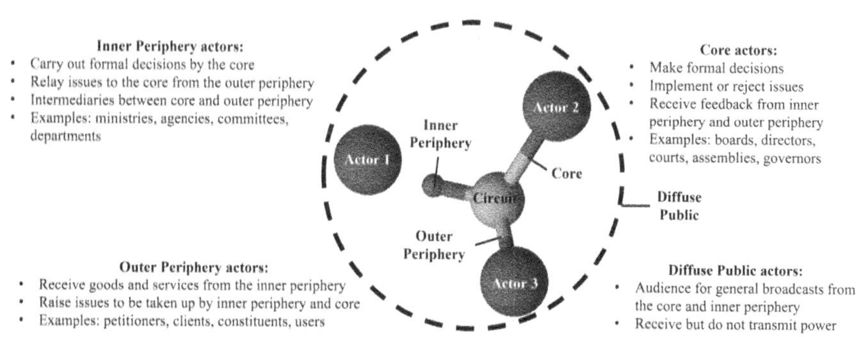

Figure 1: Positions and actors in a circuit.

As seen in Figure 1, there are four positions in a circuit: 1) core, 2) inner periphery, 3) outer periphery, and 4) diffuse public. Each position in a circuit represents a set of actors who send and receive power between one another. A circuit is a general communicative structure. This means that circuit positions are replicable and scalable across a variety of groups or organizations. A circuit structure is therefore useful for describing the configuration of different groups or organizations operating in the public sphere. It can be used to show how the circulation of power happens.

The core of a circuit represents the decision-making bodies of that circuit. Actors in the core might include a governing board, a legislature, a judicial court, a council, a director, or a voting membership. This will depend on the nature of the group or organization being described, but all groups and organizations, from small local nonprofits to the federal government, possess decision-making cores. The core is primarily responsible for transmitting formal decision power. It is the actors in the core who are responsible for making final decisions regarding issues. This is done by deciding to implement issues or not. The core takes up issues that are brought to its attention. Issues may be raised by the inner and outer periphery and transmitted inward, or issues may be raised and addressed within the core itself.

The inner periphery of a circuit works in conjunction with the core and the outer periphery. It serves as a go-between. On the one hand, the inner periphery works with the core to implement the core's formal decisions. The inner periphery therefore plays a supporting role to the core. At the same time, the inner periphery also filters and relays issues to the core that are transmitted from the outer periphery. The inner periphery therefore acts as a mediator between the core and outer periphery, both by transmitting communicative power to the core from the outer periphery and by transmitting administrative power to the outer periphery from the core. It is through communicative power that issues are raised, legitimated, and transmitted to the core, but it is through administrative power that successful issues flow back outward and are implemented. Depending on the nature of the group or organization being described, actors in the inner periphery might include governmental agencies, library administrators, or sub-committees. Just as all groups and organizations have a core, they also possess an inner periphery. Inner periphery actors have limited autonomy and some decision-making capacity, yet they don't have the final say like core actors. They are also in touch with the environment of the group or organization, yet they are not external to the circuit as are outer periphery actors.

The outer periphery is the third position in a circuit. The outer periphery represents those actors served by the core and inner periphery, in other words, actors who are affected by core and inner periphery decisions. Outer periphery actors are therefore constituents or clients of the group or organization. Examples of outer periphery actors include library patrons, voters of a certain jurisdiction, petitioners and letter writers, and libraries themselves when they are beneficiaries of foundation or government funding. Outer periphery actors receive formal decisions and administrative power from the core and inner periphery of a circuit. This power is received by them in the form of goods, services, or money. Yet, actors in the outer periphery also select issues to raise and communicate to the core and inner periphery. They advocate, rally, attend board meetings, and apply for sponsorship. Outer periphery actors legitimize those issues in the attempt to have them taken up and implemented. Issues are raised and legitimated through transmissions of communicative power.

Finally, the diffuse public position in a circuit are those actors who receive broadcasts of power from the other positions. These broadcasts can be communicative or administrative. Before a diffuse public can raise a new issue and transmit communicative power to the inner periphery or core, a public must coalesce to form an outer periphery.

The positions in a circuit are defined in part by the types of power that actors in each position send and receive. There are four types of power: communicative, social, formal decision, and administrative. Communicative power describes the type of interactions between actors where issues of common concern are raised and discussed. Issues are legitimated when they are discussed openly, inclusive of everyone involved, and actors exchange validity claims for and against an issue. Communicative power can originate from anywhere within a circuit—the outer periphery, inner periphery, or core itself. In other words, the raising and legitimating of an issue may begin anywhere. Social power is in some ways opposed to communicative power because, although it also used to influence decision making, it is a non-legitimate form of power. Social power includes both economic power—money—as well as back-channel communication and backroom deals. In other words, social power is not open, inclusive, and based on validity claims as is the case with communicative power. Formal decision power describes a vote or mandate of some sort. Formal decision emanates from core actors, be they elected officials or appointed leaders. And administrative power describes those commands and tasks whereby a decision is implemented by actors within a tessellation. These tasks might include new services, purchases, the distribution and spending of money within an infrastructure, or the creation of a new formal arrangement. The four forms of power—communicative, social, administrative, and formal decision—can be transmitted as either narrowcasts or broadcasts. Narrowcasts have a specific and discernible audience. Broadcasts, by contrast, have no single or definite recipient. Power that is broadcast is said to be transmitted to the diffuse public of a circuit. The diffuse public may be all the residents of a county, for instance, or the general readership of a news column.

What is attractive about the circuit structure is that, as a building block for describing public sphere communication, it is repeatable and scalable. More than one circuit exists in the public sphere at once, and these multiple circuits combine and interconnect to describe the entirety of public sphere communication, thus forming a larger structure with a repeated pattern. Actors across the public sphere may occupy different positions in more than one circuit simultaneously. For instance, an actor that occupies the core of one circuit might inhabit the outer periphery of another. Due to their shared actors and interconnected nature, multiple circuits link together to form a larger structure. This larger structure is called a tessellation. The scale of the tessellation is coterminous with that of the infrastructure being described. In this book, the case is the public library infrastructure in greater Pittsburgh, so the tessellation that describes that infrastructure in greater Pittsburgh is the composite of all the public sphere circuits there. Across the circuits within this larger public sphere tessellation, actors transmit and receive power between one another, forming a broader circulation of power. This set of actors, the positions they occupy, and the power they send and receive describes the overall terrain or landscape of the public sphere. The tessellation structure, composed of smaller circuits, is a consistent and reliable way to map out and illustrate this circulation of power.

Figure 2 shows a hypothetical and generalized set of circuits that connect together to form a tessellation. A tessellation is so called due to how the connected circuits form a repeated pattern. In the figure, the combined circuits appear as a larger molecule. Light nodes represent circuits. Each circuit has three positions—core, inner periphery, outer periphery—which interlink with other circuits through shared actors. Shared actors are depicted by dark nodes. All circuits share a single diffuse public.

Circuits and tessellations comprise the topographic and organizational side of the circulation of power model. On the procedural side, working within the tessellation structure, is another structural process called the decision cycle. This is the second side of the circulation of power model. Whereas circuits and tessellations describe the features and formations that exist among actors in the public sphere, allowing these positions and relationships to be mapped out in virtual space, the decision cycle describes the general sequence of events that actors carry out as power travels through the tessellation. Figure 3 depicts the decision cycle process.

A decision cycle is analogous to a traditional storyline. A decision cycle can thus be thought of as the story of how actors in a tessellation attempt to resolve a shared issue. There are four stages of a decision cycle: 1) raising of a new issue, 2) the legitimation phase, 3) the formal decision point, and 4) the implementation phase. The legitimation phase and the implementation phase correspond to rising action and falling action in a story. In a decision cycle, between and among

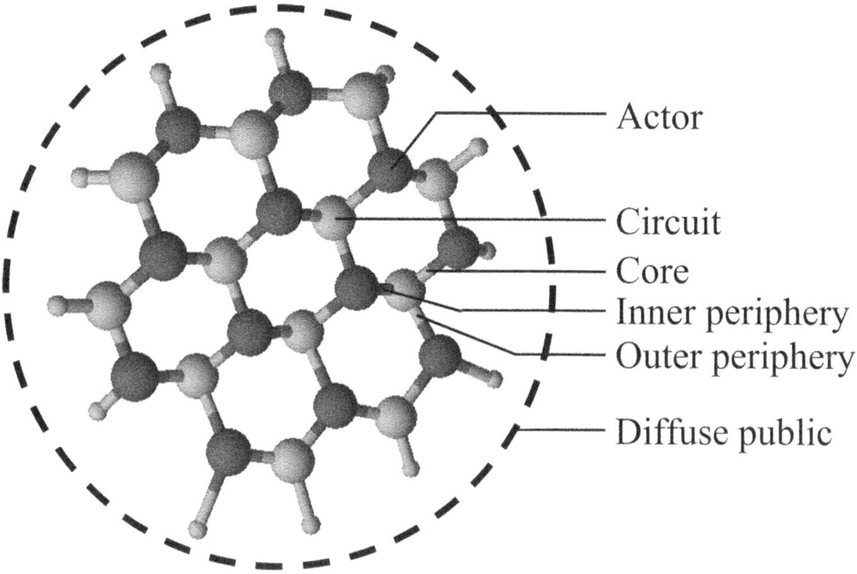

Figure 2: A tessellation.

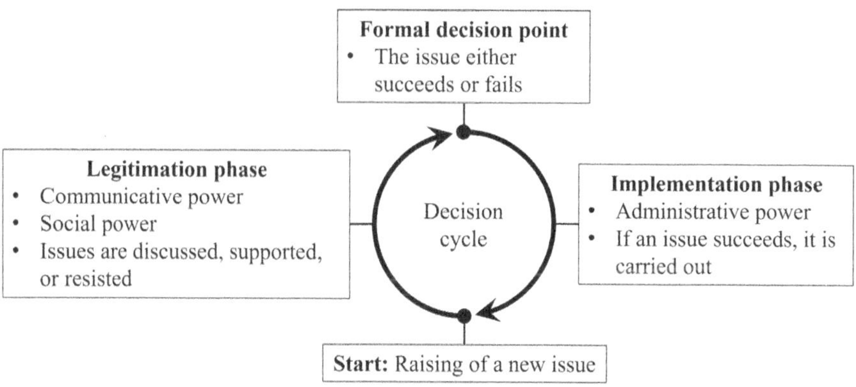

Figure 3: A decision cycle.

the actors who occupy circuits, issues are raised and legitimated. In the legitimation phase of a decision cycle, issues are discussed, supported, or resisted. The issues raised may be imported from outside a circuit, where they are relayed from the outer periphery to the inner periphery, or they may be introduced to the core from within itself. Just as in the rising action in a story, where there is some dilemma facing the characters, in the legitimation phase of a decision cycle, actors identify an issue and propose a solution to resolve it.

As an issue is discussed and a proposed solution legitimated, and as communicative power and social power are transmitted within the tessellation, issues gradually move toward a formal decision point. This is like the climax of a story. The formal decision point is where actors in the cores of a tessellation decide what will become of the issue, whether it will succeed or fail. The decision made at this point represents formal decision power.

Following the formal decision point, the final stage of the decision cycle is the implementation phase. In the implementation phase, if an issue has been successfully legitimated and decided on in at the formal decision point, it is carried out by the inner periphery through administrative power. This stage of administering the successful issue is like the denouement of a story. The outcome reached in each decision cycle influences the nature of the infrastructure. This process of administering the issue continues until a fresh issue is raised, thus starting the decision cycle anew.

The timelines of decision cycles are not standardized. Unlike other types of cycles, such as elections, the time it takes a tessellation to raise a new issue, legitimate it, reach a formal decision point, and implement the solution is indeterminate. The temporal length of each decision cycle therefore depends on the circumstances surrounding the issue, how it is legitimated, and so forth. A tessellation ends an existing decision cycle and starts a new one when a new issue is raised.

In a decision cycle, not all issues are successfully implemented. At the formal decision point, some issues fail. A set of four factors determine whether the decision cycle of the tessellation will result in a success outcome or a failure outcome, that is, whether the issue in the cycle will be implemented. Cycles that successfully implement an issue result in substantive change, while those whose issue fails result in continuity. The set of four necessary and sufficient conditions that results in a change outcome is high responsiveness, together with either high civil support, high legitimacy, or low resistance. In other words, high responsiveness, combined with at least one other condition, must be exhibited by a tessellation during a decision cycle for the tessellation to produce a positive (change) outcome.[7]

Responsiveness describes the ratio in a tessellation between legitimation and implementation. A decision cycle of a tessellation has high responsiveness when an issue that is legitimated is also implemented. In other words, decision makers in a responsive tessellation act on communicative power that is transmitted to legitimate the issue. An unresponsive tessellation, by contrast, fails to fully account for legitimation claims in its decision making. High responsiveness is a necessary condition for a change outcome in a decision cycle. Without responsive decision-making bodies, issues will not be successful. But it alone is not a sufficient one —responsiveness must be combined with at least one other condition for a cycle to result in change. The second condition, legitimacy, describes the overall volume and quality of communicative power in a tessellation that is supportive of an issue. Legitimation occurs in open, inclusive contexts where validity claims are offered to bolster an issue. High legitimacy is neither a necessary nor a sufficient condition but high legitimacy, combined with high responsiveness, results in a change outcome. Civil support is the third condition that explains positive outcomes. Civil support describes the identity of the individuals or organizations that support an issue in the decision cycle. When an issue is widely supported by civil society groups, which include any non-governmental organizations and any non-business associations, civil support is high. High civil support, like high legitimacy, is neither a necessary nor a sufficient condition for change by itself, but when it is combined with high responsiveness, the decision cycle results in a successful outcome. The last condition is low resistance. Resistance refers to objections raised against an issue over the course of a decision cycle. High resistance is not identical to low legitimacy. Legitimacy of an issue may be low not due to objections, but because discussion surrounding an issue was not open, consensus oriented, or based on the give and take of reasons. High resistance describes a specific form of low legitimacy where opposition is mobilized and targeted against an issue. A decision cycle that exhibits low resistance, together with high responsiveness, results in a successful outcome.

This four-factor theory of public library development may be abbreviated RLCr and pronounced "releaser" by combining the first letters of each condition. This theory is presented in this book and was developed considering the case of greater Pittsburgh. Releaser Theory may be described using the metaphor of a canal lock.[8] Figure 4 below illustrates this metaphor visually. The canal barge represents an issue in a decision cycle. Proponents of the issue hope to enable the barge to travel upstream, toward the core, using the lock. In the first stage, raising of an issue, supporters of the issue propose a new issue to be addressed, and the barge approaches the lock. In the second stage, the legitimation phase, the barge is in the lock. Proponents use communicative power to open the lock sluice and allow water from upstream to enter the lock, thus elevating the barge to the next level. The water entering the lock represents legitimation of the issue—the water raises the issue upward. There may be different groups of people shepherding the barge, so to speak. These passengers, who are supporters of the issue and help it to navigate through the lock, may include civil society groups, representing civil support. But the barge may encounter problems in the lock in the form of resistance. There may be a crack or hole in the lock, thus allowing water to escape, affecting the elevation of the barge to the core. Once the barge is elevated to the core level, the formal decision point is reached. At this point, the issue is transformed into goods and services when the core exhibits high responsiveness. The issue then enters the implementation phase. Using administrative power, the barge carries the resulting goods or services downstream. Overall, a successful decision cycle can be imagined as a barge (issue) that is raised (legitimated) in the lock, maneuvered by capable navigators (civil supporters), where the lock is watertight (low resistance), and the upper gate opens to allow the barge to pass and take on cargo (high responsiveness).

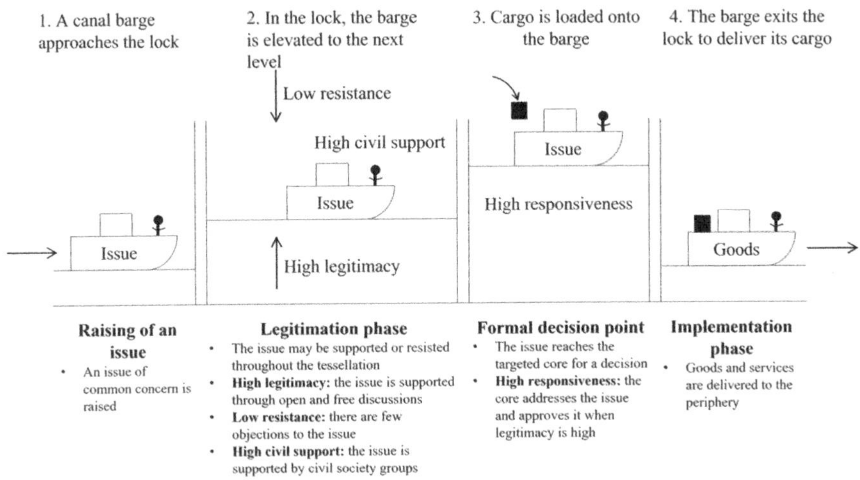

Figure 4: Releaser Theory as a canal metaphor.

The circulation of power model, described above, was used as a guiding framework for this book. Knowing and understanding this model is important because it was used to organize this book, it formed the basis for data collection, and it provides a standardized language that will be used in subsequent chapters. Knowing the model before reading the book is important. The adoption of this framework has historiographical and methodological implications, meaning that the framework determined what kind of data was collected and analyzed, how the book was organized, and what elements of the infrastructure would figure prominently in the story. Another way to say this is that the circulation of power model determined the ontology and epistemology of the research behind this book: what stuff would be talked about and how that stuff could become known. As seen in the model, the stuff of interest in this book is power. Power manifests in different forms of communication that occur between different actors who exist in configurations called circuits. Within these circuits, actors raise and legitimate issues. The story in this book therefore tracks the issues that arise and follows them as they travel between periphery and core and back again. Some issues will fail while others will succeed. Actors, circuits, power, and issues therefore represent the ontology contained in this book.

The longitudinal tracing of issues about a regional public library system in greater Pittsburgh was built piece by piece, one transmission at a time. In terms of data collection and analysis, the unit of analysis of this research was the communicative event. The data set supporting this research, its evidentiary basis, is therefore a representative sample of communicative events. A communicative event is composed of several elements: a date, type of power, sender and receiver, circuit positions for each, issue, legitimation outcome, and implementation outcome. Data from communicative events was collected and analyzed in several stages, producing both qualitative and quantitative data. The research design of this study was historical case study.[9] In the first stage, source collection and analysis, source materials were gathered to form the first layer of the evidentiary basis for the research. This project used source triangulation. Source materials for this research included archival documents, interviews, and fieldwork observations. Archival documents were the most extensive source material utilized for this study, followed by interviews and observations. Qualitative data produced during the source collection and analysis stage was used to segment the case of greater Pittsburgh into distinct decision cycles, each representing a period of the case. These periods are presented as separate body chapters in this book. The body chapters tell the story of each period's decision cycle, how issues were raised and legitimated, and whether the issues succeeded or failed. Supplementary documentation is also supplied in these chapters that relates to state and national contexts. While not part of the greater Pittsburgh case per se, it is necessary to understand state and national trends in order to see why certain issues were raised in Pittsburgh at the times they were. The second stage of the study applied a coding frame to the source materials in order to produce a set of quantitative data points. The method used during this stage was qualitative content analysis, and the coding frame that was applied to the source materials operationalized concepts from the circulation of power model presented above.[10] The third stage of the research drew from the qualitative data from stage one and the quantitative data from stage three in order to explain the outcomes of each of the nine decision cycles. It was during stage three that Releaser Theory was constructed using a method called qualitative comparative analysis.[11] The fourth stage, rule interpretation, compared the most-similar periods from the case to develop change management strategies that follow from Releaser Theory.[12] Table 1 summarizes this sequential mixed methods research design. In the end, three types of knowledge were produced using this research design: descriptive knowledge, explanatory knowledge, and technical knowledge.

Table 1: Research design.

	Stage	Method	Knowledge Product
1	Source collection and analysis	Theoretical Sampling	- Relevant source materials to use in Stage 2 - Qualitative data about each period to use in Stage 3 (description)
2	Data collection and analysis	Qualitative Content Analysis	- Circulation of Power model - Quantitative data about each period to use in Stage 3 (description)
3	Data interpretation	Qualitative Comparative Analysis	Releaser (RLCr) Theory (explanation)
4	Rule interpretation	Comparison of most-similar periods	Change management strategies (technique)

The case of the regional public library infrastructure in greater Pittsburgh that is described in this book exhibits nine decision cycles. The first decision cycle began in 1924 when the issue of a regional public library infrastructure was first imagined and proposed. The last decision cycle described in this book concludes in 2016 at the close of a re-envisioning process. The new decision cycle that begins in 2016 is outside the scope of this book. Of the nine decision cycles described here, six have change outcomes and three have continuity outcomes. A change outcome means that the decision cycle results in a substantive change, and a continuity outcome means that the infrastructure remains the same at the end of the decision cycle. Each decision cycle in the case represents an embedded unit of analysis. The inclusion of multiple embedded units, the inclusion of both positive and negative outcomes, and the variety of conditions that brought them about, make the case of greater Pittsburgh an ideal one for illustrating the Releaser Theory. Because of the case's internal variation, it is possible to isolate those causal conditions that really matter and ignore those that do not. Figure 5 shows the number of communicative events over time with key decision dates marked. Gaps in the graph indicate lulls in communicative activity for those years. Peaks in communicative activity generally indicate the formal decision dates in each decision cycle, as seen in Figure 6. Figure 6 shows the nine periods of the case, each corresponding to a decision cycle. The formal decision dates in Figure 6 match the key decision dates shown in Figure 5. Table 2 shows the periods and date ranges of the case. The titles shown are those used in the body chapters of this book.

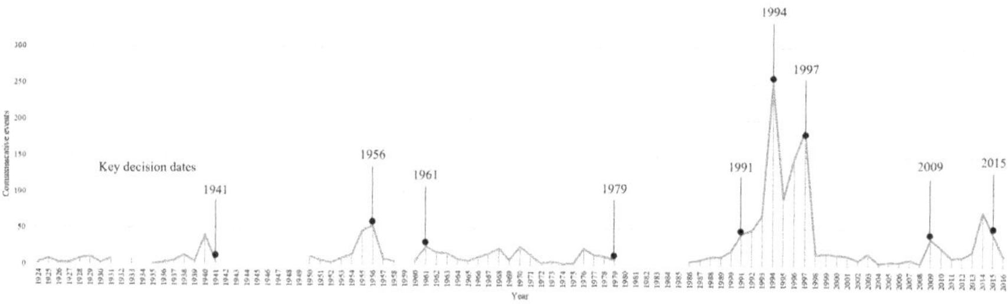

Figure 5: Communicative events over time with key decision dates.

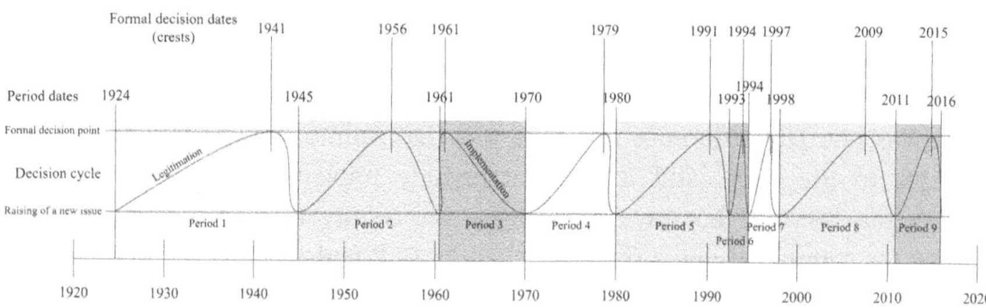

Figure 6: Decision cycles over time.

Table 2: Periods, date ranges, and titles.

Period	Year range	Period title
1	1924 – 1945	Periphery and Center
2	1945 – 1961	County Contract
3	1961 – 1970	The Rise of District Services
4	1970 – 1979	Unification Revisited
5	1980 – 1993	Fiscal Crises and Digital Visions
6	1993 – 1994	The Pursuit of RAD
7	1994 – 1997	System Dreams
8	1998 – 2011	Formula Wars
9	2011 – 2016	Compliance Culture

Each of the subsequent body chapters of this book recounts the story of a decision cycle, or period, of the case of greater Pittsburgh. During Period 1, which lasted from 1924 to 1945, the idea of a regional public library infrastructure was first imagined by the Civic Club of Allegheny County. Their earliest proposal, which was supported by other civil society groups, included a partnership between Pittsburgh's city library system, CLP, and public schools in outlying areas of the county. The proposal also included a merger between two city library systems. The exact plans for the infrastructure were modified as time went on. Despite high legitimacy, high civil support, and low resistance, county commissioners were unresponsive to the issue raised by these civil society groups, and the proposal for a regional public library infrastructure failed.

In Period 2, which spanned from 1945 to 1961, the movement for a regional public library infrastructure was reinvigorated. Some of the same civil society groups supported infrastructure as before, this time with an emphasis on a contract between Allegheny County and CLP for bookmobile services to outlying areas of the county. A new third sector organization appeared on the scene comprised of local industry leaders. Civil support and legitimacy were high, and resistance was low. The difference this time was that county commissioners were responsive to the issue, and it was successfully implemented. In 1956, county bookmobile services, borrowing privileges for county residents, and reference service for county residents were supplied by CLP and funded by the county. In that same year, the two city library systems combined.

Period 3, which ranged from 1961 to 1970, marks the advent of the Pennsylvania state plan, which included the creation of a new public library entity, the district library center. CLP became a district library for all the public libraries in Allegheny County. This led to an increase in state funding to CLP, and enhancements in services and administration of public library services. In contrast to Periods 1 and 2, where the issue of a regional public library infrastructure was homegrown, Period 3 illustrates how the issue was imported from within the state circuit and implemented locally.

In Period 4, from 1970 to 1979, due to several significant demographic, technological, and economic changes, new arrangements for the public library infrastructure in Allegheny County were proposed. There was debate about how the new structure would be funded and governed. After several preliminary proposals that failed to carry, Period 4 culminated in a countywide study and a final proposal, based on those made previously. Though there was civil support for the proposal, there was also resistance, and legitimacy was low. In the end, both CLP and county commissioners were unresponsive to the proposal, and the issue failed to be implemented.

In Period 5, which lasted from 1980 to 1993, public libraries in greater Pittsburgh struggled financially. At the same time, they were confronted with new technological opportunities, and they were urged by state planners to participate in increased resource-sharing arrangements. The years of the early 1990s marked a turning point for the development of public library infrastructure. At this time, an electronic information network began to form in Allegheny County, and the county public libraries were brought together politically to form a new organization, CLASP. These initiatives were library-led, with low resistance and high legitimacy.

In Period 6, from 1993 to 1994, the newly formed library association that included all county public libraries mobilized to pursue funding from another new county entity, the Regional Asset District. The county libraries launched a public relations campaign, invested in strategic consulting, and formed a non-profit corporation, ACLA. These efforts by the county public libraries ultimately prevailed: ACLA became recognized as a regional cultural asset, and it successfully secured funding by RAD in 1994. The mobilization for RAD funding, together with the burgeoning electronic information network, served as catalysts for bringing CLP and the county public libraries together in a new infrastructural arrangement.

During Period 7, from 1994 to 1997, CLP and the county libraries settled into their new relationship with the RAD board. RAD funding began in 1995. The electronic information network, which linked county libraries together, continued to develop. ACLA leaders proposed that the association pursue additional state aid by becoming recognized by the state library as a federated library system. The proposed formation of a federated library system became the defining issue of this period. Though the issue was legitimated, it was also met with sustained resistance. In the end, the issue was successful, and the ACLA membership voted to form a federated system in 1998.

Period 8, which lasted from 1998 to 2011, is characterized by a protracted debate about the nature of the formula used by ACLA to distribute RAD funds to member libraries. The period is marked by several failed attempts to modify the formula. Several actors were involved in the debate, including those in ACLA, RAD, and the state library. In the end, though there were some minor changes made to the formula, there was not a significant or substantive change as had been originally proposed. The issue met with high resistance, low legitimacy, and low civil support.

During Period 9, from 2011 to 2016, RAD initiated a re-envisioning process for CLP and ACLA to identify a more efficient configuration for the county public library infrastructure. The process did not result in a restructuring of the county library arrangement, as was originally intended. Instead, in the wake of this process, ACLA leadership raised the issue of system standards. Though this issue met with high resistance, a set of library standards was implemented. Period 9 is the final decision cycle described in this book.

The final chapter of this book, the conclusion, examines the changing topography of the case of greater Pittsburgh. The chapter reviews Releaser Theory and uses it to explain why some periods resulted in change whiles others resulted in continuity. Concrete examples from each of the periods in the case are used to substantiate Releaser Theory. Alternative possible theories are also presented to demonstrate why Releaser Theory is the best possible explanation. It summarizes the key themes, issues, and actors, and it analyzes changes and trends in the case's circuits and tessellation. Based on the story recounted in each of the body chapters, and in light of the findings demonstrated by Releaser Theory, the conclusion of this book looks toward the future of the public library infrastructure in greater Pittsburgh and imagines future possible directions. Several change management strategies for public library infrastructure are proposed that are based on Releaser Theory and substantiated using examples from the case of greater Pittsburgh.

This is the story of a complex infrastructure. But using the circulation of power model, which includes the simple component of a circuit, the complex arrangement and seemingly intractable behaviors can be reduced, distilled, and simplified, focusing on a few key behaviors involving the circuitry of the tessellation. Behavior of the system over time can be distilled to decisions about the arrangement and relationships within the infrastructure, the tensions that exist when the infrastructure exists as part of more than one circuit, and the initiatives to move the infrastructure into new circuits and out of old ones. The circuit is the casing or shell in which the infrastructure lives. It may draw sustenance from more than one. But from the beginning, debates about the infrastructure involved decisions concerning which inner periphery the infrastructure should be a part of. The circuitry of the infrastructure changed over time and will continue to change as new possibilities emerge. Power or control of the infrastructure has itself circulated as the circuitry of the infrastructure has evolved, as it has moved in and out of circuits within greater Pittsburgh. There has not been a stable home for the infrastructure—the infrastructure itself circulates in and out of circuits.

This book is the story of the development of public libraries in greater Pittsburgh. It is not intended as an evaluation of the current system, nor is it a prescription for its future development, though future possibilities are considered. And while the framework, theory, and story presented in this book are based on the Pittsburgh case, the findings presented here may apply to other contexts beyond the case. The circulation of power model and Releaser Theory may be applied to other similar infrastructures. Application and testing of this model and theory remain tasks for future research.

1 In the United States, a county is a governmental subdivision that is geographically smaller than a state but larger than a local municipality. Counties administer services such as health, safety, recreation, education, transportation, and law enforcement. Counties are governed by elected officials, funded by taxes, and the boundaries of existing counties and the scope of their services

are determined by each state's constitution. There are 67 counties in Pennsylvania. A municipality, as the term is used in this book, is a local government unit within a county. Municipalities include cities, towns, and boroughs. In addition to counties and municipalities, other forms of local government include school districts and special districts. School districts have their own elected boards and raise their own taxes. Like school districts, special districts raise their own taxes to support specific services, but unlike school districts, they have appointed boards. Special districts include water authorities, transportation authorities, and cultural asset districts. In this book, a prominent example of a special district is Allegheny Regional Asset District. Library districts described in later chapters in the context of Pennsylvania's state library plan are not special governmental districts in the same sense. Library districts only indicate a geographical service area. They do not have taxation power and are not governed by a board. They are administered by the State Librarian and the Governor's Advisory Council for Public Library Development.

2 An overview of previous literature related to public library development is found in Michael M. Widdersheim, "A Political Theory of Public Library Development," *Libri* 68, no. 4 (2018).

3 Discussion about a public sphere approach is found in Michael M. Widdersheim, "From Two-Track to Tessellation: A Revised Circulation of Power Model," *Journal of Documentation* 78, no. 6 (2022).

4 Preliminary models were proposed in Michael M. Widdersheim and Masanori Koizumi, "Conceptual Modelling of the Public Sphere in Public Libraries," *Journal of Documentation* 72, no. 3 (2016); Michael M. Widdersheim and Masanori Koizumi, "A Communication System Approach to the Problem of Public Library Legitimacy," *Library and Information Science Research* 39, no. 1 (2017); Michael M. Widdersheim, "Governance, Legitimation, Commons: A Public Sphere Framework and Research Agenda for the Public Library Sector," *Libri* 65, no. 4 (2015); Michael M. Widdersheim, "Late, Lost, or Renewed? A Search for the Public Sphere in Public Libraries," *Proceedings of the Ninth International Conference on Conceptions of Library and Information Science* (2017), http://informationr.net/ir/22–1/colis/colis1644.html; Michael M. Widdersheim and Masanori Koizumi, "Methodological Frameworks for Developing a Model of the Public Sphere in Public Libraries" (CoLIS 9 conference, Uppsala, Sweden, June 27–29, 2016).

5 The full modifications and enhancements are described in Widdersheim, "From Two-Track to Tessellation: A Revised Circulation of Power Model."

6 As elaborated in Widdersheim, "From Two-Track to Tessellation: A Revised Circulation of Power Model."

7 This theory is further detailed in Widdersheim, "A Political Theory of Public Library Development."

8 The circulation of power model is also known as the sluice-gate model and the Machtkreislauf model.

9 Historical case study is described in more detail as a distinctive research strategy in Michael M. Widdersheim, "Historical Case Study: A Research Strategy for Diachronic Analysis," *Library & Information Science Research* 40, no. 2 (2018).

10 For details on qualitative content analysis, see Margrit Schreier, *Qualitative Content Analysis in Practice* (Los Angeles, CA: SAGE, 2013).

11 For details on qualitative comparative analysis, see Charles C. Ragin, *The Comparative Method: Moving Beyond Qualitative and Quantitative Strategies* (Berkeley, CA: University of California Press, 1987); Benoît Rihoux and Charles C. Ragin, eds., *Configurational Comparative Methods: Qualitative Comparative Analysis (QCA) and Related Techniques* (Los Angeles, CA: SAGE, 2009).

12 Michael M. Widdersheim, Brady D. Lund, and Betty J. Kemboi, "Change Management in Public Libraries: Research-Based Political Strategies," *Journal of Library Administration* 59, no. 7 (2019).

Period 1: Periphery and Center, 1924–1945

The case of the regional public library infrastructure in greater Pittsburgh began in June of 1924 when a fixed and permanent solution to regional public library service was first envisioned by members of the Civic Club of Allegheny County. The idea was for the Carnegie Library of Pittsburgh (CLP) to establish public libraries in school buildings outside the city. Collections would be made available to students, teachers, and the community.[1] This was the first known attempt to extend public library services outside the city in a fixed building location. Earlier efforts by the Civic Club to extend public library services were either ephemeral, confined to the City of Pittsburgh, or they did not quite meet the definition of a public library. The Civic Club first attempted to broaden the City of Pittsburgh's library services beyond city limits in 1903 by establishing traveling libraries in small outlying towns.[2] But the term "libraries" was euphemistically used: traveling libraries were so called because they were wooden chests filled with books that were transported to a distribution point, usually someone's home, where the books were then lent out by a designated "librarian" to interested readers. The librarian was whoever was there or whoever signed up to receive the books. Applicants could retain the crate for a few months. CLP curated and sent traveling libraries across the city. This was a popular service at the time. At least 21 sets of "home libraries" were administered by and for children within the city from 1899 to 1918.[3] Similarly, CLP set up "station libraries" in places like fire houses where the traveling collections might attract adult borrowers.[4] The Civic Club's traveling libraries effort, which may have been supported by CLP, was successful in the neighborhood of Glenshaw, a municipality outside Pittsburgh, at least temporarily.[5] As exemplified by CLP, traveling libraries were a popular makeshift solution to library coverage in Pennsylvania in the first decades of the twentieth century. The Pennsylvania Free Library Commission (PFLC), a state agency that existed from 1899 to 1919 and was based in Harrisburg, also distributed traveling libraries across the state.[6] But this agency targeted rural counties within the state, and a list of destinations from 1910 reveals that PFLC did not send books to Allegheny County.[7] In the 1920s, the State Library of Pennsylvania inherited the traveling libraries program, but it is not likely that the program extended to Allegheny County since the state's outreach efforts emphasized rural counties. Before it focused its efforts on public libraries, the Civic Club of Allegheny County supported establishing school libraries in districts that did not have public libraries. This was in 1907.[8] It also supported sending traveling libraries to the schools. Partnering with schools was another popular strategy for coverage in Pennsylvania at this time. It was perhaps only a matter of time be-

https://doi.org/10.1515/9783111013404-003

fore the Civic Club began to consider the strengths of schools and public libraries together.

There is no evidence that state extension initiatives in the early twentieth century extended into Allegheny County. There are several reasons why the initiatives affected other areas in the state but not greater Pittsburgh. The main reason is that local municipal support was the dominant mode of public library funding in Pennsylvania at this time. Pennsylvania history and politics was firmly rooted in the principle of "home rule," or local control. Reiterating this philosophy as found in earlier laws, Pennsylvania legislation from 1917 allowed any municipality—including any county, city, borough, town, or township—to make appropriations and to levy taxes to support free, public, nonsectarian libraries.[9] It also enabled county commissioners to contract with an existing library for services. But there was no mandate to do so, and there was not yet a system of state aid that would encourage such cooperation. State legislation therefore placed both the option and the burden of public library support on local government. The state presence that did exist at the time was not noticeable in Allegheny County. The PFLC, which was a statewide agency established in 1899 and focused on rural library development, was consolidated with the State Library and Museum in 1919, thus creating a Library Extension Division.[10] This extension division continued to focus on rural library development only, thus largely overlooking greater Pittsburgh. In the 1920s, there was a statewide movement for state funding for county libraries. This resulted in legislation in 1931 that established state funding to rural county libraries. This was the first category of state funding for public libraries beyond what the state budget provided for the state library in Harrisburg. Harrisburg is the state capital of Pennsylvania. But this legislation only funded counties of the third through eighth class, according to population. Allegheny County was an urban second-class county and was therefore ineligible to receive state money for a county library.[11] Pittsburgh is the second-largest urban area in Pennsylvania after Philadelphia.

Subsequent planning efforts by the Pennsylvania Library Association (PLA) focused on the funding of rural county libraries and extension of their services.[12] In any case, Allegheny County already had a strong library system in Pittsburgh. For all these reasons, it is not difficult to see why the rural areas of Allegheny County were not the focus of state attention. In Allegheny County, therefore, in any matters related to public library services, it fell to local organizations to formulate additional means to get residents access to books. In the county, the promotion of locally taxed and locally built municipal or school libraries was one proposed solution. In sharp contrast to the dominant ideology of localism that marked Pennsylvania at this time, however, the Civic Club of Allegheny County instead proposed expanding services that already existed within the City of Pittsburgh into unserved areas. Instead of urging local investment, the Club targeted the county, which is a more regional unit of government. It is worth reiterating how noteworthy and exceptional this plan was. It flouted the parochialism of the times and instead promoted a regional way of thinking. The Civic Club's plan from 1924 is distinctive because it was the first to envision stable and sustained cooperative public library services on a regional scale. It is also notable as an example of a plan that defied localism.

But the Civic Club did not move full steam ahead with their plan immediately. In 1924, the Civic Club approached the county commissioners in Allegheny County to ask that Pittsburgh and Allegheny Libraries complete a survey "of the city and its environs to determine where there is a need of establishing further branch libraries."[13] The services were to be coordinated by a public library, not a volunteer or social library, and they would include more than just a transitory crate of books. Due to the plan's fixity and detailed consideration, though it ultimately failed, the proposal by the Civic Club for CLP to partner with schools throughout the county marks the first formal effort to create a regional public library infrastructure in greater Pittsburgh.

To be sure, there was already an infrastructure of book lending that existed in Pittsburgh at this time. In early twentieth century Pittsburgh, there was a basic level of library services available to residents within city limits and, to a limited extent, around the county. That said, these services were not coordinated in any way on a regional scale. In addition to public library collections, there were many instances of non-public or semi-public library services. Existing library services were mostly discrete and local efforts, some initiated by community clubs, social groups, and others by businesses. Though these private libraries were largely phased out by this time, some still existed.[14]

CLP was the main source of public library services at the time. The central library opened in Oakland in 1895. By 1910, the library system had three additional branches: Lawrenceville, West End, and Wylie Avenue.[15] By 1930, CLP had grown to 13 branches throughout the city's neighborhoods[16]

While it was the largest and best-coordinated library system in the area, "The Carnegie," as the public library system came to be called, was not the only public library system in the city. For much of the first half of the twentieth century, the North Side of Pittsburgh had its own public library system. There is a somewhat confusing explanation for the existence of these two library systems within the same city. In 1907, the City of Pittsburgh annexed its western neighbor, the City of Allegheny. The City of Allegheny was that area of land that occupied the western shore of the Allegheny River relative to the City of Pittsburgh. Today, the City of Allegheny roughly coincides with what is known as North Side, which it came to be called once it was annexed. At the time of annexation, most city functions of the

City of Allegheny were consolidated with those of the City of Pittsburgh. The exception to this, however, was public library service. The City of Allegheny retained its own system, and it remained separate from the CLP system.

Why did this happen? There is some speculation. The City of Allegheny had the honor of having received one of the first of Andrew Carnegie's library building donations. Opened in 1890, this library in the City of Allegheny was called the Carnegie Free Library of Allegheny. The Carnegie Free Library was a source of pride for citizens of Allegheny City. Even President Benjamin Harrison attended the opening ceremony in February of 1890.[17] The library was also distinctive because for a time it was one thing the City of Allegheny had that their rivals across the river did not. In 1881, Andrew Carnegie offered to build a library for the City of Pittsburgh, but Pittsburgh declined, saying there was no legal basis for the city to appropriate monies to maintain the library.[18] The City of Pittsburgh later recanted and accepted Andrew Carnegie's offer. The City of Pittsburgh's library became known as The Carnegie Library of Pittsburgh. It opened in Oakland in 1895, and it was part of the larger Carnegie Institute that consisted of a music hall and museums. This library was the first building in the city's CLP system, and it would become the system's central branch. But after the annexation of the City of Allegheny, the Carnegie Free library of Allegheny, together with its North Side branch, Woods Run, were administered separately from CLP. One can suppose that residents of the old City of Allegheny refused to lose their prized building to their rivals across the river. Local control and local power are recurring themes in greater Pittsburgh.

By 1930, in addition to the 13 branches of CLP, the library system also operated smaller stations, school libraries, and home libraries throughout the city. In 1930, in addition to its branches, CLP worked closely with schools throughout the city as well. It oversaw collections at 16 high schools, 86 platoon schools, and the Frick Training School for Teachers.[19] As evidenced by these micro-branches, there was some movement by the city library system to broaden its service reach, even though the coverage of CLP remained within the City of Pittsburgh.

Pre-1876, there existed in the cities of Pittsburgh and Allegheny numerous scientific or engineering society libraries, apprentices' libraries, theological seminary libraries, commercial circulating libraries, social libraries, YMCA libraries, and mercantile libraries.[20] As a boy growing up on the North Side, Andrew Carnegie famously borrowed from the private collection of Colonel James Anderson. This collection would later be expanded to become the Mechanics' and Apprentices' Library, a subscription library in the City of Allegheny.[21] These private subscription libraries were open to some members of the public for a fee, others for free, but in any case, they did not survive into the twentieth century.[22] Neighborhoods and local ethnic groups throughout Pittsburgh also maintained their own library collections, and some of these lasted into the twentieth century. One such example was the German library that for a time was housed in the abandoned athenaeum building.[23] Several YMCA and special libraries survived in Pittsburgh into the early twentieth century, according to a list compiled from 1913 – 1915.[24] By 1935, however, these private lending libraries had been largely displaced by the public library systems in the city. The earliest free public library in the area was Sewickley, which began as a Young Men's library in 1873.[25]

Besides public, school, and subscription libraries, residents of the city also had access to numerous commercial circulating libraries. These rental collections loaned books for an average of 3 cents a day. Unlike the public libraries, whose collections consisted mostly of non-fiction, the commercial circulating libraries consisted primarily of fiction. The commercial circulating libraries were therefore mainly used for leisure reading purposes, and they included fiction genres such as detective, romance, and Western, and non-fiction genres such as biographies and popular science.[26] Commercial lending was typically a side business operated out of small shops. In 1935, there were 232 such commercial lending libraries in the City of Pittsburgh. These were found in such places as bookstores and news companies, drug stores, department stores, confectionary shops, beauty shops, and cigar stores.[27]

By 1910, there were also several colleges and universities in the City of Pittsburgh. These included the University of Pittsburgh, located in Oakland; Pennsylvania College for Women, located in Shadyside and later to become Chatham University; and Carnegie Technical Schools, later renamed Carnegie Institute of Technology, then Carnegie Mellon University, now located in Oakland and Squirrel Hill. While these schools had their own collections, students also utilized materials from the Carnegie Library of Pittsburgh.[28]

As the population grew in the Pittsburgh region in the early twentieth century, independent public libraries began to appear outside the city as well. By 1920, more than half of the population of Allegheny County lived outside the city. As the population increased in the suburban ring, citizens there began to demand cultural infrastructure like what they had enjoyed in the city.[29] Many municipalities created local public libraries of some form. By 1935, there were 22 libraries in Allegheny County outside Pittsburgh's city limits.[30] This list includes Braddock, Homestead, and Duquesne—the three libraries that Andrew Carnegie built for workers at his steel plants that later became public libraries. The Carnegie Free Library in the Borough of Carnegie is distinctive because an endowment was left by Andrew Carnegie in addition to the building. These four libraries—Braddock, Homestead, Duquesne, and Carnegie Borough—are exceptional because they were supported by an Andrew Carnegie endowment rather than by public funding. The Bessemer and Lake Erie Railroad company established its own library for Bessemer workers. The building was donated by Andrew Carnegie. Most of the libraries were established independently by the municipality and were supported through local taxes. The formation of each small, independent local library followed its own path—there was no regional coordination on a countywide scale. The libraries operated independently of the city's own two systems.

By the time the Civic Club made its proposal in 1924, there was an awareness by some Civic Club members of certain shortcomings in the general library situation in the City of Pittsburgh and its environs. The population of the city and county was still headed upward, and suburban areas especially were growing. There was a perceived disparity in library service quality outside the city compared to that found inside the city, at least from the perspective of groups like the Civic Club.

Leading up to the 1920s, the Civic Club led numerous betterment projects with a view to improve living and working conditions in and around Pittsburgh. The organization's stated mission was "to promote by education and organized non-partisan effort a higher public spirit and better social order."[31] Pittsburgh was an industrial town, known in part for its soot-stained buildings from surrounding factories. It was also known for discord between labor unions and factory management, as evidenced by the Homestead Strike in 1892 at Andrew Carnegie's steel mill. The Civic Club led volunteer initiatives and legislative movements related to issues such as education, public health, public baths, smoke abatement, child labor, garbage collection, free bridges, and school inspections.[32] Some of its other notable activities included passing an anti-spitting ordinance "making it a misdemeanor to expectorate on the streets;" carrying out mental testing of high school students; introducing saluting the flag into public schools; and the assignment of a police officer devoted solely to abating unnecessary noises around the city.[33] The issue of public library services fit well within the group's scope. The club maintained a standing committee on public libraries that could mobilize at any time to address library-related issues. The function of this committee was "the extension and consolidation of adequate library service throughout the County."[34] This committee in the club was relatively active during the club's first 50 years.

Started in 1895, the Civic Club was one of several urban reform groups in greater Pittsburgh, and it was a key actor in Pittsburgh's larger progressive movement. Like most other progressive groups, and reflective of the reigning Taylorist philosophy of the time, one salient theme of the Civic Club was efficiency. The club's central committee, the Committee on City and County Government, for example, was tasked with the following: "By study and consultation with interested officials endeavor to determine the best method of improving the efficiency of governmental agencies."[35] For a group interested in efficiency, the library situation in Pittsburgh in the early twentieth century must have appeared chaotic and wasteful. The movement to consolidate the disparate library services throughout the county was thus motivated in part by this underlying drive toward more efficient service delivery.

The other philosophical tenant guiding the group's efforts toward expanded library services was its determination to raise general living conditions for those in and around the city. The group especially targeted children, workers, the poor, and, as evidenced by their library initiatives, rural residents of the county. The Civic Club viewed welfare broadly: its members combatted pollution, built public baths, established parks and playgrounds, campaigned for child labor laws, and studied refuse collection, hospitals, and housing. Education was a central concern of the club. For instance, the group organized open-air schools for youngsters and night schools for foreigners. Public libraries were viewed as educative institutions. Their efforts to extend public library services within the city and throughout the county was thus an education issue.

The Civic Club's efforts toward consolidated library services in the county was part of a larger unification movement in and around Pittsburgh. In 1920s Pittsburgh, there was a growing support for what would be known as metropolitanism. This push for legislation began as early as 1906, when the Civic Club first began to explore city-county consolidation.[36] In 1923, a commission was formed to study the possible benefits of municipal consolidation within Allegheny County. Under the proposed plan, the 122 separate units of local government within the county would be assembled under a single governmental structure whose jurisdiction was co-extensive with the county limits. The intent of this plan was to create a metropolitan government that was equipped to plan and fund regional-scale projects, thus addressing problems that were common to all municipalities in the area but were too large and complex for independent municipalities to address alone. Some of the regional problems facing greater Pittsburgh included transportation, traffic, sanitation, and health.[37] The Civic Club saw public library service as an additional infrastructural issue that could be best addressed on a countywide scale.

In 1927, legislation was passed in Pennsylvania that would allow for county consolidation. The Civic Club was involved in drafting the amendment and holding open forums on the legislation.[38] This amendment was fully adopted into Pennsylvania's constitution in 1928, and the Civic Club supported the efforts of the Metropolitan Planning Commission to develop the amendment.[39] The legislation would in effect establish county government as the legislative and administrative head of a federated city.[40]

A special election on the charter was held in Allegheny County on June 25, 1929. Perceived support for consolidation notwithstanding, and despite predictions otherwise, the charter failed to carry. Joseph T. Miller, Chairman of the Metropolitan

Planning Commission for Allegheny County, explained the charter's loss in the following way:

> The charter was not adopted, by reason of the unusual and peculiar conditions inserted in the constitutional amendment by the opponents of the idea during consideration of the constitutional amendment in the Legislature. This requirement was that in order to be accepted by the people, the charter had to be approved in a twofold way. First it had to carry in the county at large by a majority vote, and second, and in addition to that, it must carry a majority of municipalities (or sixty-two) by not less than two favorable votes to every negative vote. Notwithstanding the fact that the charter received in the county a majority of 46,834 votes, and notwithstanding the fact that it carried eighty-four municipalities 'by a majority vote, it won in only forty-nine municipalities by a vote of two to one or better. Only thirty-seven municipalities were against it by a majority vote, and in one municipality the vote was a tie.[41]

Attempts were made to immediately resubmit the charter, but these efforts failed as well, and despite many more similar initiatives throughout the twentieth century, metropolitanism never succeeded in Pittsburgh. The resulting governmental environment strongly influenced the type of library system that emerged in the Pittsburgh region. Unlike other similarly situated cities, Pittsburgh could not establish a centralized, consolidated library system.

In light of the loss for the metropolitan charter, the Civic Club and other similar groups were not dissuaded from pursuing their welfare projects. The Civic Club in particular continued its efforts to raise the quality of library services in and around Pittsburgh. The library platform of the Civic Club had three planks: 1) improve library services within the City of Pittsburgh, 2) combine the North Side libraries and the City of Pittsburgh library system into a single system, and 3) consolidate library administration throughout the county. Much of the Club's activity related to public library service was directed toward supporting greater appropriations for CLP.[42] The group targeted city council as well as the mayor, since funding for the library came from the city budget.

One of the Civic Club's key allies in their efforts toward supporting the city library system was Ralph Munn, director of CLP from 1928 to 1964. Munn often attended Civic Club board meetings, reported on library developments, and even solicited approval of the library's budget.[43] The Civic Club worked jointly with Munn to petition the mayor and city council for increased expenditures for the library.[44] In addition, the Club advocated for expanded library services in the city. The Civic Club was successful in 1924 when it petitioned the mayor and city council for appropriations for a downtown branch of the public library system. The appropriation was granted, and the business district branch was opened on the first floor of the city-county building.[45] Years later, the Civic Club and Munn worked to expand the services of the downtown branch.[46] Expanding public library services within the city was a recurring theme of the club, and the intensity of this drive increased with time as the city and county grew in size.[47] The Civic Club saw a need not only to expand services, but also to restructure the system that already existed inside the city.

The main restructuring effort carried out by the club was to merge the North Side libraries with the CLP system. The recommendation to merge the two systems was first made in 1926, at the same time the Club called for change of the library director on the North Side.[48] Merging the two systems and securing a new library director for North Side were no easy tasks. There was no immediate action taken on either proposal. In 1933, the library director was David D. Cadugan, an active member of the North Side Republicans and a North Side resident.[49] The library director position at North Side came to be seen as a political prize. Cadugan had no particular qualifications for the position. It was said that Cadugan was installed by local North Side Republicans because he was loyal to the Republican ward-chairman.[50] For a time, George M. P. Baird was considered as a candidate for the library director position. Baird was a resident of North Side and an instructor at the University of Pittsburgh with a background in English and History.[51] Ralph Munn of CLP did not know Baird but supported him as a replacement for Cadugan because Baird had at least some relevant qualifications.[52] Baird was also the Civic Club's ostensible candidate given how he supported the amalgamation of North Side libraries with the Carnegie system.[53]

By 1938, however, Cadugan remained firmly in place. Several attempts by Democratic officials to unseat him had failed.[54] The North Side libraries and the Carnegie system still operated separately. This was despite renewed efforts by the Civic Club to mobilize public support for a merger. In March of 1938, the Civic Club continued to pressure city council to combine the libraries. The Club hosted a debate on the issue in the William Penn Hotel, downtown Pittsburgh.[55]

Two years earlier, in 1936, the Civic Club reached out to the Pennsylvania Economy League in the hope that the League might help them understand the legal barriers to merging the libraries. It was also hoped that the league could assist in formulating a strong financial case for the merger.[56] The collaboration between the Civic Club and the Pennsylvania Economy League continued into 1937, though the league did not immediately assist in producing results.[57] But by mid-1937, the aid of the Economy League was confirmed. A study of the library situation was proposed to Mayor Scully and members of city council, but there is no evidence that this study led by the Economy League was ever conducted.[58]

Despite the efforts by Munn and the local civic organizations, by late 1939 the Public Libraries Committee of the Civic Club declared that the merger between Carnegie Free Library of Allegheny and CLP was no longer a live issue. The mayor had already appointed a new person to the position of Librarian of

North Side. Absent any significant protest by residents on North Side, the merger would not happen at this moment.[59] With the Civic Club alone and seemingly abandoned, the issue of the merger was set aside.

By late 1938, the Civic Club already began to shift its direction and redirect its efforts elsewhere. It shifted its focus from the city to the county, and it changed its target from the city mayor and city council to the county commissioners. The issue of public library service was reframed away from merging the two city systems and toward creating a new countywide system. From 1936 to 1940, then again from 1940 to 1944, there were three Allegheny County Commissioners: John J. Kane (Democrat), George Rankin (Democrat), and John S. Herron (Republican).[60] The Civic Club began to target these county officials.

The club found that there was renewed interest to broaden public library services throughout Allegheny County, and they found that on this issue they were not alone. Other local associations shared their interest in expanding public library services into the rural areas of the county. In late 1938, library directors in the county were contacted by Ruth Houk, Chairman of the Library Service Committee of the Allegheny County Parent-Teachers' Association. The letter expressed the view of the council that all people had the right to read books, and for this reason, everyone should have access to a library.[61] The letter went on to emphasize the importance of bookmobile services to rural residents. It was not clear from the plan how the bookmobile services would be administered, whether they would be managed by CLP or implemented by some other agency.

Meanwhile, the director of CLP, Ralph Munn, independently investigated the legal basis for county library service. In early 1938, he contacted S.L. Fuss, an attorney in Pittsburgh, regarding the possibility of forming a county library system.[62] Following an examination of the subject, Munn reported his findings back to the Civic Club. In his view, formation of a unified county library system would be nearly impossible, not just for legal reasons, but also because individual libraries would not forfeit their autonomy. Instead, Munn advanced a two-stage plan where, first, the commissioners established a library district coextensive with the county. The county drew tax support for the library system only from those municipalities without existing libraries and from municipalities who wished to join the system. The county then contracted with the city's CLP to provide library services to those municipalities in the County system. CLP therefore formed the basis of the county system. In the second stage, remaining municipalities in the county gradually joined the county system until it became a single, unified system. Munn recommended this two-stage approach not just because it was legally feasible, but also because he saw significant political challenges to establishing a single system from the outset in an area with so many governmental units.[63]

In late 1939 and early 1940, the Civic Club, the Allegheny County Parent-Teachers' Association, the Allegheny Federation of Women's Clubs, and the Carnegie Free Library of Allegheny all jointly petitioned the county commissioners for funding to support public library services in the rural areas of Allegheny County.[64] The plan was initially proposed by George Seibel, the new director of the North Side's Carnegie Free Library.[65] Under one version of the plan, the county would contract with CLP for extending services to county schools, just as the city already did for schools inside the city.[66] The basic proposal asked for $1,800 to conduct a survey of library services in the county, and $25,000 to begin county services starting in 1940. In February of that year, a delegation that included Ruth Houk and Mrs. L. G. Stone of the Parent-Teachers' Association, Marie Dermitt, Secretary of the Civic Club, and Ralph Munn, Director of CLP asked the county commissioners for appropriations for this plan. There were slight differences between the proposal by this delegation and a separate proposal made by the Federation of Women's Clubs.[67] The response of the commissioners to the proposals was not enthusiastic. The commissioners stated that they could not likely fund the plan in 1940 because the county budget was cut to avoid a tax increase.[68] The commissioners recommended that the Civic Club bring all interested groups together to decide on a definite plan to present to the commissioners in the future.[69]

In response, the groups withdrew and reconvened to consider their next move. At this point, within the groups, there were slight differences of opinion regarding what a county plan would look like. There was still no definite plan, and the groups each seemed to pull in different directions. The North Side library, the Allegheny County Parent-Teachers' Association, and the Allegheny Federation of Women's Clubs sought funding for bookmobile services only. This was the proposal that would ultimately be supported by the Civic Club. But Lars Grondahl of the Civic Club, in a letter to county commissioner John Kane, envisioned bookmobile services as only a part of a much grander countywide system structured around CLP. Grondahl's plan echoed the proposal earlier designed by Munn in 1938. CLP represented the center of the system, surrounded by regional branches, delivery stations, and book-wagon routes strategically placed, as well as classroom collections in schools.[70] Grondahl did not assign an estimated cost for this grand vision, but he did cite annual library appropriations for other large cities in the United States as a way to insist that the commissioners should appropriate more. Grondahl's letter, based on Munn's ideas, foreshadowed later debates about the countywide system that would occur throughout the following century. This was the first time Munn's ideas were presented before the county commissioners.

Following the suggestion by the commissioners that the Civic Club convene all interested parties to devise a common plan, the Civic Club met with several other groups, including the Parent-Teachers' Association, the Federation of Women's

Clubs, the Council on Adult Education, the League of Women Voters, Ralph Munn, and representatives from several schools. At their meeting on March 7, 1940, the coalition confirmed their plan of supplying collections of books to schools throughout the county. This would be funded by the county.[71] Chairman Grondahl presented this plan to the commissioners in a letter.[72] Civic Club representatives continued to discuss the proposal with the commissioners and write them letters. The commissioners considered the plan in 1940, then again in early 1941. Ralph Munn also discussed the plan with the commissioners.

In June of 1941, the Civic Club received a letter from the commissioners stating that they could not fund the survey at that time, and they did not believe they could fund the library services in the near future.[73] Needless to say, this was a disheartening response for the civic groups. By 1942, the United States entered World War II. Efforts toward extending or consolidating library services ceased at that time, as evidenced in the Civic Club's meeting minutes from late 1941 or early 1942:

> The Committee's ultimate objective is the extension of public library service throughout Allegheny County by means of a contract between the County Commissioners and the Carnegie Library of Pittsburgh. Funds for this project cannot be obtained during the war. The merging or coordination of some existing libraries is also desirable. For the present no further effort will be made in these directions.[74]

This notice marks the end of the first period to develop a regional public library infrastructure in Allegheny County. It would not be until the close of World War II that local groups once again raised the issue of regional public library service.

Period 1 is characterized by a flurry of activity by multiple civil society groups, most notably the Civic Club of Allegheny County. These groups worked cooperatively amongst themselves and with local library directors to petition two sets of decision-makers: the city, which consisted of the mayor and city council, and the county, composed of the county commissioners. These were the two core decision-making groups that were targeted by the Civic Club, which operated on the outer periphery of the city and county circuits. Other third-sector groups, including the Allegheny Federation of Women's Clubs and the Allegheny County Council of Parent-Teachers' Associations, also operated on the outer periphery and coordinated with the Civic Club to petition for public library services. The issue of an improved public library infrastructure had high legitimacy, high civil support, and low resistance. Despite a seeming abundance of support, however, the efforts to merge the city libraries and to extend library service into rural parts of the county both resulted in failure. Munn attributed the failure of the countywide extension efforts to a lack of public support in rural communities.[75] It was also the case that the decision-making bodies that the groups targeted—the city mayor, the city council, and the county commissioners—did not act on the groups' petitions. The cause of the failure of the issue in Period 1 was therefore low responsiveness on the part of the city and the county. As a result, the issue was not successfully implemented. Decision makers did not respond to legitimated concerns from civil society and could not be moved to action.

The tessellation of Period 1 shows three primary circuits: civil society, City of Pittsburgh, and Allegheny County. The civil society circuit, comprising the Civic Club and other third-sector groups, included its own core, inner periphery, and outer periphery positions and actors. But civil society groups also functioned as the outer periphery of the city and county circuits in the effort to establish a regional public infrastructure in the inner periphery of those circuits. The relationship in Period 1 between civil society groups and municipal and county government was a simple center-periphery arrangement. The formation of a regional public library infrastructure would have created a new actor in the inner periphery of the county. But these petitions to establish a new inner periphery entity in Period 1 were unsuccessful during this decision cycle.

1 Civic Club of Allegheny County, Civic Club Board Minutes from May 1924 to 1925, 11–12, Civic Club of Allegheny County Records Collection, Box 30, PITT-ASC.
2 Civic Club of Allegheny County, Fifty Years of Civic History, 1895–1945, 1945, 24, CLP-PENN.]
3 Carnegie Library of Pittsburgh, Annual Report, 1900, CLP-PENN; Carnegie Library of Pittsburgh, Annual Report, 1899, CLP-PENN; Carnegie Library of Pittsburgh, Annual Report, 1919, CLP-PENN.
4 Carnegie Library of Pittsburgh, Annual Report, 1910, CLP-PENN.
5 Traveling libraries were a popular makeshift solution in early twentieth century Pennsylvania to expand access to books. The Pennsylvania Free Library Commission, later merged with the State Library of Pennsylvania, emphasized the distribution of traveling libraries to rural areas. But traveling libraries were also used in urban areas such as Pittsburgh, as evidenced by Carnegie Library of Pittsburgh's home libraries initiative. For more on traveling libraries in rural areas and the Pennsylvania Free Library Commission, see Bernadette A. Lear, *Made Free and Thrown Open to the Public: Community Libraries in Pennsylvania from the Colonial Era through World War II* (Pittsburgh: University of Pittsburgh Press, 2021).
6 Lear, *Made Free and Thrown Open to the Public*, 86–90.
7 "Traveling Library Stations," *Pennsylvania Library Notes* 3, no. 3 (July 1910).
8 Civic Club of Allegheny County, Fifty Years of Civic History, 1895–1945, 1945, 24.
9 Act 398, Laws of the general assembly of the Commonwealth of Pennsylvania passed at the session of 1917 (1917).
10 Library Extension Division, A Library News Letter from the Library Extension Division, August 1919, SLPARSC, Pennsylvania Library Association Collection, Publications and Research Files, Box 5a PLA Bulletin and microfilm reels, Library Extension Division Newsletter, 1919, 1929.
11 "Milestones in Legislative Support of Pennsylvania Public Libraries," *PLA Bulletin* (January 1981); David L. Lawrence, 19th Biennial Budget of the Commonwealth of Pennsylvania for the Fiscal Biennium June 1, 1959 to May 31, 1961, General Fund, Submitted to the General Assembly, March 1959, 102–03, Government Documents Collection, LUL; Laws of the General Assembly of the Commonwealth of Pennsyvlania Passed at the Session of 1931, (1931); George H. Earle, *The Budget of the*

Commonwealth of Pennsylvania for the Fiscal Biennium June 1, 1935, to May 31, 1937 (February 1935), 24.

12 Sub-Committee on Library Service, Commission for the Study of Educational Problems in Pennsylvania: Report of the Sub-Committee on Library Service, 1932, 2, Pennsylvania Library Association Collection, Division Files, Box 2 County-Trustee, County-Public Division, 1934, n.d., SLPARSC; Pennsylvania Library Association Planning Committee, Preliminary Report of a Plan for Public Library Development Presented by the Pennsylvania Library Association Planning Committee to the State Planning Board, November 1934, Pennsylvania Library Association Collection, Committee Files, Box 6, O-Public Relations, Planning Committee, 1934 – 1942, SLPARSC; Margaretta A. Elder, History of the County Library Division of the Pennsylvania Library Association, 12 October 1956, Pennsylvania Library Association Collection, Division Files, Box 2 County-Trustee, County-Public Division, 1934, n.d., SLPARSC; Pennsylvania Library Association County Library Section, Standard for the Establishment and Maintenance of County and Regional Libraries in Pennsylvania, 1953, Pennsylvania Library Association Collection, Division Files, Box 2 County-Trustee, County-Public Division, 1934, n.d., SLPARSC; Planning Committee of the Pennsylvania Library Association, Library Needs of Pennsylvania, October 1939, Pennsylvania Library Association Collection, Committee Files, Box 6, O-Public Relations, Planning Committee, 1934 – 1942, SLPARSC; Harold A. Wooster, Planning Committee's Report, 11 October 1940, Pennsylvania Library Association Collection, Executive Board Files, Box 1, Minutes, Reports, Correspondence, 1940, SLPARSC.

13 Civic Club of Allegheny County, Civic Club Board Minutes from May 1924 to 1925, 5.

14 *Lear, Made Free and Thrown Open to the Public.*

15 Carnegie Library of Pittsburgh, Annual Report, 1900.

16 Carnegie Library of Pittsburgh, Annual Report, 1930, 16, CLP-PENN.

17 Carnegie Free Library, Carnegie Free Library and Carnegie Hall Opened by the President of the United States, 13 February 1890, ACLL.

18 Joseph Frazier Wall, *Andrew Carnegie* (New York, NY: Oxford University Press, 1970), 816.

19 Carnegie Library of Pittsburgh, Annual Report, 1930, 9.

20 Harold T. Shapiro, Vivian B. Shapiro, and Stephen Ferguson, *The Davies Project at Princeton University* (https://daviesproject.princeton.edu/databases/index.html, 2015).

21 Wall, *Andrew Carnegie*, 107.

22 Edward Park Anderson, "The Intellectual Life of Pittsburgh, 1786 – 1836: V: Literature," *Western Pennsylvania Historical Magazine* 14, no. 2 (1931); Donald M. Goodfellow, "Centenary of a Pittsburgh Library," *Western Pennsylvania Historical Magazine* 31, no. 1 – 2 (1948).

23 R. V. Bradshaw, "Pittsburg's First Library," *Pittsburg Dispatch*, 20 February 1898.

24 William F. Stevens, *The Keystone State Library Association, 1901 – 1915* (1916).

25 Shapiro, Shapiro, and Ferguson, *The Davies Project at Princeton University* (2015).

26 Community Councils and Group Work Division of the Federation of Social Agencies, "Resources for Leisure Time: Circulating Libraries in Pittsburgh," *The Federator* 9, no. 8 (1935): 6.

27 Community Councils and Group Work Division of the Federation of Social Agencies, "Resources for Leisure Time: Circulating Libraries in Pittsburgh," 6.

28 Carnegie Library of Pittsburgh, Annual Report, 1910.

29 Ralph Munn, Early Efforts, 1935 – 56, a Summary, 1966, 1, Movements Toward Library Services in Allegheny County, 1935 – 1961: Documents, with Connecting Narrative, compiled and annotated by Ralph Munn, Director's Office Collection, Box 21, Folder 4, CLP-OLIVER.

30 Mildred Stahl Fletcher, "Library Service in Allegheny County Outside Pittsburgh," *The Federator* 12, no. 11 (1937): 254.

31 Civic Club of Allegheny County, Civic Club Constitution, 1895, Civic Club of Allegheny County Records Collection, Box 12, Folder 2, PITT-ASC.
32 Civic Club of Allegheny County, Fifty Years of Civic History, 1895–1945, 1945; Civic Club of Allegheny County, Civic Club Pamphlets, 1930, Civic Club of Allegheny County Records Collection, Box 12, Folder 1, PITT-ASC.
33 Civic Club of Allegheny County, Civic Club Pamphlets, 1930.
34 Civic Club of Allegheny County, Civic Club Organizational Map, 1926, Civic Club of Allegheny County Records Collection, Box 12, Folder 1, PITT-ASC.
35 Civic Club of Allegheny County, Civic Club Organizational Map, 1926.
36 Civic Club of Allegheny County, Civic Club Committees, 1958, 3, Civic Club of Allegheny County Records Collection, Box 12, Folder 1, PITT-ASC.
37 Thomas H. Reed, "Pittsburgh Plans the World's Most Complete Experiment in Federated City Government," *The American City* 36 (1927).
38 Civic Club of Allegheny County, Civic Club Board Minutes from May 1925 to May 1927, 128, Civic Club of Allegheny County Records Collection, Box 30, PITT-ASC.
39 Civic Club of Allegheny County, Civic Club Board Minutes from May 1927 to May 1929, 92, Civic Club of Allegheny County Records Collection, Box 30, PITT-ASC.
40 The American City, "Why the Federated Pittsburgh Plan Failed to Carry: Defeated Plan to Be Resubmitted," *The American City* 41 (1929): 150.
41 The American City, "Why the Federated Pittsburgh Plan Failed to Carry: Defeated Plan to Be Resubmitted," 150.
42 Civic Club of Allegheny County, Civic Club Board Minutes from May 1927 to May 1929, 92.
43 Civic Club of Allegheny County, Civic Club Board Minutes from May 1929 to April 1931, Civic Club of Allegheny County Records Collection, Box 30, PITT-ASC.
44 Civic Club of Allegheny County, Civic Club Committee Roster, 1931–1932, Civic Club of Allegheny County Records Collection. Box 12, Folder 3, PITT-ASC.
45 Civic Club of Allegheny County, Fifty Years of Civic History, 1895–1945, 1945, 24.
46 Civic Club of Allegheny County, Civic Club Board Minutes from May 1929 to April 1931.
47 Civic Club of Allegheny County, Civic Club Board Minutes from May 1929 to April 1931.
48 Civic Club of Allegheny County, Civic Club Board Minutes from May 1925 to May 1927, 128.
49 Pittsburgh Post-Gazette, Change Asked for Library on Northside, 15 February 1938, Civic Club of Allegheny County Records Collection, Box 7, Folder 64, PITT-ASC.
50 Pittsburgh Post-Gazette, Change Asked for Library on Northside, 15 February 1938; George M. P. Baird, Letter from George M. P. Baird to Mr. Demmler, 6 December 1933, Civic Club of Allegheny County Records Collection, Box 16, Folder 84, PITT-ASC.
51 Baird, Letter from George M. P. Baird to Mr. Demmler, 6 December 1933.
52 Ralph Munn, Letter from Ralph Munn to Alan Bright, 21 December 1933, Civic Club of Allegheny County Records Collection, Box 16, Folder 84, PITT-ASC.
53 Baird, Letter from George M. P. Baird to Mr. Demmler, 6 December 1933.
54 Pittsburgh Post-Gazette, Change Asked for Library on Northside, 15 February 1938.
55 Pittsburgh Post-Gazette, Civic Club Renews Library Plea, 21 March 1938, Civic Club of Allegheny County Records Collection, Box 7, Folder 64, PITT-ASC; H. M. Dermitt, Please Reserve, 23 March 1938, Civic Club of Allegheny County Records Collection, Box 7, Folder 64, PITT-ASC.
56 Civic Club of Allegheny County, Civic Club Board Minutes from May 1935 to May 1937, Civic Club of Allegheny County Records Collection, Box 30, PITT-ASC.
57 Civic Club of Allegheny County, Civic Club Board Minutes from 1 May 1937 to 30 April 1939, Civic Club of Allegheny County Records Collection, Box 30, PITT-ASC.

58 Civic Club of Allegheny County, Civic Club Board Minutes from 1 May 1937 to 30 April 1939.
59 Civic Club of Allegheny County, Civic Club Board Minutes from 5 June 1939 to 16 April 1941, Civic Club of Allegheny County Records Collection, Box 30, PITT-ASC.
60 Carnegie Library of Pittsburgh, Allegheny County Commissioners, Typescript, Vertical File, CLP-PENN.
61 Ruth Houk, Letter from Ruth Houk to Library Directors, 27 October 1938, Movements Toward Library Services in Allegheny County, 1935–1961: Documents, with Connecting Narrative, compiled and annotated by Ralph Munn, Director's Office Collection, Box 21, Folder 4, CLP-OLIVER.
62 Ralph Munn, Letter from Ralph Munn to E. L. Fuss, 2 February 1938, Movements Toward Library Services in Allegheny County, 1935–1961: Documents, with Connecting Narrative, compiled and annotated by Ralph Munn, Director's Office Collection, Box 21, Folder 4, CLP-OLIVER.
63 Ralph Munn, Legal Basis for County Library Service in Allegheny County, 2 February 1938, Movements Toward Library Services in Allegheny County, 1935–1961: Documents, with Connecting Narrative, compiled and annotated by Ralph Munn, Director's Office Collection, Box 21, Folder 4, CLP-OLIVER.
64 L. O. Grondahl, Letter from L. O. Grondahl to John J. Kane, 18 December 1939, Movements toward library services in Allegheny County, 1935–1961: Documents, with Connecting Narrative, compiled and annotated by Ralph Munn, Director's Office Collection, Box 21, Folder 4, CLP-OLIVER; Pittsburgh Press, County Fund Sought for School Libraries, 1940, Movements Toward Library Services in Allegheny County, 1935–1961: Documents, with Connecting Narrative, compiled and annotated by Ralph Munn, Director's Office Collection, Box 21, Folder 4, CLP-OLIVER.
65 Pittsburgh Sun-Telegraph, County Hears P.-T.A. Plea for Rural Library, 1940, Movements Toward Library Services in Allegheny County, 1935–1961: Documents, with Connecting Narrative, compiled and annotated by Ralph Munn, Director's Office Collection, Box 21, Folder 4, CLP-OLIVER.
66 Ruth Houk, Letter from Ruth Houk to John J. Kane, 9 January 1940, Movements Toward Library Services in Allegheny County, 1935–1961: Documents, with Connecting Narrative, compiled and annotated by Ralph Munn, Director's Office Collection, Box 21, Folder 4, CLP-OLIVER.
67 Munn, Early Efforts, 1935–56, a Summary, 1966, 4.
68 Pittsburgh Post-Gazette, Lending Library Program Studied for County Schools, 1940, Movements Toward Library Services in Allegheny County, 1935–1961: Documents, with Connecting Narrative, compiled and annotated by Ralph Munn, Director's Office Collection, Box 21, Folder 4, CLP-OLIVER.
69 Pittsburgh Press, County Asked for Extension of Libraries, 1940, Movements Toward Library Services in Allegheny County, 1935–1961: Documents, with Connecting Narrative, compiled and annotated by Ralph Munn, Director's Office Collection, Box 21, Folder 4, CLP-OLIVER.
70 Grondahl, Letter from L. O. Grondahl to John J. Kane, 18 December 1939.
71 Civic Club of Allegheny County, Board Minutes, 20 March 1940, Movements Toward Library Services in Allegheny County, 1935–1961: Documents, with Connecting Narrative, compiled and annotated by Ralph Munn, Director's Office Collection, Box 21, Folder 4, CLP-OLIVER.
72 L. O. Grondahl, Letter from L. O. Grondahl to John J. Kane, 13 March 1940, Movements Toward Library Services in Allegheny County, 1935–1961: Documents, with Connecting Narrative, compiled and annotated by Ralph Munn, Director's Office Collection, Box 21, Folder 4, CLP-OLIVER.
73 L. O. Grondahl, Letter from L. O. Grondahl to Mrs. L. G. Stone, 4 June 1941, Movements Toward Library Services in Allegheny County, 1935–1961: Documents, with Connecting Narrative, compiled and annotated by Ralph Munn, Director's Office Collection, Box 21, Folder 4, CLP-OLIVER.
74 Civic Club of Allegheny County, Civic Club Board Minutes from May 8, 1941 to March 17, 1943, Civic Club of Allegheny County Records Collection, Box 30, PITT-ASC.
75 Munn, Early Efforts, 1935–56, a Summary, 1966, 1.

Period 2: County Contract, 1945 – 1961

Throughout the 1920s and 1930s, several civil society groups, most notably the Civic Club, petitioned city and county officials in the City of Pittsburgh and Allegheny County for expanded and improved public library services. The three main issues supported by the local groups were 1) improved library services within the city led by the Carnegie Library of Pittsburgh (CLP), 2) a merger between CLP and the Carnegie Free Library of Allegheny, located on the city's North Side, and 3) an extension of public library services to underserved areas of the county. In 1939, Chairman of the Civic Club's Library Committee Lars Grondahl wrote to county commissioner John J. Kane proposing a countywide system as a guiding vision for local officials. This system had CLP as its headquarters, large regional branches, many delivery stations, several book-wagon routes, and classroom collections in schools.[1] Several groups met with the county commissioners in 1940: the Allegheny County Parent-Teachers' Association, the Allegheny County Federation of Women's Clubs, the Council on Adult Education, the League of Women Voters, and CLP. Following the meeting, the groups requested that the county take the first steps to realize the vision. The groups asked that the county fund a survey of existing and required public library services.[2] This survey was never carried out by the county, and the issue of countywide public library service was tabled until after the Second World War.[3]

The failure of library reform in Allegheny County was part of a larger trend of failed regional reform attempts in greater Pittsburgh at the time. In Pennsylvania, infrastructure is designed, funded, and governed according to the premise of local control. For the most part, the state and its counties, which are essentially local extensions of state government, in many instances take a hands off approach, leaving local municipalities to plan infrastructures—or not—as they see fit. This political philosophy is especially evident in the case with public libraries, which are legally allowed by the state but not mandated or governed by it. According to their regionalist proponents, however, the comprehensive planning initiatives that failed in 1920s Pittsburgh would have made governance and funding of infrastructures more equitable and economical. The reform efforts, which symbolized a last breath of Pittsburgh's Progressive Era, clashed with a deep-seated legal and philosophical doctrine of local control.

The failure of regional reform initiatives was not limited to the library merger proposed to the city or the countywide library plan proposed to the commissioners. The same had been true for the metropolitan initiative, led by the Commission to Study Municipal Consolidation, which was appointed by the Governor of Pennsylvania in 1923. The proposal failed to carry in the referendum in 1929. Similarly, a

https://doi.org/10.1515/9783111013404-004

bond issue was ignored by the commissioners in 1924 that would have funded major highway investments. And though the Allegheny County Planning Commission was formed in 1923, it had little power to implement community-scale development plans.[4] Because greater Pittsburgh was dominated by individual municipal authorities and private interests, regional planning initiatives, while imaginable, were struck down, and they did not gain traction in the years preceding the second great war.

By 1948, the library situation in Allegheny County had not changed, but the political landscape had. Pittsburgh politics shifted in a way that was more favorable toward large-scale building projects. This changing of the political tide buoyed numerous major building projects. Following the Second World War, Pittsburgh entered a new era in regional planning and urban development now known as the Pittsburgh Renaissance.[5] The early Renaissance years were characterized by major environmental projects, including smoke control and flood control, as well as by major building projects, such as Point State Park.[6] These years marked a major turning point in the history of the city, and public library planning was of central significance. What was distinctive about this period was its emphasis not just on industrial productivity but also on quality of life. The shift in focus brought cultural institutions such as libraries into view.

The reformers of this time considered the Pittsburgh area as a single region, and with this regional view in mind, they planned accordingly. In 1948, 1.5 million people lived in Allegheny County, whose land area was 747 square miles. About 672,000 people lived within the City of Pittsburgh, which was less than half of the county population. There were 129 municipalities within Allegheny County. These municipalities included the City of Pittsburgh, 3 other cities, 74 boroughs, and 51 townships.[7] The City of Pittsburgh had gradually expanded by annexing other surrounding municipalities since the annexation of the City of Allegheny in 1907.[8] But there was still a quickly burgeoning suburban ring in the county's outskirts. By the late 1940s and early 1950s, there was a need for more comprehensive planning programs that would both acknowledge this quickly growing segment but also cut across and orchestrate the many governmental units.

Why reform, and why this moment? The central causes attributed to Pittsburgh's Renaissance at this particular time were, first, a growing feeling of crisis with respect to the region's problems, most notably traffic control, smoke control, and flood control. The Pittsburgh region was a collection of mill towns, a legacy of industrialism. The chaotic and unfeeling sprawl of mills and railroads was never designed to support life beyond labor. It was not coordinated, nor did it foresee its own unhealthy consequences. Second, there was a new generation of business elite who sought out closer relationships with those in public office. This was to safeguard the corporate interests that they had established in the Pittsburgh region.[9] Pittsburgh was known as the "Smoky City," and it was difficult to attract and retain talented workers who were willing to live in the soot-coated area. The industrial conditions of the Pittsburgh area had to change, both to support existing life and to attract new bodies.

The group at the center of Pittsburgh's post-war rebuilding efforts was the Allegheny Conference on Community Development (ACCD), led in part by Richard King Mellon. Mellon was a businessman who was the head of T. Mellon & Sons, which controlled Gulf Oil, Koppers Company, a chemical and plastics producer, and Alcoa, the Aluminum Corporation of America. The Mellon interests also influenced U.S. Steel, Westinghouse Air Brake, Pennsylvania Railroad, Pittsburgh Plate Glass, and Pittsburgh Consolidation Coal.[10] These were some of the largest corporations in the area. Mellon and others saw the need for a "comprehensive postwar planning program."[11] Other business leaders included Dr. Edward R. Weidlein, president and director of the Mellon Institute, Arthur Van Buskirk, governor of T. Mellon & Sons and board chairman of the Cleveland Federal Reserve, and Wallace Richards, who worked with the Parking Authority and was advisor to Richard K. Mellon.[12] The Allegheny Conference was composed of 50 people drawn from the fields of "industry, commerce, finance, labor, education, and public administration."[13] The Conference was led by an Executive Committee of 14 people. Dr. Weidlein was chairman of the Conference, and Park. H. Martin was Executive Director of the Conference. Martin was a planner and engineer in the County Planning Department who served as executive director of Conference from 1945 to 1958.[14] ACCD and the corporate elite who composed it in large part supplanted the earlier voluntary organizations that had taken on the tasks of civic improvement in the early twentieth century.[15] But like the Civic Club, cultural activities such as public libraries were part of ACCD's agenda.

ACCD was a key mover in Pittsburgh politics following the war. It was a small non-profit and non-partisan civic group formed by Mellon in 1943 whose mission was regional development and planning. The ACCD was distinct from other groups in terms of its composition and its power. Not only was the group composed of corporate elite, but also all members had to participate personally in the group's deliberations. What set ACCD apart from other similar groups was its personal connections and its access to resources. It worked closely with the Pittsburgh Regional Planning Association and Pennsylvania Economy League, Western Division. ACCD had access to experts who could construct detailed plans. This contrasted with existing civic groups. Another important element of the success of the ACCD was its working relationship with David Lawrence, the mayor of Pittsburgh from 1946 to 1949. ACCD worked closely with the Democratic administration and city hall in order to make improvements to the Pittsburgh cityscape. For instance, the ACCD managed opposition from the coal industry to successfully pass smoke control measures at the city and state levels. This was not something public officials could have done alone. In addition to its corporate ties, the ACCD was also successful because it built consensus on issues to implement them successfully. This meant working closely with the press and other civic groups during the planning process.[16]

The ACCD was a natural ally regarding the issue of regional public library planning. Following the war, the Civic Club renewed its efforts to establish countywide library services. By this time, their efforts reflected a national movement to establish larger regional units for libraries. Though the Civic Club may not have been aware of it at the time, there was a push for a national plan for public libraries where regional library systems were networked to form a single structure.[17] Regional public library planning was a national issue at the time the Civic Club renewed its interest in a countywide system.

In 1948, the Library Committee of the Civic Club approached the ACCD to determine if the ACCD would fund and design a countywide survey on public library service. This was at the same time that the Allegheny County Parent-Teachers' Association adopted a resolution seeking financial aid from the county to support a bookmobile.[18] But the Civic Club's effort for a survey was precisely where the civic organizations left off in their efforts from 1940. In their letter to the ACCD, Civic Club representatives noted their coalition with the Parent-Teachers Association and proposed "mobile library units to serve the outlying districts of the County."[19] By 1948, Ralph Munn, director of CLP, had joined the Civic Club and was a member of the Library Committee. Munn, together with Stanton Belfour, director of the newly formed Pittsburgh Foundation, approached the ACCD in February of 1948. They met with Mr. Martin and Mr. Stalley, director and assistant director of the Allegheny Conference, to propose a survey.[20] The Civic Club committee then met again to formalize plans for the survey. The Civic Club agreed that the ACCD should carry out the survey because the ACCD was in a strong political and financial position: it had worked successfully with city hall before and therefore had a good working relationship with city officials; it had successfully carried out other large-scale studies; and it had access to the funding necessary to implement the survey.[21]

In May 1948, F. E. Schuchman, President of the Civic Club, formally accepted the ACCD's offer to conduct the countywide survey.[22] Park H. Martin, ACCD Executive Director, requested that the Civic Club establish the scope and content of the survey. Ralph Munn then sent more detailed information to Mr. Martin to present to the ACCD's executive board for approval. An outline of the Allegheny County Library Survey established the survey's objectives which were to detail existing public library services, evaluate them, and determine how to improve them.[23] Ralph Munn was Director of CLP from 1928–1964. He was also Dean of the Carnegie Library School at Carnegie Institute of Technology from 1928 – 1962. Munn was president of the Pennsylvania Library Association (PLA) in 1930 – 31. He was then President of American Library Association (ALA) from 1939 – 1940. Given his involvement with ALA at this time, he was aware of ALA's standards for public libraries. He was also well known for his surveys of the libraries of Australia and New Zealand and his recommendations for their improvement.[24] The final survey would furnish maps and locations of all library agencies in the county, it would include a report on school libraries, and it would examine library-related laws and ordinances.[25] Most importantly, the plan would conclude by recommending a path forward.

On June 19, 1949, the ACCD announced that they would carry out the library survey and that the survey would be funded through a $3,000 grant from the Wherrett Memorial Fund of the Pittsburgh Foundation and a $2,500 grant from the Arbuckle-Jamison Foundation.[26] The study's Advisory Committee, whose composition was recommended by Munn, included members of the Civic Club, the ACCD, librarians, and teachers from around the county.[27] By the end of July 1949, ACCD formulated guidelines for potential surveyors, stating that "No thorough survey of library service in Allegheny County has ever been made. The surveyor must find the quantity and quality of service now offered in the public libraries and the extent to which service is lacking."[28] Surveyors were then advised to recommend "a plan under which an appropriate type and amount of service can be made available throughout the county with as much uniformity as is practicable."[29]

By August, 1949, two surveyors had been selected by ACCD, both from the Ohio State Library in Columbus, Ohio. Walter Brahm was Ohio State Librarian and Mildred W. Sandoe was Ohio State Library Organizer. The two surveyors were recommended to ACCD by the ALA.[30] In August, Brahm and Sandoe contacted ACCD with their proposed terms and their proposed survey methods.[31] The survey was undertaken in late 1948 and the first half of 1949. The final report was submitted to the ACCD in September of 1950.

The survey found 31 public libraries in the county.[32] These included the two city systems, Allegheny and CLP. The survey also identified 129 governmental units within Allegheny County, including the City of Pittsburgh. This count did not include school districts. In 1940, the census population for the City of Pittsburgh was 671,659, and for the county it was 1,411,539. During the time the survey was conducted, in the late 1940s, the population in the area was thought to have increased slightly. Indeed, as census data would later show, the city population peaked around 1950, and the county population continued in an upward trajectory until around 1960.

The report found that, both legally and practically, the county could raise sufficient revenue to support a library infrastructure that met the highest library standards established by the ALA.[33] The ALA had taken the lead in developing national standards for public library funding and salaries since 1921.[34] The latest in a series of ALA standards had been published in 1948, and these standards formed the basis of the surveyor's evaluations, together with other ALA recommended book purchasing lists.[35] The surveyors found that many libraries in Allegheny County had been formed in such a way that they did not obtain adequate funding. These libraries were formed by private associations on a voluntary basis. Then, later, the private libraries petitioned public government authorities for funding. These "marriages" between private groups and local governments did not result in adequate library service for the community because funders had no stake in their development from the beginning. Of the 31 libraries surveyed, the authors found 24 libraries that were under private or semi-private control, and only 7 libraries under fully public control. Of the 31 libraries surveyed, only 5 housed the recommended number of books per capita, and few libraries fared well in terms of book quality checks, due mostly to their small collection sizes. The only exception was the CLP system. Despite the shortcomings of the smaller independent libraries, the surveyors found "little evidence of inter-library loan within Allegheny County outside of the Pittsburgh system, although independent libraries could, if they chose to, borrow from their state library at Harrisburg or from Pittsburgh."[36] The CLP system fared well in terms of quantity and quality of resources, especially in light of its linked structure. The surveyors found, overall, that there was no systematic borrowing of library materials between library centers, and that "the weaker libraries were not even aware that such service might be their responsibility."[37]

In terms of personnel training, the only library system to fare well was the CLP system, whose staff accounted for 91.6 percent of all library school graduates in the county. Almost all the library salaries across the county did not meet ALA standards, and the salaries were inconsistent across similarly situated libraries. Salaries were much lower in Allegheny County than in other comparable areas in the U.S. Furthermore, library buildings were located in such a way that library service did not adequately reach the whole county population. The hilly terrain, the natural barriers such as rivers and mountains, and the many steps in the Pittsburgh area made some library locations inaccessible to residents despite a seeming proximity. Due in part to poor coverage, but also due to the recent economic depression, Allegheny County libraries fell short in terms of recommended circulation of materials per capita. CLP circulated many more books relative to other libraries in the county. Overall, service coverage in the county fell short. Public library service areas lined the river valleys, but coverage did not extend far outside the Pittsburgh city center and into more rural parts of the county, leaving 413,011 residents outside the city without library service of any kind. The "lack of money for books, for staff, for adequate and well located service outlets… and a lack of library leadership in the county" meant that the libraries of Allegheny County overall provided inadequate library services to county residents.[38] Compared to recommended standards, the surveyors found that "Allegheny County is not supporting libraries in an adequate fashion."[39] These results confirmed the suspicions long held by members of the Civic Club. The survey demonstrated through its empirical evidence the shortcomings of public library service within Allegheny County.

In light of their findings, the authors of the survey recommended a single, centralized, county-wide library system operated as a part of the county government. The library system vision, like that proposed earlier by Ralph Munn in 1938 and Lars Grondahl in 1939, was suggested as a long-term goal. Such a system would be difficult to establish in the short term given the many independent libraries in the county and the potential legal hurdles presented by forming a centralized system. These concerns echoed those voiced by Munn over a decade earlier. The interim solution recommended by the surveyors was therefore to establish two library systems in the county: a county library system and a Pittsburgh city system. This solution seemed more feasible since the CLP system was already established, leaving only the formation of a county system to achieve the plan.[40]

To establish the county library, the surveyors recommended that the county appoint a county librarian and establish a county library for all those municipalities without an existing public library. Existing public libraries could join the county system on a voluntary basis. Again, Munn had a hand in the survey's recommendation. Given CLP's size and expertise, the county library could contract with CLP for services such as reference and research, acquisitions, and cataloging. The county would also reimburse CLP for its use by county residents living outside the city.[41] Munn sided with the provisional dual-system idea, rather than advocate for a single unified system, because he feared CLP would not be adequately administered under the county commissioners.[42] Nevertheless, Brahm and Sandoe suggested that CLP and other county libraries eventually become a single administrative unit.

Surveyors recommend that the county library 1) "improve existing library facilities," and 2) "utilize bookmobile service in providing service to unserved areas."[43] In order to broaden the coverage of library service throughout the county, the surveyors recommended additional service points, including branches and bookmobiles. The plan recommended four additional county branches in West View, Mt. Lebanon, Munhall, and McKeesport. In addition to recommended ILL services, the authors especially recommended bookmobile services to the smaller schools in the county.[44] The recommendation of bookmobile services aligned with the proposed plan by civil society groups presented to the Allegheny County Commissioners in 1940.

To fund the additional branches and the purchase of bookmobiles proposed by the plan, the surveyors suggested the use of steel company endowments reserved by Carnegie-Illinois Steel Corporation for the libraries of Braddock, Duquesne, and Homestead. The endowment money, amounting to about 1 million dollars, would form the capital to build the new libraries and purchase the bookmobiles. Maintenance of the county system branches over time would then be secured through county tax support and by building library funding into the county budget. There would be a countywide tax levied. A county bond could be raised if the steel corporation did not fund the buildings.[45] The estimated annual cost of a countywide system after the second year was $783,709.[46]

The recommendations made by the surveyors were ambitious and provocative —they called for a dramatic relocation or even elimination of some libraries throughout the county, including those in the city. The plan also called for a reorganization of the administrative and funding structure. The plan even suggested enacting legislation at the state level that would prevent the further establishment of small independent libraries and to encourage the formation of larger library systems.[47]

The final report was reviewed by the Library Committee of ACCD on September 29, 1950. A motion was approved by the advisory committee to recommend the acceptance of the survey to the Executive Committee of ACCD.[48] The Library Committee suggested that ACCD recommend to the City of Pittsburgh that the two library systems in the city be merged. The Committee also suggested that ACCD recommend to the county the formation of a county library board as a first step toward the formation of a county library system. Further, the committee acknowledged that the public should be surveyed to determine the degree of popular support for a county library system.[49]

A summary of the Allegheny County Library Survey was more widely distributed in October of 1951.[50] This summary gained some media attention, though it is unclear to what degree the plan was discussed among the wider public.[51] By the end of 1951, while the Executive Committee of ACCD moved "That the Allegheny County Library Survey be received and accepted and released at once for the information of interested officials and the public pending study of the report by the Conference," it is not clear if the report was ever formally approved or adopted by ACCD.[52] There was still excitement about the survey's proposals in 1952.[53] But there is no indication that further action was taken by ACCD, the Civic Club, or any other group to push for the survey's proposals immediately after the survey was completed.

Ralph Munn and CLP nevertheless moved forward with the idea of bookmobile services within the city. In September of 1952, CLP received a bookmobile funded by the Wherrett Memorial Fund of the Pittsburgh Foundation.[54] It was said that a city bookmobile would then lead to county bookmobile service.[55] The bookmobile obtained by the city library therefore served not only to extend service within the city, but also as a pilot for countywide bookmobile service in the future. Immediately following the Allegheny County Library Survey, however, efforts toward a countywide library system stalled. The county did not immediately fund bookmobile services outside the city.

The support for a countywide public library system seemed to dwindle immediately following the publication of the Allegheny County Library Survey in 1951. There is no indication that the Civic Club, ACCD, or CLP pressed the county commissioners for the countywide system immediately following the Brahm and Sandoe report. In fact, the results and recommendations of the survey were not forgotten, but they were instead integrated into another parallel study that was underway by the time the survey had been published. This new undertaking was the Metropolitan Study Commission of Allegheny County.

The Metropolitan Study Commission of Allegheny County was formed in 1951 by the Pennsylvania General Assembly to study the problems of urban expansion in the county.[56] A memo issued by the Commission described how regional infrastructures were necessary to address the changing population.[57] Census figures in 1950 were the largest to date: the City of Pittsburgh population had risen to 676,806 and the county population was now 1,515,237. The Metropolitan Study Commission report was the latest battle in greater Pittsburgh between local autonomists and regionally minded planners. Similar studies, such as the Commission to Study Municipal Consolidation from 1923, and other similar initiatives before it, had failed to carry due to resistance from the League of Boroughs and Townships of Allegheny County, which represented local interests.[58]

The purpose of the Metropolitan Study Commission was to examine the various governmental operations in Allegheny County to determine which should be administered locally, which should be cooperatively administered, which should be administered by special districts or authorities, and which were best administered by the county.[59] The work of the Commission was divided amongst ten committees and 157 individuals.[60] The committees were tasked with studying water supply and sanitation, public health and welfare, transportation and parking, public safety, and fiscal practices and problems, among other things. One of the committees, Committee #5, was tasked with studying the functions of libraries and recreation.

The composition of the libraries and recreation committee is noteworthy. In terms of authorship, there was significant overlap between the earlier Allegheny

County Library Survey and the new library study headed by Committee #5. The Metropolitan Study Commission was chaired by Park H. Martin, then executive director of ACCD, who had sponsored the Allegheny County Library Survey in 1948 and shepherded it toward completion until 1951. Martin was charged with soliciting membership for each committee, including Committee #5. He asked Marie Davis to serve on the committee.[61] Davis was head of public relations at CLP, and it was Davis who summarized and publicized the results and recommendations of the Brahm and Sandoe report for ACCD in 1951. Ralph Munn, director of CLP, had served as a lead consultant during the Allegheny County Library Survey. With Davis, Munn was appointed by Martin to Committee #5. Also installed on the new committee by Martin was Martin Stalley, formerly assistant director of ACCD, who now worked as director of the Photographic Library at the University of Pittsburgh. In his role as assistant director of ACCD, Stalley had worked closely with Brahm and Sandoe on the Allegheny County Library Survey project. Other members of the advisory committee for Brahm and Sandoe's library survey would also serve on Committee #5: Agnes Krarup, Supervisor of Library Services for Pittsburgh City Schools, Eleanor McCann of Duquesne University libraries, and Colonel Edward H. Gallup Jr., executive secretary of the Pittsburgh Hotel Association, chairman of the Library Committee of the Civic Club, and former Civic Club president.[62] To be sure, there were many other members on the new Libraries and Recreation Committee besides this familiar cast of characters, and certainly the committee was tasked with studying not only libraries but also parks and other cultural institutions. But the composition of the Libraries and Recreation Committee of the Metropolitan Study Commission was nevertheless largely an extension of the previous library committees that supervised the county survey. This time, however, instead of only local backing from ACCD and the Civic Club, Committee #5 operated under a formal mandate from the state legislature.

The Metropolitan Study Commission began their work in September of 1953 and continued until late 1954. The final report of the commission was presented to the governor and state legislature in 1955. Study Committee #5 first met October 21, 1953.[63] Committee members focused on how best to restructure government to address their area of concern.[64] The Allegheny County Library Survey was used as a basis for the committee's findings. Committee members worked with the Pennsylvania Economy League to distribute a questionnaire designed to update and supplement the findings from the 1950 survey.[65] The commission began the data collection stage by late 1953.[66]

While the data collection stage of the Metropolitan Study Commission was underway, there were separate and unrelated talks regarding the present and future status of CLP. In early 1954, in the midst of financial difficulty, Pittsburgh's city council proposed that the city transfer responsibility for the library to the county.[67]

It was not motivated by the CLP, but some saw it as a move that could lead to a county library system.[68] Libraries throughout the county began to plan for the eventuality that the city would cede control of its library system to the county.[69] In the end, the shift from the city to the county did not take place, and the effort ultimately failed.

By June 1954, members of Committee #5, together with members of the research staff of the Pennsylvania Economy League, produced findings regarding public libraries in Allegheny County.[70] The recommendations of Committee #5 were discussed and approved by the Committee July 1954. A final report was prepared by Ralph Munn.[71]

The study found 32 public libraries in Allegheny County.[72] This is one library more than was found on the previous Allegheny County Library Survey from 1950. Ingomar, Ingram, and Sharpsburg appeared on the list from 1950 but not on the new list, and Castle Shannon, Dravosburg, North Bessemer, and Ross Township appear on the new list but not the list from 1950. The libraries on the new list were organized according to administrative and funding type. Only 7 libraries were found to be under public control, though most of the private libraries received tax support or donations of some kind.

Given the overlapping membership of the two committees, and given how the Allegheny County Library Survey formed the basis of the new report, the findings and recommendations of the Libraries and Recreation Committee largely mimicked those of the Brahm and Sandoe report from 1950. These included how CLP was the dominant library system in the area, the county public library service overall was below recommended standards, and that it was feasible for Allegheny County to fund adequate library service.[73] The recommendations of the Metropolitan Study Commission also adhere to those proposed in the 1950 Allegheny County Library Survey. The new study recommended as a long-term goal the creation of a unified county library system. Preliminarily, it recommended the creation of a county library system that would exist alongside the CLP system, at least at first, and function cooperatively. CLP would serve as the headquarters, or nucleus, of this new system. Existing libraries would gradually join the system until coverage stretched across the county. County libraries would begin cooperative services, and county residents could freely use any public library in the county.[74] The new county library system would be tax-supported. In effect, the county would become responsible for the overall planning and development of public library services.[75] The larger Metropolitan Study Commission report made other similar recommendations to shift administration to the county.[76] Given the distinctive needs of Allegheny County government, the Metropolitan Study Commission recommended that the state constitution be amended to allow for a new county government structure there.

Like the Allegheny County Library Survey in 1950, the findings and recommendations of the Libraries and Recreation Committee received some publicity immediately following the publication of the Metropolitan Study Commission report in early 1955. There was resistance to the proposals by the elected officials of 78 boroughs and townships, the Allegheny County Boroughs Association, who opposed steps toward metropolitan government and instead supported local control.[77] Again, as in the case of the 1950 survey, there is no indication that municipal leaders or county commissioners were pressed to adopt the proposals in early 1955.

While the Metropolitan Study Commission was underway and as it drew to a close in 1955, there emerged other distinct movements for improved public library services in the city and county. In contrast to the Metropolitan Study Commission, which was led primarily by public officials, these other movements were led by local civic organizations. The first movement advocated for a merger between the two city library systems. Though the Civic Club had fought for a merger of the two libraries as early as 1926, and though their efforts toward merging the libraries were particularly strong from 1936 to 1938, these efforts failed because they did not find a receptive audience in city council. In the early 1950s, this situation had changed, and interest in a merger was rekindled in 1952 when the Finance Committee of city council proposed that the Allegheny Library be administered by CLP.[78] At this time, the Allegheny Library was still controlled by city council through its Committee on Parks, Recreation, and Libraries. CLP was controlled by an 18-member board. Each library had a separate director and operated distinctly, but both libraries received funding from the city.

In March of 1952, the North Side Community Council held an open meeting to discuss the proposal. No action was taken at that time. Then in 1954, when the director of the Carnegie Free Library of Allegheny was to retire, executive committees of the Manchester Neighborhood Council on North Side and the Central North Side Neighborhood Council petitioned city council to merge the libraries.[79] A letter-writing campaign was led by the Civic Club's E. H. Gallup, Chairman of the Library Committee.[80] The campaign asked North Side residents and workers to write to the city council in support of the library. Gallup noted the potential of bookmobile service offered to suburban residents in the North Hills area: "There is now the additional advantage that Pittsburgh's new Bookmobile could be utilized to serve portions of the North Side which are remote from the main Library and its Woods Run Branch."[81] By then, CLP had successfully operated bookmobile service in other parts of the city for two years.[82] Marion Plank, a teacher in North Hills School District, wrote in support of the merger, stating how advantageous it was for her students to have such close proximity to reference materials.[83] In late 1954, the Central North Side Neighborhood Council sponsored hearings to gather opinions for and against the merger. At the first meeting, some concerns raised were that the merger would reduce library services on the North Side and that it would lose rare books in its collection. At the second meeting, Ralph Munn shared his plans for the Allegheny Library should the merger occur. The plans included an increased level of services, an improved Children's department, bookmobile service, and interlibrary loan service.[84] By the close of the letter-writing and the local hearings, the timing seemed to be right for merging the library systems. There was support from both North Side residents and workers and from city officials.

A second parallel movement occurred while the Metropolitan Study Commission was underway and as it ended. Again, this movement was from local civic groups. These groups were from outside the city and began to agitate for extended library services in the county Munn recounts that his library was "besieged by citizens groups for help in establishing local libraries in many areas at present unserved. [He] advised and assisted groups in Penn, Richland, West Deer and other townships."[85] CLP began to receive modest appropriations from the county to work with school children in outlying areas. In 1953, for example, CLP received $20,000 from the county for outreach services.[86] By 1955, this appropriation was increased to $145,000.[87] Still, this funding covered school library services but not public library services.

One group in particular became a strong voice for countywide public library services. This group was the Library Planning Committee of Allegheny County, formed on June 28, 1955, the day after the Metropolitan Study Commission published its final report. The group was an offshoot of the Allegheny County Council of Parent-Teachers' Associations, which had met in April to work toward improved county library service.[88] A new organization was then formed at a meeting for the Library Service Committee of the Allegheny County Council of Parent-Teachers' Associations. The meeting was organized by Mrs. Howard EnDean. EnDean hoped to organize support for a county library system. At the meeting, the committee formed a new organization: "Resolve that this group form a temporary organization and appoint a temporary chairman and Steering Committee, representing the four geographic areas of Allegheny County, including the chairman, to work toward improved library service in Allegheny County."[89] The group was named the Library Planning Committee of Allegheny County. Mrs. George Karnofsky was elected chairman of the group, and Mrs. Charles Strang and Mrs. Howard EnDean were included as members of the Steering Committee. Ralph Munn was also present at this first meeting. He reported on past efforts to achieve county library services and convinced the group to pursue a contract between CLP and the county.[90] Munn, then, at least in part, orchestrated the group's efforts toward county library service and directed their efforts toward his own vision for county library service.

At their second meeting in early July, the new Library Planning Committee moved to undertake actions on three fronts in order to ensure support for county library service. These were 1) promote service in unserved areas, 2) promote countywide service through newspapers, television, and the county fair, and 3) meet with county commissioners.[91] By early September, the Library Planning Committee had obtained bookmarks and informational brochures from Marie Davis, head of public relations at CLP.[92] The bookmarks included marketing language such as "Let's fill the gaps in community libraries by supporting a countywide system" and "Talk about it! Bring it up with your neighbor! Sound out community groups! Interest your local government!" These materials were distributed at the Allegheny County Fair in September 1955 alongside a bookmobile on display.[93] Following the county fair exhibit, the Library Planning Committee continued its information campaign on a county library. Thousands of bookmarks were distributed to newcomers to Pittsburgh as part of Pittsburgh Welcome Wagon service. On October 1, the Congress of Clubs hosted a radio discussion on WJAS about bookmobile service in Allegheny.[94]

Part of the motivation for the Library Planning Committee to establish a county library system was their belief that a county library would be eligible to receive state aid and provisional books while the county library was established.[95] In 1955, 22 of Pennsylvania's 67 counties had county library systems, but Allegheny County was still not eligible for state aid for a county library.[96] This was because, though legislation was revised in 1951, state funding still only applied to counties of the third through eight classes, according to population.[97]

On October 6, 1955, the Library Planning Committee met with county commissioners where they pitched the idea to extend bookmobile service into suburban areas of the county. The commissioners expressed interest in the proposal, but they requested a plan from Ralph Munn on how the plan would be carried out. Munn drafted a plan for partial library services to county residents. In the meantime, on October 24, the Library Planning Committee conducted an interview with Munn on WQED where he discussed his ideas for the countywide plan.[98]

Munn submitted his final plan to the county commissioners on November 16, 1955. Munn's proposal was based on two considerations: "(1) to bring immediate service to the greatest number of people at the least cost, and (2) to establish only such services as will fit perfectly into whatever plan is finally adopted for a complete and unified county library system."[99] Munn's plan had 4 components that could be implemented immediately without subverting the authority of existing local libraries. These components were:

(1) Providing free borrowing privileges in CLP, including its branches, to all residents of the county.
(2) Provision of bookmobile service in some areas now lacking local libraries.

(3) An ILL service which will enable existing libraries to borrow books from CLP.
(4) A centralized book purchasing and cataloguing service for those libraries which wish it.[100]

Munn offered the commissioners descriptions and cost estimates for each of these services. The total estimated cost for the first year was $280,288 and for subsequent years it was $208,801. Munn envisioned that the county would purchase two additional bookmobiles for county service. Since 1952, CLP already successfully operated one bookmobile unit for service within the city. Munn's plan received newspaper coverage the following day.[101]

No action was taken by the commissioners on Munn's proposal until a budget hearing in March, 1956.[102] At that time, Munn appeared with many of the Library Planning Committee members. The commissioners approved the plan and agreed to appropriate $225,000 for the new services. The final agreement between the county and CLP was approved on May 22, 1956.[103] Free services to county residents began on June 1, 1956. CLP became the de facto county library. There had previously been an annual $2 charge for borrowing privileges to county residents outside the city. Kenneth E. Brown was appointed to supervise all county activities.[104] On June 20 1956, county commissioners approved funding to purchase the first bookmobile.[105] The first bookmobile unit made its initial trip January 22, 1957.[106] The second unit began its services on February 22, 1958.[107] From June to December 1956, 10,241 county residents were registered as borrowers, and about 120,000 books were lent to them. In terms of ILL, nine libraries requested 108 books during the seven months of operation. The bookmobile service was also widely utilized: "at the end of the year about 30,000 books had been purchased and processed for the bookmobiles."[108] In 1957, the bookmobile ran Monday through Saturday with 9 stops: Kennedy Township, Bridgeville, Elizabeth Borough, Whitehall, Monroeville, McKnight Road in Ross Township, Hampton, Route 19 in Ross Township, and Natrona Heights.[109]

Shortly after the county commissioners approved the contract between the county and CLP, Pittsburgh City Council made preliminary movement toward a merger of the city's two library systems. City Council was predisposed to the merger since at least 1952, and the efforts by local community councils on the North Side, together with the letter-writing campaign initiated by the Civic Club in 1954, bolstered the Council's resolve. George Seibel, director of the Allegheny Library, resigned at the end of 1954, and his position had not been filled by early 1956.[110] On June 5, 1956, City Council acted to make Ralph Munn interim director of North Side library.[111] This resolution was presented by Mrs. Irma D'Ascenzo, who also served on the board of trustees for CLP.[112] The resolution was passed and Council moved forward with the merger after 30 years of discussion and despite some resistance from North Side groups, former city council members, and the former mayor.

On October 19, on petition of the City of Pittsburgh, the Orphans' Court of Allegheny County approved a combination of the library services of CLP and those of the Carnegie Free Library of the North Side. Under this order, and with an agreement between the City of Pittsburgh and the Trustees of CLP, the administration of library services of the North Side library was undertaken by CLP. The Carnegie Free Library of Allegheny came to be called the Allegheny Regional Branch Library of CLP. Anthony A. Martin, who had worked as administrative assistant to Ralph Munn, became director of North Side branches at the time of the merger.[113]

Due to residual concerns that the merger of the two city library systems would diminish the services of the Allegheny Regional branch, a local advisory committee was established as a way to guide its governance and to liaise between North Side residents and the library.[114] In a single year, then, two long-discussed issues were successfully implemented in Allegheny County: preliminary library services for county residents and a merger between the City of Pittsburgh's two library systems. The Allegheny County arrangement reflected a broader, national, and long-standing movement for library extension.[115]

For Munn, the Library Planning Committee, and other civic groups, the 1956 county contract with CLP and the merger between the two city libraries were only the first steps toward a unified county library system. The Allegheny County Library Survey from 1950 and the Metropolitan Study Commission report from 1955 both advocated unified systems, a vision that had not yet been achieved. Bookmobile service within the county, therefore, while a significant achievement, was merely part of a grander vision.

In early 1957, Munn proposed to the county commissioners a plan for a federated library system as an alternative to a completely unified county library service.[116] This plan marked a significant shift in Munn's thinking. Munn began to develop the federated system plan as early as 1956. This system could include centralized cataloging and processing, advisory services, coordinated selection and purchasing of books, and other shared services.[117] Munn worked closely with the Library Planning Committee to further strengthen library services in the county.[118] He also reached out to librarians and trustees in the county to advance a federated plan. These efforts continued into late 1957.[119] In 1958, Munn publicized his plan for a federated library system in Allegheny County.[120] Librarians within the county met to discuss federation on June 5.[121] At this meeting, librarians discussed with Munn the basic elements of Munn's federated plan.

Despite seeming interest in federation, the plan was halted in 1958 because by that time a new state plan was working its way through the state legislature.[122] The new Pennsylvania state plan would be based on the state library survey, led by Lowell A. Martin, that was begun in 1957 and completed in 1958.[123]

Legislation for the Pennsylvania State Plan took several years to be finalized. Meanwhile, the Library Planning Committee reconvened in 1960 to discuss possible improvements to county library services. The group renamed to the Citizens Committee for County Library Service.[124] Bookmobile service, interlibrary loan service, and usage by non-city residents had increased following the countywide services established in 1956, but committee members were still concerned that no other county library services had been initiated since that time. Ralph Munn and committee members met with the county commissioners on May 13, 1960. The committee expressed support for a loose federation of public libraries that would receive county support. They also petitioned for the county to establish four area libraries. The following day, the commissioners authorized Munn to present a further plan for the extension of library services in Allegheny County.[125] The Citizens Committee received favorable newspaper coverage for their proposal. At this time in the state legislature, a bill for state funding to libraries was under consideration.[126] At this time, however, no library in Allegheny County yet received state aid, and only CLP received funding from the county. The proposition that other local libraries could receive county or state aid was merely hypothetical, at least until the proposed state legislation was signed into law.

Ralph Munn nevertheless responded to the commissioners' request for a study with recommendations, which included regional reference centers, field workers, library institutes, and a new organization of public libraries based on school districts.[127] Munn's plan to fund new public libraries on school units was particularly ambitious and thoughtful. One motivation for Munn's 1961 plan was his fear that if the county established four regional reference centers, there would be little incentive for local communities to establish and fund their own libraries. This would undercut Munn's vision to extend library service coverage across the county through a federated system consisting of many public libraries established locally. Munn also feared that the county could not fund such a large undertaking as building many local libraries, and moreover that county funding would be redirected away from CLP. CLP provided significant county services that still needed to be funded by the county. In Munn's view, since there already existed 32 local libraries in the county outside the city, he thought it fair for the other municipalities without libraries to establish and fund their own rather than have the county pay. The challenge, then, was how to establish enough local libraries across the county, each with a sufficient tax base, and in such a way that they would provide sufficient public library coverage.

Munn's solution was to organize and fund local libraries using existing school districts. He grouped the county's school districts into 32 "school units." Each unit consisted of one or more school districts. Under Pennsylvania law, school districts are a type of special government unit with the power to levy taxes. They could therefore establish and fund not only schools but also public libraries. Munn organized each school unit with a large enough property tax base to support a local public library. Tax funding could also be shared between not only school districts, but also the boroughs, towns, and cities within each unit. Munn's idea was that "the school unit is recommended as the area of service, but not necessarily as the governing and financing authority of local libraries."[128] Different units might establish and fund their public library slightly differently. Startup costs could also be defrayed in other ways besides local millage: "The cost of erecting new buildings, or enlarging existing ones, must be met through bond issues or other sources. The cost of the initial book collection will also require extra financing."[129] Location was another important factor to consider. With all these considerations in mind, Munn established 32 school units. Some of these units already had existing public libraries that would serve as the library for that unit.

Munn's plan was not fully implemented by the commissioners. This was because the new state plan eventually passed the legislature in 1961. The state plan would encompass two of Munn's ideas from 1961: field workers and librarian training. Indeed, when writing his 1961 plan, Munn was cognizant of how the state plan would likely reshape the role of CLP, specifically its role as a district library. Under the state plan, as a district library, CLP would receive state funding to provide field workers and librarian training for surrounding libraries. Munn's idea of a regional reference center was not encompassed by the state plan, but it is an idea that would reappear a decade later. Finally, the idea of rearranging public library funding and basing it on 32 school units, though visionary, innovative, and ambitious, would not be part of the state plan, and it would never be realized in Allegheny County.

Period 2 is characterized by a steady flow of library studies and civic action, both of which advocated for improved public library service in Allegheny County. The Allegheny County Library Survey, sponsored by ACCD and completed in 1950, was the first substantive county library study conducted. Then followed the Metropolitan Study Commission survey in 1955, and then two studies by Munn, one in 1958 and one in 1961. Both studies by Munn were backed by a new citizens group, the Library Planning Committee, formed in 1955, later to be reactivated in 1960 under the name Citizens Committee for County Library Service.

The movement toward a regional public library infrastructure in Allegheny County during Period 2 was almost entirely locally driven. The nationally published ALA public library standards played a small role in the evaluation of library services in the county. There were other seemingly high-level national movements happening concurrently, and some of the national trends were reflected by the movement for a regional library infrastructure in Allegheny County, though these national trends were not its impetus. These national events included the formation of a Library Service Division in the U.S. Office of Education in 1937, the creation of national planning documents by the ALA in 1935, 1939, and 1948, the Public Library Inquiry that was begun in 1946, the formation of ALA's Public Libraries Division in 1944, and a movement toward federal funding of libraries that eventually culminated in the Library Services Act of 1956.[130] Local leaders in Allegheny County, such as CLP director Ralph Munn, were at the forefront of the national-level documents. Munn was involved in the committees that developed the national plan, for example. While these larger national events, publications, and laws may have operated in the background for some planners in Allegheny County, national-level developments were not significant drivers of change in Allegheny County during this time. This is evident in the local discussions. Though similar national trends were underway concurrently at a national scale, the issue of a regional infrastructure in Allegheny was itself locally driven.

Library advocates achieved two major wins in 1956. First, county commissioners agreed to fund partial public library services within the county through a contract with CLP. Second, city council acted to merge the two city library systems, making the North Side libraries branches of CLP. Munn and the members of the Citizens Committee for County Library Service later proposed more elaborate plans for county public library service, but these plans were interrupted by the new Pennsylvania state plan that was implemented in 1961.

What is significant about the second decision cycle of the greater Pittsburgh case, Period 2, is the change in the level of responsiveness by decision makers. In contrast to Period 1, where the issue of a regional public library infrastructure failed, in Period 2, the Allegheny County Commissioners and Pittsburgh City Council were receptive to the issue. Other conditions had not changed: the two periods exhibited high civil support, the issues had high legitimacy and were supported through open and inclusive dialogue, and there was not notable resistance in either period to the issue. What changed in Period 2 was the nature of the decision-making cores. There was a change in the composition of the membership of the cores, and correspondingly, there was an accompanying shift in mindset. In Period 2, decision makers were susceptible to the communicative power that targeted them, thereby opening their gates to the issue and implementing it.

In Period 2, the renewed efforts to establish the regional public library infrastructure as part of the inner periphery of the county circuit were successful. County commissioners contracted with CLP to form a county library, and thus CLP became part of the inner periphery of the Allegheny County circuit, in effect carrying out and implementing county policy at the behest of the county commissioners. The communicative power transmitted to county commissioners was transformed into administrative power, and as a result, the structure of the county circuit changed so that CLP became part of the county circuit's inner periphery.

1 L. O. Grondahl, Letter from L. O. Grondahl to John J. Kane, 18 December 1939, Movements toward library services in Allegheny County, 1935–1961: Documents, with Connecting Narrative, compiled and annotated by Ralph Munn, Director's Office Collection, Box 21, Folder 4, CLP-OLIVER.
2 L. O. Grondahl, Letter from L. O. Grondahl to John J. Kane, 13 March 1940, Movements Toward Library Services in Allegheny County, 1935–1961: Documents, with Connecting Narrative, compiled and annotated by Ralph Munn, Director's Office Collection, Box 21, Folder 4, CLP-OLIVER.
3 L. O. Grondahl, Letter from L. O. Grondahl to Mrs. L. G. Stone, 4 June 1941, Movements Toward Library Services in Allegheny County, 1935–1961: Documents, with Connecting Narrative, compiled and annotated by Ralph Munn, Director's Office Collection, Box 21, Folder 4, CLP-OLIVER; Civic Club of Allegheny County, Civic Club Board Minutes from May 8, 1941 to March 17, 1943, Civic Club of Allegheny County Records Collection, Box 30, PITT-ASC.
4 Roy Lubove, *Twentieth-Century Pittsburgh: Government, Business, and Environmental Change* (Pittsburgh: Wiley, 1969), 96.
5 David L. Lawrence, "Rebirth," in *Pittsburgh: The Story of an American City*, ed. Stefan Lorant (Garden City: Doubleday, 1964), 373.
6 Lubove, *Twentieth-Century Pittsburgh: Government, Business, and Environmental Change*, 111; Lawrence, "Rebirth."
7 Marshall Stalley, "Pittsburgh and the Allegheny Conference Effort toward a Unified Community Program for the Region," *Landscape Architecture Magazine* 38, no. 4 (1948, July).
8 Lubove, *Twentieth-Century Pittsburgh: Government, Business, and Environmental Change*, 97.
9 Lubove, *Twentieth-Century Pittsburgh: Government, Business, and Environmental Change*, 105–06.
10 Lawrence, "Rebirth," 379.
11 Lubove, *Twentieth-Century Pittsburgh: Government, Business, and Environmental Change*, 108.
12 Lawrence, "Rebirth," 380–81.
13 Stalley, "Pittsburgh and the Allegheny Conference Effort toward a Unified Community Program for the Region," 148.
14 Stalley, "Pittsburgh and the Allegheny Conference Effort toward a Unified Community Program for the Region," 148; Lawrence, "Rebirth," 403.
15 Lubove, *Twentieth-Century Pittsburgh: Government, Business, and Environmental Change*, 106.
16 Lubove, *Twentieth-Century Pittsburgh: Government, Business, and Environmental Change*, 110–11.
17 Lowell A. Martin, "A Plan for Public Library Service in America," *ALA Bulletin* 40, no. 8 (1946); Committee on Post-War Planning, *A National Plan for Public Library Service: Prepared for the Committee on Postwar Planning of the American Library Association* (Chicago: American Library Association, 1948).
18 "County Notes," *Pennsylvania Library Association Bulletin* 4, no. 3–4 (Autumn-Winter 1948).
19 Fred E. Schuchman, Letter from F. E. Schuchman to Park H. Martin, 8 May 1948, Movements Toward Library Services in Allegheny County, 1935–1961: Documents, with Connecting Narrative, compiled and annotated by Ralph Munn, Director's Office Collection, Box 21, Folder 4, CLP-OLIVER.

20 Ralph Munn, Early Efforts, 1935 – 56, a Summary, 1966, 34, Movements Toward Library Services in Allegheny County, 1935 – 1961: Documents, with Connecting Narrative, compiled and annotated by Ralph Munn, Director's Office Collection, Box 21, Folder 4, CLP-OLIVER.
21 Munn, Early Efforts, 1935 – 56, a Summary, 1966, 34 – 35.
22 Schuchman, Letter from F. E. Schuchman to Park H. Martin, 8 May 1948.
23 Ralph Munn, Letter from Ralph Munn to Park H. Martin, 27 July 1948, Movements Toward Library Services in Allegheny County, 1935 – 1961: Documents, with Connecting Narrative, compiled and annotated by Ralph Munn, Director's Office Collection, Box 21, Folder 4, CLP-OLIVER.
24 Ralph Munn and John Barr, *New Zealand Libraries: A Survey of Conditions and Suggestions for Their Improvement* (Christchurch: Libraries Association of New Zealand, 1934); Ralph Munn and Ernest Roland Pitt, *Australian Libraries: A Survey of Conditions and Suggestions for Their Improvement* (Melbourne: Australian Council for Education Research, 1935).
25 Munn, Early Efforts, 1935 – 56, a Summary, 1966, 38.
26 Allegheny Conference on Community Development, Press Release on Allegheny County Library Survey, 19 June 1949, Movements Toward Library Services in Allegheny County, 1935 – 1961: Documents, with Connecting Narrative, compiled and annotated by Ralph Munn, Director's Office Collection, Box 21, Folder 4, CLP-OLIVER.
27 Ralph Munn, Letter from Ralph Munn to Park H. Martin, 1949, May 27, Movements toward library services in Allegheny County, 1935 – 1961: Documents, with connecting narrative; Allegheny Conference on Community Development, Press Release on Allegheny County Library Survey, 19 June 1949; Walter T. Brahm and Mildred W. Sandoe, *Allegheny County Library Survey* (1950).
28 Munn, Early Efforts, 1935 – 56, a Summary, 1966, 57.
29 Munn, Early Efforts, 1935 – 56, a Summary, 1966, 57.
30 Munn, Early Efforts, 1935 – 56, a Summary, 1966, 33.
31 Mildred W. Sandoe, Letter from Mildred W. Sandoe to Marshall Stalley, 18 August 1949, Movements Toward Library Services in Allegheny County, 1935 – 1961: Documents, with Connecting Narrative, compiled and annotated by Ralph Munn, Director's Office Collection, Box 21, Folder 4, CLP-OLIVER.
32 Brahm and Sandoe, *Allegheny County Library Survey* (1950), iv, 14 – 27.
33 Brahm and Sandoe, *Allegheny County Library Survey* (1950), 6 – 7.
34 ALA Council adopted resolutions on public library revenues and public library salaries at the Mid-Winter Meetings in 1921 and 1922.
35 American Library Association, "Standards for Public Libraries," *Bulletin of the American Library Association* 27, no. 11 (1933); Committee on Post-War Planning, *Post-War Standards for Public Libraries* (Chicago: American Library Association, 1943).
36 Brahm and Sandoe, *Allegheny County Library Survey* (1950), 41.
37 Brahm and Sandoe, *Allegheny County Library Survey* (1950).
38 Brahm and Sandoe, *Allegheny County Library Survey* (1950), 98.
39 Brahm and Sandoe, *Allegheny County Library Survey* (1950), 120.
40 Brahm and Sandoe, *Allegheny County Library Survey* (1950), 155 – 56.
41 Brahm and Sandoe, *Allegheny County Library Survey* (1950), 165.
42 Munn, Early Efforts, 1935 – 56, a Summary, 1966, 65.
43 Brahm and Sandoe, *Allegheny County Library Survey* (1950), 157.
44 Brahm and Sandoe, *Allegheny County Library Survey* (1950), 168.
45 Marie A. Davis, *Libraries in Allegheny County Today and Tomorrow: A Summary of the Allegheny County Library Survey by Walter T. Braham and Mildred W. Sandoe* (1951), 9; Brahm and Sandoe, *Allegheny County Library Survey* (1950), 159 – 60.

46 Brahm and Sandoe, *Allegheny County Library Survey* (1950), 162.
47 Brahm and Sandoe, *Allegheny County Library Survey* (1950), 167.
48 Marshall Stalley, Minutes of the Meeting of the Library Committee of the Allegheny Conference on Community Development, 29 September 1950, 4, Movements Toward Library Services in Allegheny County, 1935–1961: Documents, with Connecting Narrative, compiled and annotated by Ralph Munn, Director's Office Collection, Box 21, Folder 4, CLP-OLIVER.
49 Stalley, Minutes of the Meeting of the Library Committee of the Allegheny Conference on Community Development, 29 September 1950.
50 Davis, *Libraries in Allegheny County Today and Tomorrow: A Summary of the Allegheny County Library Survey by Walter T. Braham and Mildred W. Sandoe* (1951).
51 "Combined Library System Needed in County, Specialists Contend," *Pittsburgh Press*, 4 October 1951.
52 Davis, *Libraries in Allegheny County Today and Tomorrow: A Summary of the Allegheny County Library Survey by Walter T. Braham and Mildred W. Sandoe* (1951), i.
53 Ella English Daub, "The Allegheny County Library Survey," *Carnegie Magazine* 26, no. 1 (1952).
54 Dorothy Kenneweg, "Here Comes Your Library on Wheels," *Carnegie Magazine*, 1952.
55 Kenneweg, "Here Comes Your Library on Wheels," 274.
56 Munn, Early Efforts, 1935–56, a Summary, 1966, 90; Metropolitan Study Commission of Allegheny County, *An Urban Home Rule Charter for Allegheny County: A Report of the Metropolitan Study Commission of Allegheny County* (Pittsburgh, 1955).
57 Metropolitan Study Commission of Allegheny County, A Prospectus for the Study of Metropolitan Problems of Allegheny County, 17 September 1953, District Services Collection, Series 2, Box 4, Folder 6, CLP-OLIVER.
58 Lubove, *Twentieth-Century Pittsburgh: Government, Business, and Environmental Change*, 97–100.
59 Metropolitan Study Commission of Allegheny County, A Prospectus for the Study of Metropolitan Problems of Allegheny County, 17 September 1953.
60 David H. Kurtzman, Present Status and Future Plans of the Metropolitan Study Commission of Allegheny County, 7 December 1953, District Services Collection, Series 2, Box 4, Folder 6, CLP-OLIVER.
61 Park H. Martin, Letter from Park H. Martin to Marie A. Davis, 8 October 1953, District Services Collection, Series 2, Box 4, Folder 6, CLP-OLIVER; Martin, Letter from Park H. Martin to Marie A. Davis, 8 October 1953; Marie A. Davis, Letter from Marie A. Davis to Park H. Martin, 6 October 1953, District Services Collection, Series 2, Box 4, Folder 6, CLP-OLIVER.
62 Brahm and Sandoe, *Allegheny County Library Survey* (1950); Allegheny Conference on Community Development, Press Release on Allegheny County Library Survey, 19 June 1949; Stalley, Minutes of the Meeting of the Library Committee of the Allegheny Conference on Community Development, 29 September 1950; Martin, Letter from Park H. Martin to Marie A. Davis, 8 October 1953; Schuchman, Letter from F. E. Schuchman to Park H. Martin, 8 May 1948; Pennsylvania Economy League, "The League Gets a Big New Job: Research for the Metropolitan Study Commission of Allegheny County," *P.E.L. Newsletter for Western Pennsylvania*, September-October 1953.
63 Martin, Letter from Park H. Martin to Marie A. Davis, 8 October 1953.
64 Metropolitan Study Commission of Allegheny County, Metropolitan Study Commission of Allegheny County, Tentative Check List for Field Study, 8 October 1953, District Services Collection, Series 2, Box 4, Folder 6, CLP-OLIVER.
65 Ralph Munn, Letter from Ralph Munn to David H. Kurtzman, 18 November 1953, District Services Collection, Series 2, Box 4, Folder 6, CLP-OLIVER; The Study Committee on Libraries Parks and

Recreation Facilities of the Metropolitan Study Commission, Findings of Study of Public Libraries in Allegheny County: For the Study Committee on Libraries, Parks, and Recreation Facilities of the Metropolitan Study Commission, June 1954, District Services Collection, Series 2, Box 6, Manual 3, CLP-OLIVER.

66 Kurtzman, Present Status and Future Plans of the Metropolitan Study Commission of Allegheny County, 7 December 1953.

67 Munn, Early Efforts, 1935–56, a Summary, 1966, 84.

68 "Library Shift May Be Illegal," *Pittsburgh Sun-Telegraph*, 18 February 1954.

69 Ralph Munn, Letter from Ralph Munn to Librarians of Allegheny County, 2 April 1954, District Services Collection, Series 2, Box 4, Folder 6, CLP-OLIVER.

70 The Study Committee on Libraries Parks and Recreation Facilities of the Metropolitan Study Commission, Findings of Study of Public Libraries in Allegheny County, June 1954.

71 R. Otto Amann, Letter from R. Otto Amann to Members of Study Committee #5, 5 August 1954, District Services Collection, Series 2, Box 4, Folder 6, CLP-OLIVER.

72 The Study Committee on Libraries Parks and Recreation Facilities of the Metropolitan Study Commission, Findings of Study of Public Libraries in Allegheny County, June 1954.

73 The Study Committee on Libraries Parks and Recreation Facilities of the Metropolitan Study Commission, Findings of Study of Public Libraries in Allegheny County, June 1954, Appendix A.

74 Amann, Letter from R. Otto Amann to Members of Study Committee #5, 5 August 1954.

75 The Study Committee on Libraries Parks and Recreation Facilities of the Metropolitan Study Commission, Findings of Study of Public Libraries in Allegheny County, June 1954., Appendix B.

76 Metropolitan Study Commission of Allegheny County, *An Urban Home Rule Charter for Allegheny County: A Report of the Metropolitan Study Commission of Allegheny County* (1955).

77 "New Report Lists Many Objections to Charter Plan," *Pittsburgh Press*, 1 October 1955; Munn, Early Efforts, 1935–56, a Summary, 1966, 90.

78 Central North Side Neighborhood Council, Report of Temporary Library Committee, 18 January 1955, District Services Collection, Series 2, Box 4, Folder 6, CLP-OLIVER.

79 Central North Side Neighborhood Council, Report of Temporary Library Committee, 18 January 1955.

80 E. H. Gallup, Letter from E. H. Gallup to Fabian C. Mcintosh, 8 July 1954, Civic Club of Allegheny County Records Collection, Box 19, Folder 134, PITT-ASC.

81 Gallup, Letter from E. H. Gallup to Fabian C. Mcintosh, 8 July 1954.

82 Donald B. Hirsch, "Community Center on Wheels," *Carnegie Magazine*, 1954; Kenneweg, "Here Comes Your Library on Wheels."

83 Marion S. Plank, Letter from Marion S. Plank to Civic Club of Allegheny County, 30 July 1954, Civic Club of Allegheny County Records Collection, Box 19, Folder 134, PITT-ASC.

84 Central North Side Neighborhood Council, Report of Temporary Library Committee, 18 January 1955.

85 Munn, Early Efforts, 1935–56, a Summary, 1966, 94.

86 Allegheny County Controller, Controller's 93rd Annual Report of the Fiscal Affairs of Allegheny County: For the Year Ending December 31, 1953, DETRE.

87 County Controller, Controller's Annual Report for the Fiscall Affairs of Allegheny County: For the Year Ending December 31, 1955, DETRE.

88 Ralph Munn, Larger Units of Library Administration: What Has Happened in Allegheny County, October 1955, District Services Collection, Series 2, Box 4, Folder 6, CLP-OLIVER.

89 Florence Karnofsky, Minutes of Library Meeting, 28 June 1955, Movements Toward Library Services in Allegheny County, 1935 – 1961: Documents, with Connecting Narrative, compiled and annotated by Ralph Munn, Director's Office Collection, Box 21, Folder 4, CLP-OLIVER.
90 Munn, Early Efforts, 1935 – 56, a Summary, 1966, 97.
91 Dorothea S. EnDean, Minutes of Library Meeting, 11 July 1955, Movements Toward Library Services in Allegheny County, 1935 – 1961: Documents, with Connecting Narrative, compiled and annotated by Ralph Munn, Director's Office Collection, Box 21, Folder 4, CLP-OLIVER.
92 Marie A. Davis, Letter from Marie A. Davis to Mrs. George Karnofsky, 9 September 1955, District Services Collection, Series 2, Box 4, Folder 6, CLP-OLIVER.
93 Dorothea S. EnDean, Minutes of the Library Meeting, 1 December 1955, District Services Collection, Series 2, Box 4, Folder 6, CLP-OLIVER.
94 Library Planning Committee of Allegheny County, Newsletter of the Library Planning Committee of Allegheny County, September 1955, Movements Toward Library Services in Allegheny County, 1935 – 1961: Documents, with Connecting Narrative, compiled and annotated by Ralph Munn, Director's Office Collection, Box 21, Folder 4, CLP-OLIVER.
95 Pennsylvania State Library Extension Division, Why Our County Should Have a County Library, 1955, District Services Collection, Series 2, Box 4, Folder 6, CLP-OLIVER.
96 Karnofsky, Minutes of Library Meeting, 28 June 1955.
97 "Milestones in Legislative Support of Pennsylvania Public Libraries," *PLA Bulletin* (January 1981); No. 621, Laws of the General Assembly of the Commonwealth of Pennsylvania passed at the Session of 1951 (1951).
98 Library Planning Committee of Allegheny County, Newsletter of the Library Planning Committee of Allegheny County, September 1955; "Book-Mobile Service May Be Extended," *Pittsburgh Press*, 7 October 1955.
99 Ralph Munn, Letter from Ralph Munn to County Commissioners, 16 November 1955, Movements Toward Library Services in Allegheny County, 1935 – 1961: Documents, with Connecting Narrative, compiled and annotated by Ralph Munn, Director's Office Collection, Box 21, Folder 4, CLP-OLIVER.
100 Munn, Letter from Ralph Munn to County Commissioners, 16 November 1955.
101 "County-Wide Library Plan Advocated," *Pittsburgh Post-Gazette*, 17 November 1955; "New Library Plan," *Pittsburgh Press*, 17 November 1955.
102 Munn, Early Efforts, 1935 – 56, a Summary, 1966, 116.
103 County Commissioners Meeting Minutes, 2 January 1956 to 27 December 1956 Inc., Vol. 37, 1956, May 22, DETRE.
104 Carnegie Library of Pittsburgh, Annual Report, 1956, CLP-PENN.
105 "Bookmobile for County Service Ok'd," *Pittsburgh Post-Gazette* (Pittsburgh, PA), 20 June 1956.
106 Kenneth Brown, "Books on Wheels," *Carnegie Magazine*, 1958, 45.
107 Munn, Early Efforts, 1935 – 56, a Summary, 1966, 116.
108 Carnegie Library of Pittsburgh, Annual Report, 1956, 11.
109 "Allegheny County Bookmobile Schedule," *Carnegie Magazine*, 1957.
110 "Step to Merge Libraries Due in Council," *Pittsburgh Press*, 11 March 1956.
111 "Council Acts for Merger of Libraries," *Pittsburgh Post-Gazette*, 5 June 1956.
112 Carnegie Library of Pittsburgh, Annual Report, 1956.
113 Carnegie Library of Pittsburgh, Annual Report, 1956, 18.
114 Carnegie Library of Pittsburgh, Annual Report, 1956, 10.
115 Carleton Bruns Joeckel, *Library Extension: Problems and Solutions: Papers Presented before the Library Institute at the University of Chicago, August 21 – 26, 1944* (Chicago: University of Chicago Press, 1946); Gretchen Knief Shenk, *County and Regional Library Development* (Chicago: American

Library Association, 1954); Nelson Associates, *Public Library Systems in the United States: A Survey of Multijurisdictional Systems* (Chicago: American Library Association, 1969).

116 "Library Expansion, Unification Planned," *Pittsburgh Post-Gazette*, 7 February 1957.

117 Kenneth Brown, "The New Library Service Plan," *Carnegie Magazine*, 1956, 228 – 29.

118 Steering Committee of the Library Planning Committee, Letter from the Steering Committee to Library Planning Committee of Allegheny County, 29 March 1957, District Services Collection, Series 2, Box 4, Folder 5, CLP-OLIVER.

119 Ralph Munn, Letter from Ralph Munn to Librarians of Allegheny County, 1957, District Services Collection, Series 2, Box 4, Folder 5, CLP-OLIVER; Munn, Letter from Ralph Munn to Librarians of Allegheny County, 1957.

120 Ralph Munn, *A Plan for the Federation of Libraries in Allegheny County*, Carnegie Library of Pittsburgh (Pittsburgh, 21 April 1958); Ralph Munn, "Libraries for Allegheny County," *Carnegie Magazine*, 1958.

121 Munn, Early Efforts, 1935 – 56, a Summary, 1966, 133.

122 Carnegie Library of Pittsburgh, Annual Report, 1958, 9, CLP-PENN.

123 Pennsylvania State Library, Summary of Recommendations of the Pennsylvania Library Survey, 1958, District Services Collection, Series 2, Box 4, Folder 5, CLP-OLIVER.

124 Dorothea M. Strang, Citizens Committee for County Library Service, Minutes of Meeting, 19 April 1960, Movements Toward Library Services in Allegheny County, 1935 – 1961: Documents, with Connecting Narrative, compiled and annotated by Ralph Munn, Director's Office Collection, Box 21, Folder 4, CLP-OLIVER.

125 M. W. Snyder, Letter to the Members of the Citizens Committee for County Library Service, 14 May 1960, Movements Toward Library Services in Allegheny County, 1935 – 1961: Documents, with Connecting Narrative, compiled and annotated by Ralph Munn, Director's Office Collection, Box 21, Folder 4, CLP-OLIVER.

126 "Unmet Library Needs," *Pittsburgh Post-Gazette*, 20 May 1960.

127 Ralph Munn, Public Library Service in Allegheny County: A Study, with Recommendations, Made at the Request of the Board of County Commissioners, 1961, District Services Collection, Series 2, Box 2, Folder 6, CLP-OLIVER.

128 Munn, Public Library Service in Allegheny County, 1961, 6.

129 Munn, Public Library Service in Allegheny County, 1961, 6.

130 American Library Association, "Standards for Public Libraries."; Committee on Post-War Planning, *Post-War Standards for Public Libraries*; American Library Association, "A National Plan for Libraries," *Bulletin of the American Library Association* 29, no. 2 (1935): 91; American Library Association, "A National Plan for Libraries," *ALA Bulletin* 33, no. 2 (1939): 146; Committee on Post-War Planning, *A National Plan for Public Library Service*, 19; Douglas Raber, *Librarianship and Legitimacy: The Ideology of the Public Library Inquiry* (Westport: Greenwood, 1997); Robert D. Leigh, *The Public Library in the United States: The General Report of the Public Library Inquiry* (New York: Columbia University Press, 1950); American Library Association, *Public Library Service: A Guide to Evaluation, with Minimum Standards* (Chicago: ALA, 1956); Lowell A. Martin, "Standards for Public Libraries," *Library Trends* 21, no. 2 (1972); Peggy Sullivan, *Carl H. Milam and the American Library Association* (New York: H. W. Wilson, 1976); Redmond Kathleen Molz, *National Planning for Library Service, 1935 – 1975* (Chicago: American Library Association, 1984).

Period 3: The Rise of District Services, 1961–1970

By 1961, a regional public library infrastructure in greater Pittsburgh had begun to take shape. As seen in Period 2, the two city library systems merged to form a single system administered by the Carnegie Library of Pittsburgh (CLP). Then, with a new contract established between CLP and the Allegheny County Commissioners, all Allegheny County residents enjoyed free borrowing privileges from CLP. There was interlibrary loan between CLP and the smaller local libraries outside the City of Pittsburgh, making it easier for users to acquire more specialized materials from their local libraries. CLP provided bookmobile services to some areas in the county that did not have a local public library. This service improved the overall coverage within the county. And CLP offered centralized book purchasing and cataloging services for other area libraries. This allowed the smaller county libraries to save money on staffing and training, and by saving time on technical tasks, they could spend more time on front-line services. These were significant first steps toward a regional public library infrastructure.

In the eyes of some, however, these achievements in library service were only preliminary steps toward a much grander vision. Their vision was a more robust countywide system. Since Munn's and Grondahl's proposals to the county commissioners in 1938 and 1939, both local civic groups and CLP had called for a system of public libraries in the county. Several county studies since that time, including the one by Brahm and Sandoe in 1950, had called for a countywide system. In 1961, Ralph Munn submitted a detailed plan to the county commissioners outlining how a countywide system could be formed yet remain funded and administered locally. All of the proposed plans varied in their details. Some proposed a unified system while others imagined a federated one. Regardless, a library system did not yet exist in greater Pittsburgh in 1961. Certainly, the libraries in greater Pittsburgh were not yet part of a state system of libraries. This was because there was not yet a formalized public library administrative entity at the state level. In Period 3, that new state entity would emerge, resulting in lasting effects for the public library infrastructure in Allegheny County.

As demonstrated in Period 1 and Period 2, the movement toward a regional public library infrastructure in greater Pittsburgh was locally driven. There was little to no state or federal influence on the local scene. Any impact from relevant state organizations and state laws was largely absent in Allegheny County during that time. Prior to the emergence of a strong state agency presence in the late 1950s and early 1960s, the Pennsylvania State Library was still a fledgling agency whose efforts focused on the maintenance of the state library proper. Statewide planning and initiatives for public library services were not led by some other stable gov-

https://doi.org/10.1515/9783111013404-005

ernment bureau, either. Instead, statewide initiatives were spearheaded by adjacent organizations, most notably the Pennsylvania Library Association (or PLA, formerly Keystone State Library Association, now PaLA) and the Pennsylvania Free Library Commission (PFLC), which was a standalone state agency. But the work of these organizations, together with the legislation they proposed, supported, or were associated with, focused primarily on rural public library development, not urban libraries such as those in greater Pittsburgh. The PFLC, founded in 1899, operated as an autonomous state agency. Its work was distributing traveling book collections across the state, which it did from 1899 to 1919. It also helped to found more than 300 public libraries.[1] But the traveling library stations and the PFLC's advisory services focused on rural areas of Pennsylvania. A list of traveling libraries from PFLC from 1910 shows that Allegheny County did not receive traveling libraries from PFLC.[2] This was because Allegheny was not considered a rural county. When PFLC was disbanded in 1919, its work was integrated into the newly-formed State Library and Museum.[3] The Pennsylvania State Library therefore carried on extension and advisory services, yet these were advisory services only, and there was never any mandate in Pennsylvania for local or regional public libraries.[4] The PFLC, superseded by the State Library and Museum, was not involved in planning and administration across the state.

State legislation related to libraries also did not touch Allegheny County. The Keystone State Library Association (KSLA) was founded in 1901 and it was later renamed the Pennsylvania Library Association (PLA) in 1922.[5] PLA and the state library worked closely in several ways. One notable statewide initiative that involved the state library was the county library movement that gained momentum throughout Pennsylvania in the 1920s. This rural and agricultural movement, which exhibited quasi-religious tones, was led by the State Federation of Pennsylvania Women in conjunction with the Pennsylvania State Grange, the state library, and other local clubs.[6] The movement emphasized the creation of county libraries, and it also called for state funding for them. In response to county library petitioners, the state passed legislation in 1931 to provide funding to county libraries.[7] Throughout the 1930s, 1940s, and 1950s, Pennsylvania State Library continued to promote county libraries as a means to reach rural areas of the state.[8] But the 1931 statute that established state aid for county libraries only applied to counties of the third through eighth classes, not Allegheny, which was a second-class county in terms of population.[9] Even when CLP became the county library for Allegheny County in 1956, neither the county or CLP received state aid for the library.[10]

The first high-level planning efforts for public libraries across the state of Pennsylvania were initiated in 1932 under the auspices of the state's Commission for the Study of Educational Problems in Pennsylvania. Ralph Munn chaired the

Commission's Sub-Committee on Library Service. According to the Committee, and with rural areas as the plan's focus, the ultimate goals for public library service across the state were 1) county libraries in every county, 2) mandatory tax support for them, 3) public libraries in all of the larger municipalities across the state, and 4) adequate libraries in every school.[11] The plan also proposed that county managers contract with existing local libraries for library extension services throughout the county. This was what civic groups petitioned the Allegheny County Commissioners to do in Period 1 and Period 2. It is not surprising that aspects of Munn's plans from Allegheny County were reflected in the state documents he helped to create around the same time. Nevertheless, the state plan was designed to address the lack of library service in rural areas, not urban ones. Allegheny County did not have a county contract until 1956, which illustrates how state-level documents from as early as 1932 did not orient efforts in Allegheny County at that time.

State planning efforts by PLA continued with its effort to establish standards and certification for librarians. This coincided with the release of public library standards by ALA in 1933.[12] A Standards Committee was formed by PLA in 1933, and standards were accepted in 1937.[13] PLA formed its county division in 1940. [14] Standards for regional and county libraries and librarians were adopted in the early 1940s, in 1945, and in 1953.[15] Though these standards represent significant planning efforts at the state level, it is difficult to assess how they impacted the situation in greater Pittsburgh since PLA's standards were only recommendations, not mandates, and the certification of librarians remained voluntary.

In 1934, PLA proposed its own statewide plan for public libraries, derived from its own Planning Committee. The plan echoed that proposed in 1932. Munn participated in this committee as well, and the plan largely reiterated the main points from the plan created by Munn's committee in 1932. The plan emphasized larger administrative units, particularly counties, and stressed the importance of library extension to those areas without library service. In 1934, the state library still operated the system of traveling libraries that were supplied to small libraries or communities without libraires. Only five or six county libraries had been established across the state out of a total of 67 counties. This was despite the state aid to county libraries that had been in effect since 1931.[16] Later planning documents from PLA from 1939, 1940, and 1944 grappled with the failure of the 1931 act to promote county library development as leaders predicted it would. Planning documents also opined the absence of any legislation mandating the formation of public libraries across the state.[17] With planning architects hamstrung, state planning efforts did not approach the development of urban libraries like that underway in Allegheny County during Period 1 and Period 2.

Things began to change in 1952. The Pennsylvania State Library began to be seen as a decision-making body that should oversee public library development across the state, not just the library proper in Harrisburg. PLA itself saw the state library as the most appropriate agency for promoting plans and administering them, especially with respect to regional units.[18] The event that solidified the state library's leadership role in public library development came in 1956. Undoubtedly, this timing was due to events at the federal level. In 1956, the U.S. Congress passed the Library Services Act (LSA), which was signed into law by President Eisenhower. This landmark legislation established the first stable federal funding for public libraries. LSA was designed to stimulate states to form stronger library programs, especially in rural areas. By 1956, there were 28 county libraries across Pennsylvania, forming the basis of a statewide system. Additional federal aid from LSA, flowing to the states, would stimulate new developments in the state system. [19] This is what happened in Pennsylvania and in Allegheny County.

The 1930s, 1940s, and 1950s witnessed a wave of developments on the American national stage related to the purpose and role of public libraries. Ideological, administrative, and legislative changes within the country leading up to, during, and after the Second World War affected regional, state, and local public library policies in the 1960s. To understand why public libraries in greater Pittsburgh developed the way they did in the 1960s, therefore, it is necessary to understand these national developments.

In the 1930s and 1940s, the professional library association in America, the American Library Association (ALA), issued several key studies and policy documents related to public libraries. Some planning and standards documents were created by ALA committees, a prominent one being the Post-War Planning Committee of the ALA.[20] Other studies were issued by the Public Library Inquiry, a social-scientific research project whose focus was public libraries in America.[21] The Public Library Inquiry was initiated by leaders in ALA, but it was completed by independent researchers from outside the profession, resulting in several publications.[22] Amid the activity in the 1930s and 1940s by the Post-War Planning Committee, the Public Library Inquiry, and other committees in ALA, a division in the federal Office of Education was formed. Started in 1937, the Library Service Division was dedicated to federal library planning.[23] Lastly, another important development during this time at the national level was federal library aid. Federal aid was a controversial issue within ALA in the earlier twentieth century, but it gradually gained support, and it was eventually taken up and voted on by the U.S. Congress. This legislative movement culminated in several library-funding laws, most notably LSA of 1956, which provided federal funding to libraries and accelerated state planning efforts. Taken together, these activities on the national level, whether legislative, administrative, or ideological, established the context for public library development in greater Pittsburgh in the 1960s.[24]

ALA issued a new set of public library standards in 1956. The chair of the standards committee was Lowell A. Martin, who previously served as a consultant during the development of the 1943 standards and wrote part of the 1948 national plan. The 1956 standards were incorporated ideas that had been laid out during the previous two decades.[25] Significantly, the 1956 standards addressed library systems instead of individual libraries. The concept of a library system was a central tenet of the 1956 standards. While systems were mentioned in the national plans of 1935 and 1939, the standards of 1956 adopted systems as a central concern. A system was defined as a cooperative group of interdependent libraries.[26] Libraries that operated in systems shared resources and made a greater number of better materials available to their patrons. The 1956 standards thus served as a clarion call for libraries to form and maintain such systems, even if the exact structure of each system varied.

A necessary component of a library system as described in the 1956 standards was the state library agency. The importance and responsibilities of state library agencies was elaborated at length in the national plan of 1939, much more so than in the plan of 1935. The standards of 1956 reiterated the importance of the state's role and accorded states a central role within library systems.[27] These 1956 standards influenced ongoing efforts at the state, regional, and local levels to draw up plans for library systems. Central to the 1956 standards was the idea of a larger unit, or system, of libraries. This concept had been built up in ALA's planning documents, including the standards from 1943, the national plan from 1948, and the Public Library Inquiry thereafter.

The publication of national policies and the formation of a federal library office occurred concurrently as part of a larger movement to pass federal legislation to fund public libraries. The efforts by the ALA to develop standards, approve a Library Bill of Rights, create a national library plan, establish a Library Service Division in the Office of Education, and implement the Public Library Inquiry—all these efforts paved the way for establishing federal library funding. They did this by establishing the role and purpose of the public library and by legitimating the public library in the eyes of the American public. Furthermore, the ALA conducted surveys of public library locations and use that further demonstrated to lawmakers the deficiencies of existing public library services.[28] During the first half of the twentieth century, many library planners looked to private funding for support. Andrew Carnegie from 1886 to 1911, then the Carnegie Corporation from 1911 onward, greatly funded library buildings, services, and education.[29] This funding notwithstanding, there was still a need for additional public library support in the 1930s and 1940s, funding that was beyond the scope of local tax income.[30] The campaign for federal funding by the ALA during the 1920s, 1930s, and 1940s influenced the creation of library bills that eventually resulted in successful legislation.[31]

The Library Services Act (LSA) was not the only library funding legislation at this time. Other, similar national laws were also implemented in the 1950s and 1960s.[32] These other funding sources notwithstanding, LSA was a milestone for public library funding. On June 19, U.S. President Dwight D. Eisenhower signed LSA into law.[33] This act, administered under the U.S. Office of Education, amended Title 45 of the Code of Federal Regulations, which governed public welfare. LSA offered federal funding to states to extend public library services to rural areas. States with state library plans were eligible to receive federal funding for public library programs. States were responsible for distributing the funding and overseeing the locally administered programs. The funding was limited to "salaries and wages, the purchase of books, other library materials, and equipment, and operational costs, applied to the further extension of public library services to rural areas." Funding did not apply to library construction projects.[34] LSA was a 5-year act. It was later extended in 1960 by Congress and President Eisenhower. Five years later, in 1964, it was extended and renamed to the Library Services and Construction Act (LSCA).[35] LSCA was then amended in 1966 to include new funding areas.[36]

Passage of LSA in 1956, and the public library standards published by ALA that same year, directly impacted public library development in Pennsylvania, including greater Pittsburgh. The new standards and funding led to a wholesale restructuring of the state library system in Pennsylvania. CLP and the libraries in Allegheny County became essential components of that system.

As the bills for federal library funding worked their way through U.S. Congress in the 1940s and 1950s, PLA was already working toward improving the state library. Before their efforts to transform the state library in 1955, in 1951 the Public Library Service Committee of PLA published preliminary results of its survey of Pennsylvania libraries.[37] This survey used the most recent standards established in ALA's national plan of 1948 and the recommendations from the Public Library Inquiry from 1949 and 1950 to assess the quality of Pennsylvania libraries, not including those in the state's two major cities, Philadelphia and Pittsburgh. The survey found that many Pennsylvanians lacked public library service and that many existing public libraries were inadequate.

Work done on the national stage in the form of planning, standards, and funding made a state system in Pennsylvania more urgent. The turning point was the passage of LSA in 1956. In order to receive federal money from the U.S. Office of Education to develop public library services in rural areas, LSA first required that states possess a state plan. Given the opportunity to receive federal money, Pennsylvania acted immediately to create a state library plan.[38]

By June 30, 1957, 35 states and Hawaii submitted state plans to the U.S. Office of Education. An additional 12 plans from other states and territories were added in fiscal year 1958, bringing the total to 48.[39] Pennsylvania was among this second round of plans.[40] The initial state plan from 1957–1958 was vague: it focused on strengthening the state library agency, aimed toward conducting a study of Pennsylvania libraries, and provided county library grants.[41] Though Pennsylvania did not have a detailed state plan in place in 1957, federal funds from the newly passed LSA were acquired by Pennsylvania. Pennsylvania used its initial LSA funds to strengthen rural county libraries by augmenting the state library's collections, conducting workshops, improving interlibrary loan (ILL) service, and publishing public library statistics.[42] ILL services involved the reciprocal sharing of books and other materials between libraries. In the 1961/62 budget request, about forty percent of the state library's operations were carried by federal funds.[43] The second major project was to conduct a comprehensive study of the state's public libraries. This study would form the basis of any subsequent state plan. The study began in October 1957.[44] By December 31, 1960, $652,262 in federal LSA funds were received by Pennsylvania.[45]

Lowell A. Martin was asked to lead the study. Martin was a significant figure in the library world and on the national stage at that time. His accomplishments by that point were numerous. Martin had studied under the prominent Carleton B. Joeckel in the Graduate Library School at the University of Chicago. Martin completed his master's thesis at Chicago in 1940 and his dissertation there in 1945.[46] He co-authored a book on library administration in 1941 with Arnold Miles.[47] Some of Martin's most significant and lasting achievements included his contributions to national library planning documents. Due to his close student-mentor relationship with Joeckel, he worked as a consultant to the Post-War Planning Committee while Joeckel served as its chair. In his role as consultant to the committee, Martin helped to develop the national public library standards published in 1943.[48] Also while working as a consultant to the committee, he wrote the first chapter of the National Plan for Public Library Service which was published in 1948.[49] After the Post-War Planning Committee was discontinued in 1948, Martin then led the Co-ordinating Committee on Revision of Public Library Standards, which published new public library standards in 1956. In addition to his work on national planning documents, Martin worked as Instructor and Assistant Professor at the Graduate Library School at the University of Chicago, he was appointed Professor in the School of Library Service at Columbia University, and he served as Dean of the Graduate School of Library Service at Rutgers University. Martin was contacted by the Pennsylvania State Librarian, Ralph Blasingame, to lead the Pennsylvania study in 1957. The study was commissioned by the state library and overseen by the Library Development Committee of PLA. [50]

The Martin study consisted of two main parts. The first part of the study was a survey of the current state of public libraries in Pennsylvania. The second part was a plan for a state library system. In the survey portion, Martin and his team found that millions of Pennsylvanians had no local public library of any kind, grossly inadequate facilities, or they lacked reasonable access to specialized research materials. Surveyors found a total of 367 local public libraries around the state. Despite this seemingly large number, there were large gaps in service.[51]

The survey found that Pennsylvania libraries did not meet quality standards. For a set of baseline standards, the surveyors used the national public library standards of 1956, which Martin had authored. But the surveyors also used a tailored set of standards that they developed specifically for Pennsylvania as a basis for evaluation. Results of the evaluation found that most libraries did not receive adequate per capita support or sufficient millage, and few libraries had adequate book collections, staffing, services, or buildings. The libraries in Philadelphia and Pittsburgh, despite their impressive services, also showed room for improvement. It was speculated that several causes contributed to the relatively low public library service levels in Pennsylvania. This included lack of funding, the outlook of citizens and government officials, the governance structure of public libraries, a lack of adequately trained library personnel, and passive leadership at the state level.[52]

Special criticism was reserved by the Martin study for county libraries in Pennsylvania. The study noted the county library movement in Pennsylvania from 1930 to 1950 and now county libraries were the only type of public library to receive state aid, up to $4,000. Despite the enthusiasm for county libraries within the state, and despite their state support, county libraries actually constituted "one of the weakest parts of Pennsylvania library service," according to Martin.[53] And despite the state library's promotion of county libraries in the 1930s and 1940s, no integrated library system developed across the state. County libraries did not figure prominently into Martin's vision of a statewide public library system, though they were included in the end anyway, perhaps given how the state had already invested so much into them and there were some already established.

Following the survey and analysis, Martin proposed a detailed plan intended to improve public library service across the state. The stated goal of the Martin plan was to propose "a coordinated system of public library service which would remove these deficiencies and provide Pennsylvania with the reading resources and information needed to realize its potentialities."[54] This language of a coordinated public library system was taken from Martin's own 1956 standards. The total cost of Martin's proposed state plan was $24 million annually. The plan incorporated the ideas from the ALA national plan from 1948, where Martin served as a consultant and co-author, the recommendations of the Public Library Inquiry from 1949 – 1950, and the most recent standards from 1956, which Martin had authored.

Central to the Martin plan of 1958 was the idea of a coordinated, statewide library system. This was an idea that had been built up in national planning documents and was especially prevalent in the 1956 standards.[55] The statewide system proposed by Martin had three service levels. The guiding mantra for his system was:

- a day-to-day library within 15 to 20 minutes of one's home;
- a subject and professional library within an hour;
- a research library within a day.[56]

In Martin's plan, each of the three levels of service was delivered by a distinct library role. Martin named these roles local units, district library centers, and regional resource centers.[57] This tripartite structure formed the backbone of the plan. The service levels operated together to form a single library system with coverage across the state. State aid was structured to support each of the three library roles. Any particular library might play one, two, or all three roles. Because state aid recognized three distinct service levels and three corresponding service roles in any library, a single library would receive state aid for all the roles it played.

Formative ideas for the tripartite service level structure were first articulated by Martin in his dissertation in 1945. There, he distinguished three kinds of reading: diversional reading, purposeful reading, and research reading.[58] Each type of reading required a different type of library collection. The corresponding collection types were recreational materials, educational materials, and scholarly materials. Purposeful reading, which was the focus of Martin's study at that time, required educational materials, but it also drew from recreational materials and scholarly materials (see Figure 7). Martin states:

> Purposeful reading... is broader than purely educational reading. It includes at one end scholarly materials... and at the other end it includes recreational materials... Yet it is a selective area, for it excludes voluminous diversional materials and specialized research materials. And it is not an area that becomes confused by the fiction—nonfiction distinction employed in libraries, for many fiction items are of use in purposeful reading.[59]

In his dissertation study, Martin argued that because purposeful reading and purposeful readers have increased, the minimum public library size should be redefined to accommodate purposeful reading.

A similar 3-level service structure reappeared in chapter 1 of the national plan of 1948, which Martin authored. He said:

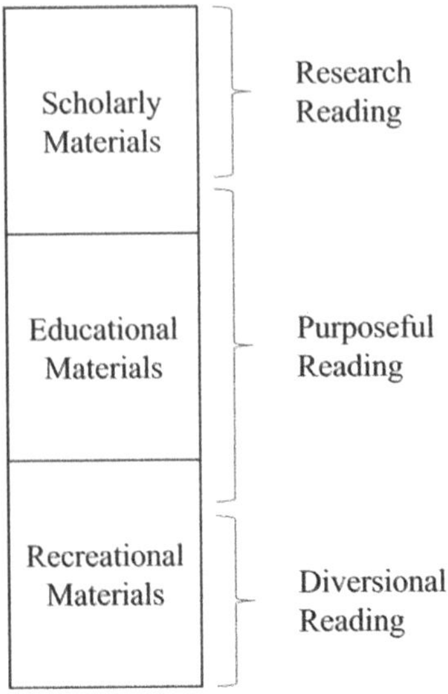

Figure 7: Relation of purposeful reading to educational materials.

> In planning for library service to America it is useful to note three levels of required materials, distinguished by scope and quantity of use.
>
> There are, first, the materials needed in practically every American community, sources of information on topics of current general interest (national affairs, for example) and the significant books of contemporary literature.
>
> A second group is composed of materials not required in every community but which, when needed at all, are in sufficient demand locally to justify the purchase of at least one copy by a library serving ten or twenty thousand people. The "core" materials and the "local interest" materials should be accessible to every citizen in his own neighborhood.
>
> A third group is composed of materials which, while not research or scientific in nature, are sufficiently specialized that only one or relatively few readers may want them. These may be on the special hobbies or interests which people develop, the personal problems they face, the unusual vocational interests they follow. Only the larger libraries can afford such materials; the smaller libraries can neither afford nor justify their purchase. This is not a serious handicap if local libraries have a "second line of defense," in the form of state and region-wide libraries to which they can turn when special items are needed. The American library system will be incomplete until materials at this, as well as more popular levels, are available on a nation-wide basis.[60]

The 3-part structure proposed by Martin did not appear elsewhere in the national plan of 1948. It was distinctively Martin's idea. It is striking how such an idiosyncratic idea would enjoy such broad and long-lasting support in Pennsylvania.

The plan downplayed the role of the county library and emphasized coordination between local and district levels. Martin's plan included state funding for county libraries, but county libraries did not fit neatly into the plan because the plan did not specify what type of reading or what type of collection county libraries should support. The plan gave district library centers additional leadership, consulting, and reporting responsibilities. It included academic and a state library as regional resource centers. Figure 8 shows how the state plan was built out from Martin's 3-part structure of collection types and reading types.

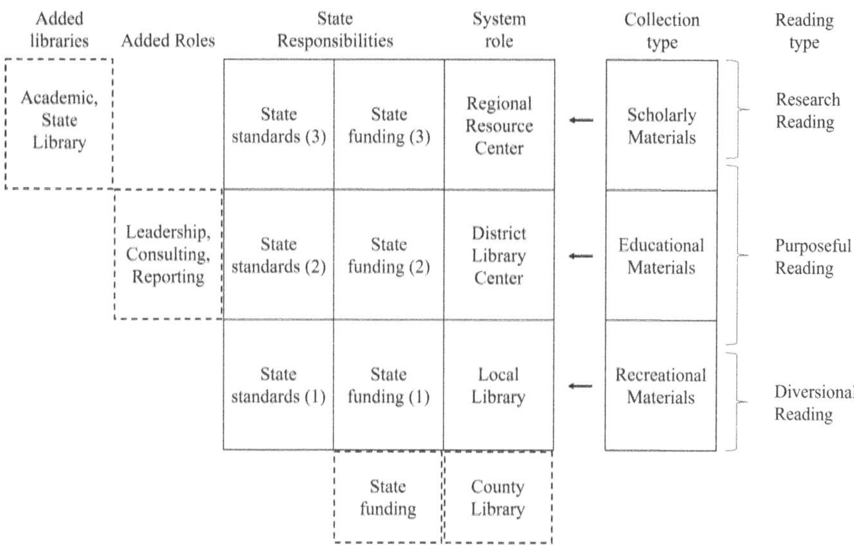

Figure 8: Expansion of Martin's 3-level structure into the state plan.

The first step in the plan was to retain and strengthen local library units. The second step in the plan was to establish district library centers. These district centers would function alongside local library units to strengthen them. Each local library would be affiliated with a district library center. There would be up to 30 district library centers established across the state. District centers would house additional resources and provide training and guidance to affiliated local libraries. District libraries would be selected from already-strong local libraries. These libraries would extend their services beyond their existing service areas. District library centers would be allocated state aid for their district services. The third step in the plan was to create four regional resource centers, in addition to local libraries and district library centers. Regional resource centers would collect and provide specialized materials to supplement existing collections.[61]

There were several implications of the state plan to the regional public library infrastructure in greater Pittsburgh. First, all county libraries, including the CLP, would receive state funding so long as they maintained adequate local support. Support for local libraries would flow from the state so long as minimum local money was raised, and there would be additional state funding available for constructing library buildings.[62] For new libraries, state aid would only be available if those local outlets joined the district system.

Second, CLP was identified as the district library center for the areas of Allegheny and Beaver Counties and the southern half of Butler County. CLP was one of 27 district library centers stated in the plan. CLP took on additional roles and responsibilities as district library center. It would form service contracts and guidance contracts with surrounding local libraries. District library centers provided additional services to readers and local libraries:

Direct Service to Readers

- full access for persons in the whole district to consultation of resources in the central library.
- Full availability of the reference service provided by the central library.
- Free circulation of books to individuals throughout the district.
- Consultation by readers throughout the district with the librarian and specialized staff members available in the central library.
- Interlibrary loan service to get resources from over the state and beyond as needed by readers.

Services to Readers through Local Libraries

- Loans of titles from the district center to local libraries for use as specifically needed by individual readers.
- Placement of groups of books in local libraries on a rotating basis to supplement resources available in the locality.
- Reference service for quick information by telephone call from the local library.
- Extensive reference and advisory service on request for readers who cannot or prefer not to go to the central library.

Guidance and Coordination Functions for Libraries

- Guidance to communities on the establishment of local facilities.
- Consultation through regularly scheduled visits with personnel responsible for local libraries.
- Program to aid in the selection of materials throughout the district by means of establishing standards and preparing buying lists in the light of these standards.
- Arrangements for centralized book ordering, so that orders are pooled and maximum discounts obtained.

- Experimentation with centralized cataloguing to relieve the burden of duplication of this work in each small library.[63]

Then, in addition to being a designated local library and district library center, CLP was designated as one of four regional resource centers across the state due to its strong collections in science and technology. The Free Library of Philadelphia, Pennsylvania State Library, and Pennsylvania State University Library were the other three designated regional resource centers. [64]

The total projected annual cost of the proposed plan was about $24 million, supported by local and state funding. State aid to libraries was to be determined by a formula and standards for each service level:

- $.25 per capita to local libraries.
- up to an additional $.25 to tax districts definitely below average wealth for the state.
- $.25 per capita to district library centers for service over their districts.
- $100,000 per year to each of four major libraries responsible for specialized service over a region of several million people.[65]

To receive funding, local libraries were required to raise a "fair share" of local funds, comply with standards for local library service, and participate in the local district library system. Equalization aid would be given to local libraries that struggled to fund basic services,[66] and additional money would fund district centers, regional resource centers, the state library, and capital expenditures for library buildings.

Finally, the plan proposed new legislation that would establish the state system. Legislative revisions proposed by the new state plan included clarifying the role of the public library as part of the educational program of the state, removing maximum funding restrictions, and specifying guidance and supervisory functions of the state library.[67]

The next challenge in Pennsylvania was to get Martin's plan passed in the legislature. This effort consisted of educating and informing local librarians and advocates about the plan. In Pittsburgh, local library leaders with knowledge of the plan explained its significance and impact on the regional library system there. Before the plan had passed the legislature, the CLP began to plan for its imminent adoption. CLP continued to strategize to ensure it secured the largest share possible of state aid. The library would receive aid for each level of service it offered: local, district library center, and regional resource center. The library would also receive funding as its role as a county library.[68]

Legislation to implement the Martin plan passed the House of Representatives in the 1959 session, but it failed to move through the State Senate. Addressing the

Pennsylvania Library Association in 1961, and reflecting on the Library Development Plan's failure in the Senate, Governor Lawrence, the former mayor of Pittsburgh, insisted on the importance of building public support:

> Our foremost task...is to mobilize public opinion behind the primary need for good library service if we are to have good education. For my own part, I pledge that this State Administration will do everything within its power to press for passage of the Library Development Program and to encourage local interest in your libraries. We cannot let our experiences of this past two years discourage or deter us from that effort.[69]

Another attempt was made to pass the legislation in the 1961 session.[70] Act 188 passed the Pennsylvania General Assembly in 1961. Public Law 324 was signed June 14, 1961, by Governor David L. Lawrence, the former mayor of Pittsburgh, who several years before in 1956 had facilitated the merger of the city library systems. Though the legislation passed in 1961, several details still needed to be worked out, including the exact boundaries and definitions of library districts. These were further studied in 1962.[71]

State funding for other types of public libraries, other than for rural county libraries, was not established until 1961 when the new Pennsylvania library code was passed, though municipalities, school districts, and counties had the authority to establish and appropriate monies toward public libraries.[72] Five categories of state funding for public libraries were established with the new library code in 1961.[73] Each category of funding had its own formula for determining how much public libraries of that type would receive. The five categories of state aid were local libraries, county libraries, district library centers, regional library resource centers, and equalization aid.[74] Local libraries that made the minimum financial effort could receive a maximum of twenty-five cents for each person residing in its direct service area.[75] Libraries would receive additional aid if they also served in other roles. To receive state aid, libraries had to submit annual plans to the State Librarian. The new Library Code of 1961 designated the Advisory Council on Library Development as the authority for issuing rules and regulations for the State Librarian to use to distribute state funds to local libraries and to approve or disapprove plans for use of state funds by local libraries.[76]

Thus marked the beginning of a robust state aid system to public libraries. In 1962, 230 public libraires in 60 counties received state aid. There were 430 public libraries in Pennsylvania in 1962.[77] In Allegheny County, 16 libraries received state aid.[78] The Governor's Advisory Council (GAC) had been brought into existence by the 1961 legislation. At their first meeting in 1962, members discussed the shortage of staff at the state library required to carry out the new library plan and the approval process for regulations for the distribution of state aid.[79] The statute gives the State Librarian the authority to approve or disapprove of library plans for the use of state aid. But the statute does not specify what goes in the plans, or what criteria would be used to evaluate them. The regulations speak to the criteria for approving plans. These regulations were released by GAC in 1962. Some of the main components that were required to be in plans were system participation, achievement of standards of service, library services to the whole community, standards of library expenditures, and economical use of public funds. [80] Importantly, public libraries had to maintain their own adequate level of local support to be eligible for state support. Slight amendments to the regulations appeared in 1965, and standards would begin to appear in 1968 and 1969.[81] Funding for LSA, then LSCA, continued into the mid-1960s.[82]

In 1962, the board of trustees for CLP formally accepted its designation as a district library and regional resource center, "with the provision that no City of Pittsburgh tax funds be used for extending library service beyond the city. As a result, this library became eligible for financial assistance from the state."[83] The district center for the area included all of Allegheny County, the ten southern townships in Butler County, and the northern section of Westmoreland County including the cities of Greensburg, Latrobe, and New Kensington. The population of the district, not including Pittsburgh, was 1,339,100.

In 1962, CLP anticipated receiving $218,055.16 in state aid for the following year. These were for its roles as a local library, county library, district library center, regional resource center, and library for the blind. CLP first received state funding in 1962 in the amount of $19,792 for planning and book purchases.[84] A field representative was selected for district services, and preliminary consulting work was begun in 1962.[85] District library center services included reference services through local libraries and directly to individuals on the same basis as Pittsburgh residents, interlibrary loans to all local libraries, field workers to local libraries, a book exchange pool, and professional development for local libraries.[86]

The state system began to function in greater Pittsburgh on January 1, 1963. At the start, there were 50 libraries in the Pittsburgh District.[87] By 1965, the number of libraries in the district increased to 52.[88] By 1963, the overall number of library district centers across Pennsylvania had been increased from 27 to 30.[89] During its first year as district library center, the field representative visited local libraries in the district and held meetings with library personnel. Interlibrary loan requests were made by 27 libraries in the district. In the second year, interlibrary loan requests increased, with 39 libraries requesting materials. Two new libraries, Monroeville and New Florence, joined the system. Consulting and personnel meetings continued. In 1964, a special telephone line was installed in the reference department that allowed local libraries to call the department directly without cost to the library.[90] District services continued to expand in 1965. The number of libraries increased to 53. Consulting was provided to local citizens groups interested in starting libraries in their communities. A public relations specialist was employed in 1965 and began to serve in the library districts around Pittsburgh.[91]

Regional resource center allocations were not yet assigned by 1963, but CLP did receive funding for its role as a county library and for providing services to the blind. Funding to the library was gradually increased as the annual state funds for implementing the state plan were increased. The full funding for the state plan was met by the state for the first time in 1966.[92] CLP continued to expand its bookmobile services that reached into the county.

The years 1955 to 1965 was a time of dramatic financial growth in CLP. The first major growth point occurred in 1956 when it merged with the North Side libraries and contracted with the county commissioners. These changes added branches and county funding. The number of total registered borrowers more than doubled from 1955 to 1960. The next major growth point occurred in 1962 and 1963 when the library began receiving state aid for its added roles of county library, district library center, and regional resource center (see Table 3).

Table 3: Growth of the Carnegie Library of Pittsburgh, 1955–1965.

Year	Branches	Circulation (including bookmobile)	Registered borrowers		Funding		
			City	County	City	County	State
1955	14	2,760,265		114,924	$1,335,882	0	0
1956	17	2,967,109		128,776	$1,428,998	$125,000	0
1957	17	3,072,403		185,573	$1,591,985	$206,520	0
1958	17	3,395,516		209,495	$1,694,285	$254,925	0
1959	17	3,569,370		240,508	$1,769,070	$365,949	0
1960	17	3,757,094		269,593	$1,833,937	$384,377	0
1961	17	4,499,847	145,937	139,659	$1,884,584	$384,786	0
1962	17	4,683,711	157,804	144,800	$1,935,607	$414,063	$19,792
1963	17	4,848,281	172,963	137,458	$1,924,425	$450,918	$218,295
1964	16	4,784,954	185,152	143,298	$1,854,512	$749,805	$313,892
1965	16	4,748,308	196,390	149,311	$1,898,995	$857,050	$408,994

There were several public library building projects in Allegheny County at this time, including a regional reference center in Monroeville and public library buildings throughout the county.[93] Some of this activity was stimulated in part by the passage of LSCA in 1964, which allowed states to award federal money for library building projects. This contrasted with LSA from 1956, whose federal grants were reserved only for the extension of library services into rural areas. In 1964, there were 36 public libraries in Allegheny County, not including CLP.[94] But the largest and most striking growth was in the authority of CLP. As a result of the new state code of 1961, CLP became a data-collecting site for county and district libraries. The library collected data about the number of libraries in the county, the county's population, the total and itemized income of libraries in the county, library expenditures on salaries and books, and the service populations of each library.[95] It tracked who used ILL, who requested advisory services, who attended classes, and who received state aid, and who did not.[96] The district collected data on how libraries in the county were funded, how much they received in support from municipalities and school districts.[97] CLP became the gateway for other local libraries to receive state aid. Following the Martin survey in 1958, libraries began to file detailed annual reports with the district library center and state library.[98] This created data that the state library had never possessed. This new administrative function of CLP was completely new. Becoming the county library in 1956 meant that it contracted with the county to provide services to county libraries. These services included bookmobile routes, free borrowing for county residents, and reference services for county residents. But as a county library, the CLP did not administer other libraries in the county. It was only with the additional role of district library center that CLP took on additional administrative functions associated with the state. It became an extension of the state government. In that role, it began to monitor the activities of the other libraries in its district. This served the purposes of the state library because, under the new state code, the state library was charged with distributing state aid and ensuring that libraries achieved established standards. As local libraries within the district, the libraries became required to provide data to the district library. The district services office began to rank-order Allegheny County libraries as acceptable, possibly acceptable, and other.[99] The standards for district library centers would be further discussed in the 1960s. A district library center study committee, whose members included district coordinators from around the state, including CLP director Keith Doms, as well as the state librarian, Ernest E. Doerschuck, met in 1966 to discuss what would count toward defining high-quality district centers.[100]

In 1967, a decade after publication of the initial study, Lowell A. Martin was again commissioned by the Pennsylvania State Library to conduct a re-survey of public libraries in the state. The purpose of the re-survey was to "determine the extent to which the 1958 Pennsylvania library program has been achieved" and to "propose a revised or new program to meet present and emerging library needs in the Commonwealth."[101] Results of the re-survey identified growth and improvement. Nearly a million more people had library service, and there was better resource sharing between libraries.[102] Proposed lines of action that resulted from the re-survey included mandating county libraries in the 28 counties lacking total library coverage; establishing minimum funding levels for county libraries; and the establishment of inter-county units in sparsely populated areas of the state, corresponding with intermediate school units.[103] Libraries would receive new funding as a result of the plan. CLP calculated that the library would receive $1,918,664 in state aid under the new proposed Martin plan in 1967, as a district center, local library, county library, and a regional resource center.[104]

Discussions followed the Martin re-survey. Library district centers around the state discussed its recommendations. CLP, as a district library, was asked by the Library Development Committee of PLA to sponsor discussions on the study.[105] There was some discussion of mandating that school districts provide for all public library funding, as they have taxing authority.[106] This plan was rejected by Martin as too controversial to succeed in the legislature.[107] In light of the discussions of the Martin re-survey, several legislative changes occurred to the Pennsylvania state library code in 1967, 1968, and 1970.[108] Following that, in 1970, LSCA was renewed, which prompted the State Library of Pennsylvania to transition to develop a new 5-year plan. In 1969 and 1970, CLP continued in its role as district library center. As part of its responsibilities, it continued to monitor and record data about county libraries.[109]

Period 3 is significant because a new circuit and decision-making center emerged in the Pennsylvania State Library. This was a result of developments beyond the boundaries of the greater Pittsburgh case. The advent of federal funding led to the construction of a new state plan, which included state funding for various public library roles. CLP opted into this new administrative structure, and with the acceptance of its new administrative and service responsibilities, it grew in influence, becoming not just a county library, but a district library center and a regional resource center, alongside its local role. Within the state circuit, CLP became part of the inner periphery in how it implemented and administered decisions made in Harrisburg while at the same time collecting and relaying data from local libraries in Allegheny County.

What is also significant about Period 3 was how the issue of a reorganization of the public library infrastructure during this period was imported from the outside. This contrasts with the issues raised in Period 1 and Period 2, which were wholly locally driven. Period 3 shows how decision cycles can take up issues from the outside, leading to new developments locally. The decision cycle in Period 3 exhibited low legitimacy and low civil support. But there was also low resistance to the state plan, and the relatively few instances of communicative power were taken up and acted on by the decision-making cores and inner peripheries, including CLP, which agreed to participate in the state plan as a district library center. This decision was a major turning point for the development of public library infrastructure in greater Pittsburgh.

In Period 3, CLP became part of the state circuit, located in the inner periphery of the state. This was because CLP now, at least in part, carried out not just local operations but also state plans. By the end of Period 3, CLP was nested within the inner peripheries of the city, county, and state circuits, implementing decisions made by the core decision making bodies of those circuits.

1 Robert Bray Wingate, "The Pennsylvania State Library: 225 Years of Service," *PLA Bulletin* 25, no. 6 (1970): 341.

2 "Traveling Library Stations," *Pennsylvania Library Notes* 3, no. 3 (July 1910): 7.

3 Library Extension Division, A Library News Letter from the Library Extension Division, August 1919, SLPARSC, Pennsylvania Library Association Collection, Publications and Research Files, Box 5a PLA Bulletin and microfilm reels, Library Extension Division Newsletter, 1919, 1929.

4 Bernadette A. Lear, *Made Free and Thrown Open to the Public: Community Libraries in Pennsylvania from the Colonial Era through World War II* (Pittsburgh: University of Pittsburgh Press, 2021), 90.

5 William F. Stevens, *The Keystone State Library Association, 1901 – 1915* (1916), 10; William F. Stevens, *The Keystone State Library Association, 1916 – 1922: Supplement to 1901 – 1915* (Altoona: Tribune Print, 1923), 9.

6 Anna A. MacDonald, "The County Library Movement in Pennsylvania," *Pennsylvania Library Notes* 12, no. 2 (1929).

7 Bernadette A. Lear, "A State Library Transformed: Pennsylvania, 1878 – 1921," *Information & Culture: A Journal of History* 48, no. 1 (2013): 41.

8 Lowell A. Martin et al., *Library Service in Pennsylvania, Present and Proposed: A Survey Commissioned by the Pennsylvania State Librarian at the Request of the Honorable George M. Leader, Governor of the Commonwealth of Pennsylvania*, Pennsylvania State Library (Harrisburg, 1958), 29.

9 Laws of the General Assembly of the Commonwealth of Pennsyvlania Passed at the Session of 1931, (1931). 447 Laws of the General Assembly of the Commonwealth of Pennsylvania passed at the Session of 1929 (1929).Commonwealth of Pennsylvania, 1955, LCLL.

10 "Milestones in Legislative Support of Pennsylvania Public Libraries," *PLA Bulletin* (January 1981); David L. Lawrence, 19th Biennial Budget of the Commonwealth of Pennsylvania for the Fiscal Biennium June 1, 1959 to May 31, 1961, General Fund, Submitted to the General Assembly, March 1959, 102 – 03, Government Documents Collection, LUL; *Laws of the General Assembly of the Commonwealth of Pennsyvlania Passed at the Session of 1931*; George H. Earle, *The Budget of the Commonwealth of Pennsylvania for the Fiscal Biennium June 1, 1935, to May 31, 1937* (February 1935), 24; Pennsylvania Library Association Planning Committee, Preliminary Report of a Plan for Public Library Development Presented by the Pennsylvania Library Association Planning Committee to the State Planning Board, November 1934, Pennsylvania Library Association Collection, Division Files, Box 2 County-Trustee, County-Public Division, 1934, n.d., SLPARSC.

11 Sub-Committee on Library Service, Commission for the Study of Educational Problems in Pennsylvania: Report of the Sub-Committee on Library Service, 1932, 2, Pennsylvania Library Association Collection, Division Files, Box 2 County-Trustee, County-Public Division, 1934, n.d., SLPARSC.

12 American Library Association, "Standards for Public Libraries," *Bulletin of the American Library Association* 27, no. 11 (1933); Joseph L. Rafter, "Standards and Certification for the Public Libraries and Librarians of Pennsylvania," *Pennsylvania Library and Museum Notes* 16, no. 1 (1937); "Committee on Standards and Certification," *Pennsylvania Library and Museum Notes* 16, no. 5 (1938).
13 Katharine Shorey, "Milestones on the Path to Success with H.B. 132, 1933–1961," *PLA Bulletin* 17, no. 1 (1961): 186; Committee on Standards and Certification for the Public Libraries and Librarians of Pennsylvania, "Standards and Certification for the Public Libraries and Librarians of Pennsylvania," *Pennsylvania Library and Museum Notes* 16, no. 1 (January 1937).
14 Margaretta A. Elder, History of the County Library Division of the Pennsylvania Library Association, 12 October 1956, Pennsylvania Library Association Collection, Division Files, Box 2 County-Trustee, County-Public Division, 1934, n.d., SLPARSC.
15 Pennsylvania Library Association County Library Section, Standard for the Establishment and Maintenance of County and Regional Libraries in Pennsylvania, 1953, Pennsylvania Library Association Collection, Division Files, Box 2 County-Trustee, County-Public Division, 1934, n.d., SLPARSC.
16 Pennsylvania Library Association Planning Committee, Preliminary Report of a Plan for Public Library Development Presented by the Pennsylvania Library Association Planning Committee to the State Planning Board, November 1934.
17 Planning Committee of the Pennsylvania Library Association, Library Needs of Pennsylvania, October 1939, Pennsylvania Library Association Collection, Committee Files, Box 6, O-Public Relations, Planning Committee, 1934–1942, SLPARSC; Harold A. Wooster, Planning Committee's Report, 11 October 1940, Pennsylvania Library Association Collection, Executive Board Files, Box 1, Minutes, Reports, Correspondence, 1940, SLPARSC; Harold A. Wooster, "Planning Committee, PLA, Report," *Pennsylvania Library and Museum Notes* 17, no. 7 (July 1940); Committee on Public Library Service, Plans for Post-War Library Service in Pennsylvania, 1944, Pennsylvania Library Association Collection, Executive Board Files, Box 1, 1901–1960, Minute Book, 1944–1954, Part 1, 1944, SLPARSC.
18 Executive Committee of the Pennsylvania Library Association, Pennsylvania State-Wide Library Program for Developing Adequate Public Library Service to All Citizens: Part II Long Term Program, 1952, Pennsylvania Library Association Collection, Division Files, Box 2 County-Trustee, County-Public Division, 1934, n.d., SLPARSC.
19 "Library Services Bill," *Pennsylvania Library Association Bulletin* 11, no. 3 (Winter 1956): 2.
20 Committee on Post-War Planning, *Post-War Standards for Public Libraries* (Chicago: American Library Association, 1943); Carleton Bruns Joeckel and Helen A. Ridgway, "Postwar Planning Committee," *ALA Bulletin* 42, no. 11 (1948); Carleton Bruns Joeckel, "Postwar Planning," *ALA Bulletin* 41, no. 11 (1947); Lowell A. Martin, "A Plan for Public Library Service in America," *ALA Bulletin* 40, no. 8 (1946): 276; Committee on Post-War Planning, *A National Plan for Public Library Service: Prepared for the Committee on Postwar Planning of the American Library Association* (Chicago: American Library Association, 1948), 19.
21 Joeckel, "Postwar Planning," 352; Oliver Garceau, *The Public Library in the Political Process* (New York: Columbia University Press, 1949)., preface.Robert D. Leigh, "The Public Library Inquiry," *Library Journal* (1947, May 1): 722; Douglas Raber, *Librarianship and Legitimacy: The Ideology of the Public Library Inquiry* (Westport: Greenwood, 1997), 51; Bernard Berelson, *The Library's Public: A Report of the Public Library Inquiry* (New York: Columbia University Press, 1949), 18.
22 Raber, *Librarianship and Legitimacy: The Ideology of the Public Library Inquiry*, 81–82. Books: Bernard Berelson, *The Library's Public* (New York: Columbia University Press, 1949) Alice I. Bryan, *The Public Librarian* (New York: Columbia University Press, 1952) Oliver Garceau, *The Public Library in the Political Process* (New York, Columbia University Press, 1949) Robert D. Leigh, *The*

Public Library in the United States: The General Report of the Public Library Inquiry (New York: Columbia University Press, 1950) James L. McCamy, *Government Publications for the Citizen* (New York: Columbia University Press, 1949) William Miller, *The Book Industry* (New York: Columbia University Press, 1949) Gloria Waldren, *The Information Film* (New York: Columbia University Press, 1949) Supplementary reports: Charles M. Armstrong, *Money for Libraries: A Report on Library Finance* (New York: Social Science Research Council, 1951) Angus Campbell and C. A. Metzner, *Public Use of the Library and Other Sources of Information* (Ann Arbor, MI: Survey Research Center, Institute for Social Research, University of Michigan, 1950) Joseph T. Klapper, *The Effects of the Mass Media* (New York: Bureau of Applied Social Research, Columbia University, 1949) Otto Luening, *Music Materials and the Public Library: An Analysis of the Role of the Public Library in the Field of Music* (New York: New York Public Library Inquiry, Columbia University, 1949) Watson O'D. Pierce, *Work Measurement in Public Libraries* (New York: Social Science Research Council, 1949)

23 Carleton Bruns Joeckel, *Library Service* (Washington, DC: US GPO, 1938), 46; Office of the Federal Register, *United States Government Organization Manual, 1972/73* (Washington, DC: US GPO, 1972), 221; Charles H. Compton, "Looking toward Greater Federal Participation in Library Development," *Bulletin of the American Library Association* 29, no. 2 (1935): 57–58; American Library Association, "A National Plan for Libraries," *ALA Bulletin* 33, no. 2 (1939): 146.

24 American Library Association, "Standards for Public Libraries," 513; American Library Association, "Midwinter News," *ALA Bulletin* 36, no. 1 (1942): 17; American Library Association, "Midwinter Council Minutes," *ALA Bulletin* 36, no. 2 (1942): 150; American Library Association, "Council," *ALA Bulletin* 36, no. 10 (1942): P-39; Carleton Bruns Joeckel, "Postwar Planning," *ALA Bulletin* 36, no. 12 (1942); Redmond Kathleen Molz, *National Planning for Library Service, 1935 – 1975* (Chicago: American Library Association, 1984), 44–47; Committee on Post-War Planning, *Post-War Standards for Public Libraries*; Louise S. Robbins, *Censorship and the American Library: The American Library Association's Response to Threats to Intellectual Freedom, 1939 – 1969* (Westport: Greenwood, 1996), 12–13; American Library Association, "The Library's Bill of Rights: Minutes of the Council of the American Library Association, June 19, 1939," *ALA Bulletin* 33, no. 11 (1939): P-60-P-61; Michael H. Harris, *History of Libraries in the Western World*, 4th ed. (Lanham: Scarecrow, 1999), 248; Molz, *National Planning for Library Service, 1935 – 1975*, 20; American Library Association, "A National Plan for Libraries," *Bulletin of the American Library Association* 29, no. 2 (1935): 91.

25 Lowell A. Martin, "Standards for Public Libraries," *Library Trends* 21, no. 2 (1972): 166; American Library Association, *Public Library Service: A Guide to Evaluation, with Minimum Standards* (Chicago: ALA, 1956), xvi.

26 American Library Association, *Public Library Service: A Guide to Evaluation, with Minimum Standards*, 7.

27 American Library Association, *Public Library Service: A Guide to Evaluation, with Minimum Standards*, 8.

28 James W. Fry, "LSA and LSCA, 1956 – 1973: A Legislative History," *Library Trends* 24, no. 1 (1975): 7–8.

29 George S. Bobinski, *Carnegie Libraries: Their History and Impact on American Public Library Development* (Chicago: American Library Association, 1969), 196 – 97.

30 *Fry, "LSA and LSCA, 1956 – 1973: A Legislative History."; Bobinski, Carnegie Libraries: Their History and Impact on American Public Library Development.*

31 Peggy Sullivan, *Carl H. Milam and the American Library Association* (New York: H. W. Wilson, 1976); Molz, *National Planning for Library Service, 1935 – 1975*, 75; Compton, "Looking toward Greater Federal Participation in Library Development," 57–58; Edward A. Chapman, "Theory and Prac-

tice in the Organization and Operation of WPA Library Service Projects," in *NASL Proceedings and Papers, 1938* (Springfield: National Association of State Libraries, 1939).
32 Harris, *History of Libraries in the Western World*, 264–65.
33 Fry, "LSA and LSCA, 1956–1973: A Legislative History," 9.
34 Library Services Act, 45 Code of Federal Regulations 130, (1956).
35 Fry, "LSA and LSCA, 1956–1973: A Legislative History."
36 Alex Ladenson, "Overhauling the Library Services & Construction Act," *Library Journal* 103, no. 10 (1978): 1026–27.
37 Pennsylvania Library Association Public Library Survey Committee, *How Good Are Pennsylvania's Public Libraries? (Exclusive of Pittsburgh and Philadelphia)* (1951, April).
38 *Library Services Act.*
39 U.S. Department of Health Education and Welfare Office of Education, *State Plans under the Library Services Act: A Summary of Plans and Programs for Fiscal 1957 Submitted under Public Law 597, 84th Congress*, US GPO (Washington, DC, 1958).
40 U.S. Department of Health Education and Welfare Office of Education, *State Plans under the Library Services Act, Supplement 1: A Summary of Programs for Fiscal 1958 Submitted under Public Law 597, 84th Congress*, US GPO (Washington, DC, 1959).
41 U.S. Department of Health Education and Welfare Office of Education, *State Plans under the Library Services Act, Supplement 1* (1959), 54–58.
42 Ernest E. Doerschuk, "Library Services Act in Pennsylvania," *PLA Bulletin* 16, no. 3 (1961); Ralph Upshaw Blasingame, "Pennsylvania Library Trends," *Pennsylvania Library Association Bulletin* 13, no. 2 (Fall 1957): 14–15.
43 Committee on the State Library Governor's Commission on Public Library Development in Pennsylvania, Minutes of First Meeting, 23 September 1960, Department of Education, Commonwealth Libraries, Minutes of the Governor's Advisory Council on Library Development 1962–2009 collection, Carton 1, Folder 1, PASA; Doerschuk, "Library Services Act in Pennsylvania," 142.
44 Pennsylvania State Library, Summary of Recommendations of the Pennsylvania Library Survey, 1958, District Services Collection, Series 2, Box 4, Folder 5, CLP-OLIVER.
45 Doerschuk, "Library Services Act in Pennsylvania."
46 Lowell A. Martin, "The Desireable Minimum Size of Public Library Units" (Ph.D. University of Chicago, 1945).
47 Arnold Miles and Lowell A. Martin, *Public Administration and the Library* (Chicago: University of Chicago Press, 1941).
48 *Committee on Post-War Planning, Post-War Standards for Public Libraries.*
49 *Committee on Post-War Planning, A National Plan for Public Library Service.*
50 Martin et al., *Library Service in Pennsylvania* (1958), iii.
51 Martin et al., *Library Service in Pennsylvania* (1958), 15.
52 Martin et al., *Library Service in Pennsylvania* (1958), 75–80.
53 Martin et al., *Library Service in Pennsylvania* (1958), 128.
54 Martin et al., *Library Service in Pennsylvania* (1958), viii.
55 Martin et al., *Library Service in Pennsylvania* (1958), 28.
56 Martin et al., *Library Service in Pennsylvania* (1958), 87.
57 Martin et al., *Library Service in Pennsylvania* (1958), 4.
58 Martin, "The Desireable Minimum Size of Public Library Units," 19.
59 Martin, "The Desireable Minimum Size of Public Library Units," 20.
60 Committee on Post-War Planning, *A National Plan for Public Library Service*, 5–6.
61 Martin et al., *Library Service in Pennsylvania* (1958), 87.

62 Martin et al., *Library Service in Pennsylvania* (1958), 89.
63 Martin et al., *Library Service in Pennsylvania* (1958), 105 – 06.
64 Martin et al., *Library Service in Pennsylvania* (1958), 109.
65 Martin et al., *Library Service in Pennsylvania* (1958), 129.
66 Martin et al., *Library Service in Pennsylvania* (1958), 130 – 31.
67 Martin et al., *Library Service in Pennsylvania* (1958), 165 – 66.
68 Kenneth Brown, State Aid to Libraries of Allegheny County, 1960, District Services Collection, Series 2, Box 4, Folder 5, CLP-OLIVER.
69 David L. Lawrence, "Address of Governor David L. Lawrence," *PLA Bulletin* 16, no. 2 (1960).
70 Carnegie Library of Pittsburgh, Annual Report, 1959, CLP-PENN.
71 Kenneth E. Beasley and Carl E. Robinson, *A Study and Recommendations of Library Districts for Pennsylvania*, Institute of Public Administration, The Pennsylvania State University (1962).
72 Lawrence, 19th Biennial Budget, March 1959, 103; David L. Lawrence, 1961–1962 Budget of the Commonwealth of Pennsylvania for the Fiscal Period June 1, 1961 to June 30, 1962, General Fund, Submitted to the General Assembly, January 1961, Government Documents Collection, LUL; David L. Lawrence, 1962 – 1963 Budget of the Commonwealth of Pennsylvania for the Fiscal Year July 1, 1962 to June 30, 1963, General Fund, Submitted to the General Assembly, January 1962, Government Documents Collection, LUL.
73 The Library Code, Pub. L. No. Act No. 188 (14 June 1961).
74 Lawrence, 1962 – 1963 Budget, January 1962, 108.
75 *The Library Code.*
76 *The Library Code.*
77 Ron Fink, *Libraries: The Story of State Aid 1962* (Harrisburg: Pennsylvania State Library, Public Information Office, May 8, 1963), 2.
78 Fink, *Libraries: The Story of State Aid 1962*, 7; Lawrence, 1962 – 1963 Budget, January 1962; William W. Scranton, 1963 – 1964 Budget of the Commonwealth of Pennsylvania for the Fiscal Year July 1, 1963 to June 30, 1964, General Fund, Submitted to the General Assembly, April 1963, Government Documents Collection, LUL.
79 Governor's Advisory Council on Library Development, Meeting Minutes, 11 January 1962, Department of Education, Commonwealth Libraries, Minutes of the Governor's Advisory Council on Library Development 1962 – 2009 collection, Carton 1, Folder 1, PASA; "Governor Appoints Twenty-Five Member Commission to Study Pennsylvania Libraries," *Pennsylvania Library Association Bulletin* 13, no. 4 (Spring 1958); Ron Fink, "State Library News," *PLA Bulletin* 20, no. 1 (August 1964).
80 Advisory Council on Library Development, *Rules and Regulations Governing Payment of State-Aid to Public Libraries and Approval of Plans for Its Use* (Harrisburg: Commonwealth of Pennsylvania, April 1962), 16 – 17.
81 Ernest E. Doerschuk, *Amendments to Rules and Regulations Governing Payment of State Aid to Public Libraries and Approval of Plans for Its Use* (Harrisburg: State Library, June 24, 1965); Donald C. Potter, "New Directions for the State Library's Bureau of Library Development," *PLA Bulletin* 23, no. 5 (August 1968): 296 – 97; Donald C. Potter, "Catching up with Yesterday," *PLA Bulletin* 24, no. 5 (July 1969).
82 Ernest E. Doerschuk, "View from the State Library," *PLA Bulletin* 30, no. 4 (July 1975); "Milestones in Legislative Support of Pennsylvania Public Libraries."; Joseph H. Myers, "The Library Services and Construction Act: Title II, Construction, in Pennsylvania, 1966," *PLA Bulletin* 22, no. 2 (November 1966).
83 Carnegie Library of Pittsburgh, Annual Report, 1962, 6, CLP-PENN.
84 Carnegie Library of Pittsburgh, Annual Report, 1962, 9.

85 Carnegie Library of Pittsburgh, District Library Center, Carnegie Library of Pittsburgh: A Brief History, 1963–1965, 1965, 1, District Services Collection, Series 2, Box 4, Folder 5, CLP-OLIVER; Carnegie Library of Pittsburgh, Annual Report, 1963, 10, CLP-PENN.
86 Carnegie Library of Pittsburgh, District Library Center, Carnegie Library of Pittsburgh: A Brief History, 1963–1965, 1965, 2.
87 Carnegie Library of Pittsburgh, Annual Report, 1963, 10.
88 Carnegie Library of Pittsburgh, Annual Report, 1963, 10.
89 Carnegie Library of Pittsburgh, Annual Report, 1963, 9.
90 Carnegie Library of Pittsburgh, District Library Center, Carnegie Library of Pittsburgh: A Brief History, 1963–1965, 1965, 3.
91 Carnegie Library of Pittsburgh, District Library Center, Carnegie Library of Pittsburgh: A Brief History, 1963–1965, 1965, 5.
92 Lowell A. Martin, *Progress and Problems of Pennsylvania Libraries: A Re-Survey*, Pennsylvania State Library (Harrisburg, 1967), 2; Carnegie Library of Pittsburgh, Annual Report, 1963, 11.
93 Keith Doms, A Public Library for the Churchill Area, 1963, District Services Collection, Series 2, Box 2, Folder 7, CLP-OLIVER; Keith Doms, A Public Library for Penn Hills, Pennsylvania, 1963, District Services Collection, Series 2, Box 2, Folder 7, CLP-OLIVER; Keith Doms, The Northland Public Library: A Cooperative Library for Ross Township, West View Borough, Mccandless Township, Franklin Park Borough, Pine Township, Bradford Woods Borough, Marshall Township, 1963, District Services Collection, Series 2, Box 2, Folder 7, CLP-OLIVER; Mary Lou Sebastian, Library History of Northland Public Library, 1 June 1968, PITT-HILL; Ralph Munn, Special Report: 1928–1964, 1964, CLP-PENN; Carnegie Library of Pittsburgh, Annual Report, 1964, CLP-PENN; Keith Doms, Regional Reference Center Proposal, 10 October 1966, District Services Collection, Series 2, Box 4, Folder 7, CLP-OLIVER; Ralph Munn, Public Library Service in Allegheny County: A Study, with Recommendations, Made at the Request of the Board of County Commissioners, 1961, District Services Collection, Series 2, Box 2, Folder 6, CLP-OLIVER; Doms, Regional Reference Center Proposal, 10 October 1966, 1; *Library Services Act*; Library Services and Construction Act, 20 USC Pub. L. No. 88–269 (1964); Keith Doms, Letter from Keith Doms to Library Directors, 1968, District Services Collection, Series 2, Box 4, Folder 7. CLP-OLIVER; Joseph F. Falgione, Letter from Joseph F. Falgione to Keith Doms, 8 May 1968, District Services Collection, Series 2, Box 4, Folder 7. CLP-OLIVER; Donald C. Potter, Statement from Donald C. Potter Re: Regional Reference Center, November 1968, District Services Collection, Series 2, Box 4, Folder 7. CLP-OLIVER; Keith Doms, Letter from Keith Doms to Donald C. Potter, 29 November 1968, District Services Collection, Series 2, Box 4, Folder 7. CLP-OLIVER; Carnegie Library of Pittsburgh, Annual Report, 1970, 16, CLP-PENN.
94 Carnegie Library of Pittsburgh, Financial Support of Public Libraries in Allegheny County, 23 March 1964, District Services Collection, Series 2, Box 4, Folder 5, CLP-OLIVER; Carnegie Library of Pittsburgh, Statistics for Allegheny County Libraries, 1965, District Services Collection, Series 2, Box 4, Folder 5, CLP-OLIVER; Carnegie Library of Pittsburgh, Public Libraries of Allegheny County (Exc. Pgh), June 1969, District Services Collection, Series 2, Box 4, Folder 5, CLP-OLIVER.
95 Carnegie Library of Pittsburgh, Fact Sheet on Allegheny County Libraries, December 1966, District Services Collection, Series 2, Box 4, Folder 5, CLP-OLIVER; Carnegie Library of Pittsburgh, Expenditures by Allegheny County (Outside Pittsburgh) Public Libraries for Last Complete Year (1966 or 1967), 1967, District Services Collection, Series 2, Box 4, Folder 5, CLP-OLIVER.
96 Carnegie Library of Pittsburgh, District Library Center, Carnegie Library of Pittsburgh: A Brief History, 1963–1965, 1965., Appendix III
97 Carnegie Library of Pittsburgh, Financial Support of Allegheny County Public Libraries, April 1966, District Services Collection, Series 2, Box 4, Folder 5, CLP-OLIVER; Joseph F. Falgione, Letter

from Joseph F. Falgione to James T. Golden, 19 August 1966, District Services Collection, Series 2, Box 4, Folder 5, CLP-OLIVER.

98 Martin, *Progress and Problems of Pennsylvania Libraries* (1967), 1.

99 Carnegie Library of Pittsburgh, Acceptable, Possibly Acceptable, and Other Libraries, 1966, District Services Collection, Series 2, Box 4, Folder 5, CLP-OLIVER.

100 Dorothy Bendix, Letter from Dorothy Bendix to Members of the District Center Standards Committee, 9 September 1966, District Services Collection, Series 2, Box 4, Folder 1, CLP-OLIVER; Committee on Standards for District Library Center Libraries, Committee on Standards for District Library Center Libraries, Summary of Discussion, 25 February 1966, District Services Collection, Series 2, Box 4, Folder 1, CLP-OLIVER.

101 Martin, *Progress and Problems of Pennsylvania Libraries* (1967), 1.

102 Martin, *Progress and Problems of Pennsylvania Libraries* (1967), 3.

103 Martin, *Progress and Problems of Pennsylvania Libraries* (1967), 16.

104 Joseph F. Falgione, State Funds for C.L.P. Under New Martin Plan, August 1967, District Services Collection, Series 2, Box 4, Folder 1, CLP-OLIVER.

105 Keith Doms, Letter from Keith Doms to Mrs. Franklin Malick, 11 December 1967, District Services Collection, Series 2, Box 4, Folder 1, CLP-OLIVER; Edmond J. Doherty, Letter from Edmond J. Doherty to Library Development Committee, 27 March 1968, District Services Collection, Series 2, Box 4, Folder 1, CLP-OLIVER; Pennsylvania Library Association, Library Development Committee Meeting of the Public Library and Trustee Sub-Committee, 19 April 1968, District Services Collection, Series 2, Box 4, Folder 1, CLP-OLIVER.

106 Dean C. Gross, Letter from Dean C. Gross to Ernest K. Doerschuk, 20 November 1967, District Services Collection, Series 2, Box 4, Folder 1, CLP-OLIVER.

107 Ernest E. Doerschuk, Agenda of Library Legislation Considerations in Martin Resurvey, 9 November 1967, 4, District Services Collection, Series 2, Box 4, Folder 1, CLP-OLIVER.

108 State Library of Pennsylvania, A Five-Year Plan: Pennsylvania Library Development, 1971–1976, Library Services and Construction Act, 1971, 62, District Services Collection, Series 2, Box 9, Manual 2, CLP-OLIVER.

109 Carnegie Library of Pittsburgh, Fact Sheet for Allegheny County Public Libraries (Excluding City of Pittsburgh), 4 June 1969, District Services Collection, Series 2, Box 4, Folder 5, CLP-OLIVER; Carnegie Library of Pittsburgh, Statistics for Allegheny County Libraries, May 1969, District Services Collection, Series 2, Box 2, Folder 2, CLP-OLIVER.

Period 4: Federation Revisited, 1970 – 1979

In 1970, the public library infrastructure in greater Pittsburgh continued to operate under the aegis of the state plan. The district library center, a role fulfilled by the Carnegie Library of Pittsburgh (CLP), completed its eighth full year. It served 50 local public libraries in the Pittsburgh district. As a regional resource center under the state plan, CLP continued to receive ILL requests from across the state. The library received $100,000 annually in state funds for this service over the previous four years (1967–1970). Other state-supported regional services supported by CLP included the Regional Library for the Blind and Physically Handicapped and the Regional Film Center.[1]

In addition to state-sponsored regional services, the public library system in greater Pittsburgh also continued to operate according to the county arrangement. The county bookmobile service, started in 1956 at CLP, was still going strong. There were 2 city bookmobiles, 3 county bookmobiles, and 1 "sidewalk van."[2] There were 42 stops around the county in places without local library service. The Board of County Commissioners approved a request for funding to replace a bookmobile that had been in constant use for more than 13 years. By 1970, there were 38 independent public libraries in operation in Allegheny County besides CLP.[3]

Some regional service initiatives were funded using a mix of county and federal funds. The year 1970 was the second year of the Allegheny County Regional Reference library, located in the Monroeville Mall. It was financed by a $140,000 grant from Allegheny County for books and materials and a federal grant of $125,000 under LSCA for operating expenses.[4] The year 1970 was also the third year of operation of Northland Public Library, which was a multi-municipal library located in the North Hills of Greater Pittsburgh. Resource-sharing networks between libraries had begun to emerge, and libraries also began to connect through electronic networks. The Pittsburgh Regional Library Center (PRLC), of which CLP was a member, maintained connections to the Ohio College Library Center (OCLC), establishing the first OCLC terminal outside of Ohio.[5] By 1977, 40 OCLC terminals were operational in 31 PRLC member libraries, including CLP, which was the first public library to join OCLC. [6]

Though a stable foundation had been established in the 1960s for regional services in the county, there were significant political, economic, and cultural changes underway since the state plan had been implemented. In 1970, the Public Affairs Room at CLP reported that subjects in great demand included "Drugs; Environment; Women's Rights, Black History and Culture; Race Relations; Vietnam; Middle East; Poverty; Urban Problems; Education Problems; Youth; Protest; Generation Gap; and Crime."[7] On the social and political fronts in the U.S., the civil rights

https://doi.org/10.1515/9783111013404-006

movement in the 1950s and early 1960s led to the Civil Rights Acts of 1957, 1960, and 1964. In his State of the Union address in 1964, President Johnson announced a "War on Poverty." The Economic Opportunity Act was later passed in 1964, and the Department of Housing and Urban Development was formed in 1965. President Kennedy was assassinated in 1963, Malcolm X was assassinated in 1965, and Martin Luther King, Jr. was assassinated in 1968. Following the assassination of King, riots occurred across the U.S. Robert F. Kennedy, the brother of the late President, was assassinated later in 1968. Neil Armstrong became the first person to set foot on the moon in 1969. Abroad, the beginning of 1970 marked the fourteenth year of the war in Vietnam. President Richard Nixon had begun troop withdrawals amidst American deaths abroad and anti-war protests at home. By all accounts, the 1960s and 1970s were an extraordinary and tumultuous period in American history.

By 1970, public library leaders in greater Pittsburgh were beginning to grapple with what would become major cultural, social, and economic changes. The first major change underway was economic—a shift from a primarily manufacturing economy to a primarily service- and technology-based one. One catchphrase of the day used to capture the state of the economy was post-industrial.[8] Deindustrialization was occurring nationwide as heavy industry and manufacturing jobs moved overseas and as international competition increased. Between 1950 and 1960, the service sector in the U.S. surpassed the manufacturing sector. This change was reflected in part by changes in occupation categories.[9] The economy of greater Pittsburgh followed this larger pattern. Greater Pittsburgh was directly affected by this economic change given how industrialized the region had been.

Pittsburgh was an industrial city throughout the first half of the twentieth century. Public policy in the region concentrated on maintaining this industrial base and managing its social and environmental costs. In the 1950s, for example, urban policy in the City of Pittsburgh concentrated on revitalizing the city's downtown and central business district. The 1950s in Pittsburgh became known as part of the city's first Renaissance period (1945 – 1970), characterized by pollution and flood control as well as downtown revitalization projects.[10] There were efforts to revitalize the dilapidated and deteriorating housing, led by ACTION-Housing from 1957 onward.[11] Throughout Renaissance I years, however, the economic base of greater Pittsburgh went unchallenged—it remained centered on coal, iron, steel, heavy electrical machinery, glass, clay, and stone products.[12] The years 1970 to 1980 were therefore a time of economic transition away from manufacturing and toward services. In the early 1970s, greater Pittsburgh was on the cusp of this significant economic change.[13] The unemployment rate rose in greater Pittsburgh in the mid-1970s, and it continued to stay high in the early 1980s as the national economy entered a recession.[14] The shift away from manufacturing caused substantial upheaval. This was due to how the cultural and social life of greater Pittsburgh for a century developed around mills and industrial towns. The energy crisis in the early and late 1970s, combined with rising inflation, further exacerbated the economic situation in greater Pittsburgh.

The second major change in greater Pittsburgh around 1970 was demographic. Despite public and private investment in the city center, and due in part to economic changes, improvements in transportation, and advances in communication technology, Allegheny County and the City of Pittsburgh faced population decreases between 1950 and 1970. This was counterintuitive given how overall the population of the United States was steadily increasing. The population of the City of Pittsburgh showed a decrease from 1950 to 1960. Similarly, the population of Allegheny County decreased from 1960 to 1970. Likewise, all four cities in Allegheny County exhibited population decreases from 1950 to 1960.[15] This exodus from the cities in greater Pittsburgh reflected a larger pattern across the country. Other industrial cities in the East and Midwest also witnessed population declines at the same time. The same pattern did not apply to city centers in the South and West, however, because they were younger, had different economic bases, and were still destinations of western expansion. Because greater Pittsburgh was a relatively older, industrial-based area, it fit the pattern of population decline.

Given the steadily growing national population, where were people locating themselves within the country if not industrial city centers? A further demographic change occurring in greater Pittsburgh around 1970 was an increase in the metropolitan population. As the city centers were decreasing in population, the population of urban peripheries was increasing. In other words, people gathered outside of the older city centers, forming more dispersed and distributed urban and suburban areas around the cities. One word used to describe this population shift was suburbanization. Urban fringe areas were growing while city centers and rural towns were declining.[16] By 1950, more people in the U.S. lived in metropolitan areas than in nonmetropolitan areas. By 1970, the metropolitan population increased by 60 percent compared to 1950. In Allegheny County, though the overall population began to decline from 1960 to 1970, several urban areas outside the city grew between 1960 and 1967.[17]

In addition to a changing economy, a shrinking urban core, and an expanding urban periphery, there was also a shift in the racial demographics of greater Pittsburgh. Pittsburgh was a destination for African Americans who traveled northward from the rural southern U.S. during the Great Migration. The Great Migration was a time when African Americans moved out of the South to escape economic hardship and oppressive racist conditions known as Jim Crow. In the second phase of the Great Migration, which lasted from 1940 to 1970, there was a slight but steady climb in the African American population in the U.S. overall.[18] The change in the City of Pittsburgh was more pronounced than what was occurring nationally or even regionally. [19] In 1940, the percentage of African Americans living in the City of Pittsburgh was below the national percentage, at about 9 percent. But by 1950, the percentage of African Americans in the city exceeded the national percentage—it was over 12 percent. The city's African American population grew further in 1960, becoming nearly 17 percent. By 1970, the percentage of African Americans in the City of Pittsburgh was over 20 percent. This meant that 1 person in 5 residing in the city was African American. This percentage was well above the national level, and it was more than double the proportion it had been in the city in 1940. The City of Pittsburgh therefore had a distinct demographic composition, one that was not generally reflected in other areas of Allegheny County outside the city.

The Civil Rights movement in the US during the 1950s and early 1960s led to significant civil rights legislation, including the Civil Rights Acts of 1957, 1960, and 1964. The 1964 legislation, in part, prohibited racial segregation in schools and public housing. Nevertheless, greater Pittsburgh remained a largely segregated city by 1970. Though the Great Migration brought many African Americans from the South to Pittsburgh's city center, African Americans were segregated there by neighborhoods. Following the Great Depression, there were large increases in the African American populations in neighborhoods such as the Hill District, East Liberty, Homewood-Brushton, Beltzhoover, North Side, Manchester, Homestead, Rankin, and Braddock.[20] This segregation occurred partly due to the building and planning efforts of the City of Pittsburgh, guided by federal policies. The City of Pittsburgh faced a housing crisis in the 1960s.[21] There was a shortage of housing available for low-income families. Public housing projects, led by ACTION-Housing, were created in neighborhoods such as East Hills, Sheraden Park, Hill District, Homewood-Brushton, Hazelwood-Glenwood, and Perry Hilltop. African Americans were located there. This geographical segregation caused segregated schooling. By 1965, as the proportion of African American children attending school in the City of Pittsburgh increased, the segregation of the district became more pronounced. Of 87 total elementary schools, 15 (17.2 percent) had a student population that was 80 percent or more African American, and of 15 total secondary schools, 4 (16 percent) had a student population that was 80 percent or more African American.[22] Pittsburgh formed white and black neighborhoods as well as white and black schools.

In addition to geographical and educational segregation, African Americans in Pittsburgh were disadvantaged in terms of economic opportunity and decision-making power. There was unequal distribution of wealth and resources. Further, African Americans suffered forced displacement by city planners. In the 1940s, 1950s, and 1960s, many African American families and businesses in Pittsburgh were displaced due to revitalization projects, implemented under the mantra of "renewal" and "rebirth." One such project was in the Lower Hill, part of the historic and predominantly African American Hill District. Led by the City Planning

Commission and the Urban Redevelopment Authority, demolition in 1956 displaced 1,551 mostly African American families and 413 businesses.[23] This led communities in the 1960s to form Action Committees to inform public housing projects.

Given such dramatic social and economic changes, the new and emerging "urban library" was a salient topic in library literature in the late 1960s and early 1970s. The changing nature of metropolitan public library systems was not just a national but an international topic. Referring to urban metropolitan libraries internationally, researchers asked how libraries might adapt.[24] In the U.S. specifically, library planners struggled to understand how urban public library systems might continue to function despite declining populations in the city centers, the migration of white middle-class residents to the suburbs, a shifting racial demographic within the urban core, and economic changes that included deindustrialization, inflation, and recurring energy crises. These changes affected public libraries on a national level.[25] A series of national studies in the 1970s explored how to restructure public library financing in a way that would increase and distribute funding in a more equitable way.[26]

At the same time, while also continuing to serve a white middle-class demographic in the peripheries, public libraries had to address new concerns associated with race and poverty within the city.[27] The social, economic, and cultural changes in the 1970s presented new problems to urban library planners.[28] The changing economic base and population of the metropolitan area presented a dilemma: how could urban public library systems serve both affluent, largely white segments of users in the urban peripheries and also serve poorer, mostly African-American users in the city center?

Commenters came down on both sides of this dilemma. Some argued for the importance of library services for the urban core. There was an increased sensitivity to economic and educational disparities between whites and African Americans. Public libraries were seen to be critical for providing low-income families in urban areas with enriching childhood experiences. Libraries were positioned to offer early interventions in intellectual development.[29] Some argued for focusing on the urban core not just for benevolent or professional reasons, but also for strategic and financial reasons.[30] In contrast to the views of those who emphasized public library services to the urban core, other commentators recommended emphasizing library services that targeted specialized research areas. Lowell Martin, who conducted the Pennsylvania survey in 1958, and who designed the Pennsylvania state plan, weighed in on the debate following an extensive study of the Chicago Public Library system. [31] The Chicago system faced even more exaggerated urban challenges than CLP.[32] Martin's recommendation echoed a report two decades earlier by Bereleson of the Public Library Inquiry.[33] In his 1949 report, Berelson found that, despite pretensions that public libraries serve the poor and uneducated class, the primary clientele of the public library was in fact the white educated elite. Martin suggested that public libraries continue to maintain close relationships with this primary clientele.

To be sure, library services to people of color within the City of Pittsburgh were far from non-existent by 1970. CLP maintained enhanced library services to the Hill District, for example, and services were informed by a local advisory committee there.[34] In light of the changing circumstances within the county, however, in 1969, the Board of County Commissioners of Allegheny County funded a survey of the public libraries in the county. The final report was published in October 1970.[35] The charge to the director of the study was "To prepare a plan for the provision of adequate public library service to all residents of Allegheny County utilizing existing public libraries, and particularly the Carnegie Library of Pittsburgh, as a base."[36]

The commissioners hired Ralph Upshaw Blasingame, Jr. to lead the study. Blasingame was a leading voice in the literature related to the demographic changes underway that affected urban libraries. By this time, Blasingame was well-known in Pennsylvania and nationally. From 1957 to 1964, he served as Pennsylvania State Librarian. There, he oversaw the passage and early implementation of the state plan that was adopted in 1961. It was Blasingame who hired Lowell Martin to conduct the study that led to the state plan. He was a student of Lowell Martin while working on his dissertation.[37] His dissertation was titled *The Public Library as an Urban Phenomenon.*[38] The Blasingame survey from 1970 was titled *Organization for Public Library Services in an Urban Field.* The concept of an urban field was central to Blasingame's understanding of public library infrastructure. Though the concept was established and applied in the 1970 study, it was further elaborated in Blasingame's dissertation in 1973. The urban field referred to social and economic conditions associated with a declining population in city centers and a growth in a distributed network of urban areas around the city.[39] The concept of an urban field meant that there was no longer a center and periphery—urban and rural, city and suburb—but instead there existed a network of multiple centers that were broadly distributed across a wide geographical area. The urban field was thought to accompany not only demographic and economic change but also cultural change. According to Blasingame, public libraries were products of the industrial era. It was not social and mechanics libraries that led to or created public libraries in the mid-nineteenth century. Rather, it was industrialization—the development of an urban-industrial society at that time. Public libraries were the products of cities. Public libraries served a function: they supplied specialized knowledge and recreational reading materials to industrial workers. Cities of the mid-nineteenth century were markedly different than those previously—with respect to urbanization and the volume of population that resided there—and that is why public libraries developed in conjunction with modern industrial cities instead of earlier cities. With the decay of the physical and spatial structures exhibited by industrialization, Blasingame thought old city centers in their traditional form had come to an end. The population had shifted to the urban periphery—people commuted in to work in the cities but resided outside of them. People could take advantage of the urban cultural and educational institutions of the old cities, but not live there. Cities saw an influx of low-income populations, and the cities became less able to support public services. These changes coincided with a shift in the economy from manufacturing to services, which characterized a post-industrial society.[40]

Some vestigial structures that developed during the industrial era lived on in the post-industrial era. Blasingame thought that public libraries were one such structure—the basic features of public libraries had not changed. Blasingame called the antiquated structure of public libraries the "prototype library." Prototype libraries had several characteristics: they were local agencies, governed by an elite board and administered by a bureaucracy; their institutional research focused on procedures, not goals; they had a central library; their administrators were not specially trained and their staff were generalists; they had few connections to state and federal government; they had competing service commitments to serious study and leisure reading; and the need for their services was not as apparent as it was for other infrastructures.[41] There was a need for a fundamental change in the public library paradigm.[42] The industrial-era structure and purpose of public libraries, supported by a specific demographic, had to adapt to new circumstances.

Blasingame warned against public libraries retaining their traditional structure given the significant societal changes.[43] While some observers predicted that city centers would soon revert to their previous economic and social composition, it was generally acknowledged that urban public library systems had to restructure in fundamental ways to accommodate new clientele and a changing demographic.[44] One recommendation was an urban library network that crossed municipal and other boundaries.[45] Other general solutions included either the federation or consolidation of the city library system with the regional authorities, together with a two-tier administrative structure that coordinated regional services and accommodated local autonomy.[46]

In their 1970 survey of Allegheny County public libraries, Blasingame and his team examined library reports and census data, projected future growth and changes in the county, and conducted field visits and interviews with librarians, trustees, and other officials. The district services of CLP provided the surveyors with fact sheets about the bookmobile and service areas of the county libraries. From 1963, the data-gathering practices of CLP increased as its role as district library center began. The study surveyed 31 of the 36 libraries in Allegheny County.[47]

The survey reported several empirical findings. In general, the collections of the county libraries were not keeping pace with publishing trends: "The most reasonable conclusion one can draw from the data is that, in view of the rapidly growing output of the publishing houses, these collections are gradually losing ground."[48] Salaries and wages at county libraries were also generally low, making it difficult for the libraries to attract and retain skilled workers. Moreover, library buildings were generally old and small. Though funding to CLP for city and county services had risen substantially, there was still a noticeable gap in public library services for residents in the city compared to those residents in the county but outside the city. Surveyors found that the number of people in the county who were served by a local library was 486,099, while the number of people with no local library service was 578,901.[49]

The views of county librarians were mixed, but most believed their problems stemmed from a lack of financial support, particularly from the county. Most county librarians viewed the district services of CLP positively, and most supported the idea of contracting between either CLP or an Allegheny County Library in order to provide county library services. Librarians were resistant, however, to the idea of a county library established as part of the county government that was distinct from CLP and would place governance of the system outside libraries' hands.[50]

The surveyors identified several trends in the county. The first trends were demographic: people were continuing to move away from the strong, central library in the area, CLP, they were moving to areas that did not operate local public libraries, and there was a need for new services, both formal and informal, that could help people transition to new occupations. Other trends were educational. Students of all ages were increasingly traveling farther from their homes to seek library services, and any student who commutes to college also seeks library service near home, which will overburden local libraries around the county that are not well equipped to assist college-level students.[51] Without some county-wide plan, the problems presented by these trends would continue to become more pronounced.

To address the growing challenges of public library service in the county, surveyors focused on the question of what roles CLP should play in a larger county system and what that larger system might look like. CLP was already contracting with the county, and it was receiving state aid for its roles of district library center and regional resource center. Beyond these cooperative and system-level efforts, surveyors found no other cooperative efforts among the other libraries in the county. Except for the services received by county libraries from CLP, there was not any other visible way that county libraries shared materials, services, staff, or planning responsibilities. CLP was already overextended, the surveyors felt, and therefore the surveyors did not recommend that county libraries extend contracts to CLP for additional improvements.[52]

Blasingame and his team did not recommend the establishment of a new county department responsible for library development in Allegheny County. The appointment of a county librarian would solve the problem of countywide planning, which CLP was not able to do. It would place the responsibility of countywide planning in the hands of a single individual or a single department. It would also allow the independent county libraries some degree of flexibility and independence—they would each still be governed by their own boards. The downside of this option, however, was that the county librarian or county department would not be accountable to any policy-making group of citizens.[53] Because of problematic governance issues, the establishment of a new county government department did not appear to be a viable option.

Rather than establish a new county government department to oversee public library planning in Allegheny County, the Blasingame study recommended that there be a headquarters for an Allegheny County Library System established at CLP in Oakland. Independent county libraries could opt-in to the new system, but if they did, they would still retain their autonomy and remain governed locally by their own boards. The new system would be governed by a representative board composed of members from different county constituencies.[54] The surveyors recommended this new restructuring despite potential complications related to policies, funding, board composition, administration, and staffing.[55]

Blasingame and his co-authors recommended an 11-member governance board. Ex-officio members were thought to include county commissioners and the mayor of the City of Pittsburgh. Members at large could be selected by county commissioners, library directors, and trustees. Several board powers were recommended, included setting objectives for the system, establishing a budgeting procedure, establishing policies, and approving professional appointments.[56] Several other recommendations were made regarding programming and services, organization, and personnel.

In terms of funding, the consultants saw the current situation as unsatisfactory. One problem was that the present funding did not enable complete coverage of public library service throughout the county. Another problem was that CLP did not adequately utilize its county appropriation to provide county services. CLP "continued to grow while public library services in nearly all other parts of the County have grown only slightly, if at all."[57] In order to address the funding issue, the consultants recommended funding to all county library services on a per capita basis according to a distribution formula: the onus would rest on CLP to restructure and form a new county library headquarters. The county commissioners would fund this new office. The office would be responsible for overseeing county library development and distributing county funds amongst the county libraries to support county library services. The distribution would occur according to a formula that was created by the board. Funding distribution would be implemented by the new county library office, but decisions about the composition of the formula and the nature of the countywide planning would be made by the new county library board.

Some recommendations of the Blasingame study resembled those of earlier studies. Like the plan proposed by Ralph Munn in 1958, the Blasingame plan proposed an opt-in, federated system of county libraries. This federated structure differed significantly from what had been proposed by Brahm and Sandoe in 1950 and the Metropolitan Study Commission of Allegheny County in 1955.[58] These studies recommended a two-stage process leading to a unified system administered by the county government. In the first stage of unification, a dual system would exist in the county: one system, CLP, would operate alongside a second system consisting of all other public libraries in the county. In the second stage, these two systems would merge to form a single system administered and funded by the county government. While the unified system would have subsumed the administration and funding of all libraries under county government, the plan proposed by Blasingame, reflecting the earlier proposal by Munn in 1958, called for a federated system with a mix of local and centralized control. Under the federated system, county libraries could opt in to receive the benefits of centralized services, such as purchasing and cataloging. Decisions related to services and funding would be made by a representative board. At the same time, independent libraries would retain board governance to decide local issues. System funding was also mixed: funding would flow from the county to support countywide service, but local municipal funding would also remain under the control of local libraries. County funding would not replace local funding, only supplement it in such a way to support county services.

Following the completion of the Blasingame report in 1970, there followed countywide discussions about the report's proposals. These discussions were organized by CLP and held at the request of the Board of County Commissioners.[59] Copies of the report were distributed to the Board of Trustees of CLP. Copies were also distributed to the 38 independent library boards in Allegheny County. A discussion session was held with library boards on October 16, 1970. At least 32 people attended the October 16 meeting, including representatives from libraries from around the county and representatives from the state library.[60] Follow-up discussions were held at county libraries, including Clairton, Mt. Lebanon, Upper St. Clair, Dormont, and Northland.[61] Senior staff of CLP endorsed the Blasingame plan in principle,[62] but the county discussions revealed a number of concerns regarding local funding, governance, and legality. An administrator at CLP noted concerns from county library boards.[63] In response to these concerns, CLP reached out to other county library systems, for example in New York and California, to determine how they solved problems related to funding.[64] Representative from CLP were in contact with the county commissioners.[65] In light of discussions with the commissioners, CLP created a funding formula for a grant system for the independent libraries in Allegheny County.[66] This formula was created with the vision that any county library program should be a grant program.[67]

While discussions at the county level occurred between CLP, the independent libraries in the county, and the county commissioners in response to the Blasingame study, other discussions happened simultaneously between citizens groups in the county and at the state level. A representative from CLP was in close communication with Dorothea M. Strang as early as 1971. Strang had been a member of the Library Planning Committee, formed in 1955, which petitioned the county commissioners in 1955 for countywide services.[68] The effort for countywide library service at that time was successful. Strang had also been a member of the Citizens Committee for County Library Service, formed in 1960 to petition the commissioners for countywide services.[69] This effort, coordinated with Ralph Munn, was sidelined when it became apparent that the state plan would be implemented in 1961. Though the state plan had been implemented since that time, resulting in the formation of the district library center role for CLP, Strang still actively advocated for improved county library services.

In February 1971, CLP kept Strang updated on developments at the county level with respect to the Blasingame discussions.[70] Separately and using different channels, Strang advocated to the county commissioners and library trustees for the implementation of the recommendations proposed in the Blasingame study. At the same time, there was a movement by the Trustee's Section of the Pennsylvania Library Association which called for a mandating of county libraries by state law. Strang noted that the legislative action was being taken up at the state level by the Pennsylvania Library Association.[71]

The Pennsylvania State Library created a long-range plan in response to the 1970/71 passage of LSCA.[72] The plan suggested continuing state aid to metropolitan or other libraries serving as regional resource centers, also state aid to district library centers, local library units, and outreach grants using federal funds, and development of state union catalog.[73] In Allegheny county, Mt. Lebanon, Penn Hills, and Northland were identified as grant recipients for library building projects.[74] Then, in 1973, the State Library of Pennsylvania continued to refine the statewide system.[75] The efforts to mandate county libraries statewide were unsuccessful. Meanwhile, talks at the county level to restructure CLP and establish a county library system stalled on all sides.

By 1973, a new County Government Study Commission was underway. The "question to create a Government Study Commission for Allegheny County was placed on the November, 1972 ballot by unanimous action of the County Commissioners."[76] This effort was similar to the Metropolitan Study Commission of Allegheny County of 1955. The charges of the 1973 commission were similar. The purpose of the study was to determine the most economical way to structure county government.[77] If the resulting Home Rule Charter were adopted, it would replace the Second Class County Code as the law of the county, thus placing the primary responsibility for county decision-making with the people of the county, not legislators in the State Assembly.[78] The similar, earlier efforts at Home Rule from 1955 had failed to carry in the subsequent referendum.

The Government Study Commission offered a new opportunity for county library advocates to present their case. Dorothea Strang and others presented arguments for a county library before the commission. Strang presented on behalf of an Ad-Hoc Committee to Improve Public Library Service in Allegheny County. She noted the consistent themes of county service. Bee Hirschl also petitioned the commission.[79] In the end, there was no interest by the commission in the idea of a county library system.[80] In response to Strang's and other's arguments, an editorial ran in Pittsburgh Post-Gazette that sided against the formation of a county system, siding instead with the idea of shared cooperative arrangements between local libraries.[81]

In the resulting home rule charter, a new Department of Community Services was mentioned. The duties of the department were said to include the coordination of county participation in and funding of library service.[82] This change would have implemented a new county office to oversee county library development, precisely what the Blasingame report had advocated against. In the charter, there was no mention of the formation of any unified public library system in the county. In the end, the home rule charter was defeated in 1974 by majority vote, with a similar charter defeated again in 1978.[83] Table 4 compares features of the county library system proposals from the 1970s with those from earlier and later studies.

Table 4: Comparison of county library system proposals, 1924 – 1991.

Date	Author	Proposal	Stages	Federation/ unification	Funding	Governance
1924	Civic Club	extend CLP services to county schools	unknown	unknown	unknown	unknown

Table 4: Comparison of county library system proposals, 1924 – 1991. *(Continued)*

Date	Author	Proposal	Stages	Federation/ unification	Funding	Governance
1938	Ralph Munn	first, extend bookmobile services to areas in the county not served by a local library, then, local libraries gradually join unified county system	2-stage	unification	First, county provides funding to CLP for bookmobile services. Then, libraries gradually join single system	county libraries gradually merge into unified system headed by CLP
1939	Lars Grondahl, Civic Club	bookmobile services to rural areas, plus a countywide system headquartered at CLP	unknown	unification	county	county libraries form unified system headed by CLP
1940	Various civil society groups	Extend bookmobile services to the county	unknown	unknown	county	unknown
1950	Brahm and Sandoe	county library under county government, with county board; city and county exist side-by-side at first, then merge	gradual	unification	county eventually funds entire system	county library system, organized under county government, governed by library board, single administrator
1955	Metropolitan Study Commission	county system form with CLP as nucleus, incorporate interested county libraries	immediate	unification	county	CLP transferred to county, city board becomes county board with county representatives
1955	Library Planning Committee	bookmobile services	immediate	unknown	county	unknown
1955	Ralph Munn	bookmobile services, free borrowing privi-	2-stage	unification	county	contract between county and CLP for CLP

Table 4: Comparison of county library system proposals, 1924 – 1991. *(Continued)*

Date	Author	Proposal	Stages	Federation/ unification	Funding	Governance
		leges, interlibrary loan, centralized purchasing and cataloging; vision of eventual unification still in place				to provide partial county services
1956	Allegheny County, Ralph Munn	county contract with CLP for county services; partial acceptance of Munn's 1955 plan	immediate	semi-federated	county	CLP provides ILL, borrowing privileges, reference, bookmobile service; funded by county
1958	Ralph Munn	opt-in system with centralized services	immediate	federation with unification as distant goal	county provides financial aid to those libraries that join federation	local autonomy, centralized processing at CLP, some distribution to local libraries using formula
1958	Lowell Martin	Pennsylvania state plan, tiered service levels for roles of local, county, district, and regional research libraries	immediate	federation	state funding for local libraries, county libraries, district centers, and regional research centers	state aid contingent on meeting minimum standards for each role
1961	Ralph Munn	create regional reference centers, restructure library service areas on school districts	immediate	federation	municipalities and/or school districts; county funds CLP county services	service areas determined by school districts; funding and governance decided by municipalities and school districts
1970	Ralph Blasingame	federated system headquartered at CLP, re-	immediate	federation	municipal funding with supple-	representative city-county CLP

Table 4: Comparison of county library system proposals, 1924–1991. *(Continued)*

Date	Author	Proposal	Stages	Federation/ unification	Funding	Governance
		constitution of CLP's board to include reps from county			mentary funding from county	board; local boards
1974	Allegheny County Government Study Commission	establish county department to oversee library service	immediate	federation	local funding remains; county funding used for county services	county department, county board contracts with CLP and other libraries to develop county services
1978	Citizens Study Committee on Libraries; Ed and May Dowlin	independent county board, advisory council, executive director	immediate	federation	county funding to be distributed by county board; continuance of local funding	independent county board (citizens), advisory council (librarians), executive director
1990	CLP President's Advisory Commission	proposed 3 options for county library system; 4 service tiers; CLP oversees system	immediate	hybrid	county funding for county services; maintain local and state funding	unclear: CLP directs 4-level system but libraries somehow remain autonomous
1991	Frank Lucchino, *A Quiet Crisis*	county funding to independent libraries; formation of county government office; appointment of county library director	phased over a 5-year period	federated	county funding for county services; maintain local and state funding	county library director in county government office oversees federated system

Despite some initial inertia, in the years immediately followed the Blasingame study, the movement toward a change in the county library system in Allegheny County lost momentum. Joseph H. Myers of the Pennsylvania State Library was in contact with Dorothea Strang of Allegheny County, who was a proponent of a county system. In a memo, Myers summarized two options for how a county library system could be formed.[84] First, the existing contract that CLP had with the county could be enlarged to further extend service where none yet existed. CLP would also contract with the other local libraries for certain services. Second, a county library board could be formed by the county commissioners, and that board could contract with CLP and the other libraries for certain countywide services. A paid county coordinator would oversee operations and state aid payments. Neither option was adopted. The State Library continued to promote county libraries.[85] Strang continued to try to petition for a county library system by approaching the Pennsylvania State Library and the county commissioners.[86] But by 1975, the issue of county library formation no longer had substantive form.

Then, in 1976, a budget shortfall within Allegheny County renewed the issue. In the background were dramatic social and economic changes. Unemployment and inflation spiked nationwide and in greater Pittsburgh in 1975. The county commissioners had also changed.[87] County government was a significant source of public library funding across Pennsylvania, and Allegheny County had the largest appropriation. [88] In 1976, the county reduced its appropriation to CLP by $422,000, from $2,222,000 to $1,800,000.[89] The library was aware of the potential cut and attempted a campaign to try to avert it.[90] The funding cut went through, and some county services were forced to close.[91]

In light of the budget cut from the county, CLP drew up a detailed plan for county services.[92] This definite, detailed proposal for a county library system balanced the library's responsibilities to the state and those to county residents.[93] The plan addressed two primary areas of concern raised by the Blasingame discussions: governance and funding. To solve the governance problem, the plan proposed that the board of CLP be reconstituted to include not only city but also county representation. The resulting 9-member board would be composed of 3 members from the city of Pittsburgh and 6 from the county.[94] The county members would be employees or trustees from other libraries. In addition to restructuring the board, the director position of CLP would be retitled Director of the CLP and Allegheny County. An additional administrative position would be created to oversee countywide development. In early 1978, this governance proposal would become a reality. On January 1, 1978, CLP board was restructured with permission of the Orphans' Court to include four county representatives (the three county commissioners or their appointees and a high-level county official.[95] This change made it possible for the CLP board to govern any future county library system.

To solve the funding problem, the plan proposed a funding formula modeled on the formulas used by other county libraries across the state.[96] Funding for libraries across the county would use a combination of local and county funds on a per capita basis. Local libraries meeting state standards would qualify for county aid. Some county funding would go directly to local libraries, and local library boards would determine how to use that county aid. Other county aid would be distributed by the board of CLP and Allegheny County for countywide services. The countywide money might be used for such things as bookmobile services, county services provided by local libraries, shipping, storing, and interlibrary loan, processing, and administrative costs.[97] In effect, this plan drafted by CLP worked out the implications of the recommendations proposed by Blasingame.

Meanwhile, citizens groups in Pittsburgh worked directly with the state library to try to obtain support from the state for a county system led by CLP.[98] One citizen, Dorothea Strang, mediated between the state library and the county commissioners. The state librarian was supportive of a new agreement between Allegheny County and CLP, one that would improve library services across the county and CLP in light of recent setbacks.[99] The state library recommended that the county commissioners restore funding to CLP, appoint county residents to the CLP board, and create a study group that would study the library situation in the county and make recommendations.[100] CLP had anticipated some of these points when it changed the composition of its board. It was left to the commissioners to form a study group and implement any resulting recommendations. CLP and the county began to jointly plan and carry out a new study of library services, one that would assess current needs in the county and recommend improvements to library services.[101] Notably, the study's grant application named a County Librarian position to oversee the project. [102] The grant was funded in 1976, and the study group was formed in 1977. [103]

The study project envisioned by the county and overseen by the state library included both a citizens' committee and a research team, led by consultants. The composition and structure of this new study differed from previous county library surveys that used consultants only, such as Brahm and Sandoe and Blasingame.[104] These previous studies did not include the committee component, and the research and recommendations were authored entirely by the consultants. By contrast, in the new study, the decision-making was to be performed by a citizen-led committee vested with the power to hire consultants, then develop and implement a plan for improving countywide library service. The purpose of the committee was "to improve citizen input into the provision of library services by working with a staff to develop a plan, approving the plan and submitting it to the county commissioners for action."[105] Research undertaken by the consultants and their assistance was to form the basis for the committee's plan. The inclusion of an advisory council or a citizens' planning group reflected new federal requirements and contemporary trends in project planning.[106]

Around the time the county's study project was beginning, the League of Women Voters began to study the county libraries on their own.[107] This parallel study began with the belief that independent libraries could join together in order to claim additional state aid as a county system.[108] Members were also interested in documenting how inflation had affected the purchasing power of libraries, and how this decrease in purchasing power necessitated increases in state aid.[109] It was thought that the state's ceilings on local aid, district aid, county aid, and regional resource center should be removed.[110] In the end, the league's work did not produce an independent study, but their fact-finding efforts contributed to the final report issued by the Citizens Study Committee.[111]

The citizens study faced problems of staffing and membership. Staff and members were difficult to find or quickly resigned.[112] This made the composition of the group not truly representative of various constituencies around the county. By August 1977, two consultants were hired to work with the Citizens Study Committee: Charles Edwin "Ed" Dowlin, who previously worked at the State Library of Ohio and was the former State Librarian of New Mexico from May 1970–July 1977, and his wife May.[113]

It is notable that the Dowlins were hired as consultants for the Citizens Study Committee project. This is because though Ed Dowlin had served on several committees in ALA, he was a relatively unknown figure in the public library world. Ed was the brother of Ken E. Dowlin, also a librarian.[114] It is not clear what if any library or research experience May had. Compared to Lowell Martin or Ralph Blasingame, for instance, the Dowlins had no previous research experience. It is not evident that the Dowlins were familiar with contemporary thinking on consulting work, public library standards, not to mention public libraries in greater Pittsburgh. By contrast, Martin had approached his consulting work with extensive previous research experience: his perspectives on libraries were informed by his time at the Chicago Public Library, his research with Carleton Joeckel, and his development of national library standards. Blasingame served as the State Librarian in Pennsylvania, he had complex views on the nature, history, and purpose of urban libraries, and he had conducted library surveys before. Both Martin and Blasingame approached their consulting work with some preformed theoretical focus in mind and extensive experience. The Dowlins, by contrast, did not present a clear theoretical focus, they had no research background, and they had no familiarity with the Pittsburgh situation. It was perhaps the Dowlins' personal network that led to their hiring.[115]

By February of 1978, the Dowlins circulated to committee members a written draft of preliminary observations regarding the current state of library services in the county. The draft alluded to policy conflicts within the county that hindered further library development. The draft also established the consultants' position that they would "create awareness of the clinical process of identifying, comparing, and modifying basic assumptions which allows the participants to develop joint policies."[116] The consultants contrasted their approach with superimposition and negotiation.[117] The consultants believed that past developments in the county library system had been superimposed by either the county or the state, that additional money led libraries to comply, but that additional funding would not be forthcoming in the future. The consultants saw themselves as discussion facilitators who would assist the study group to identify a solution of their own design.[118]

In March, April, May, and June of 1978, the consultants and members of the committee visited libraries across the county, and they collected statistical information about the county's population and libraries' circulation activities.[119] By July 1978, the consultants and the committee completed some preliminary research and discussions.[120] The winter had delayed travel to libraries and affected attendance at meetings due to energy shortages and coal strikes.[121] Research and writing were rushed. By July 1978, the Citizens Study Committee created a draft of an action plan for public libraries in Allegheny County.[122] The goal of the draft plan was "Equal access to all library resources for the citizens of Allegheny County." The mission of the draft plan was to "Create a workable, affordable pattern of library service in Allegheny County leading toward reduction of inequities in services and improvement of the services which currently exist." The draft plan included 5 objectives, such as the creation of a county library board and improvement of the ILL system.

This draft of the study was roundly criticized by CLP and the state library. CLP criticized the draft as vague and ill-informed.[123] CLP observed that the plan did not accomplish what the study group had been charged to do: design an incremental plan for developing a county library system.[124] The state library questioned how the consultants arrived at the conclusions they did and how objectives would be met.[125]

In September, the committee and consultants held 5 community hearings at libraries around the county to discuss a revised county plan.[126] The final plan was issued September 30, 1978.[127] The committee approved the plan by a 6 to 1 vote. Presumably, 5 of the 12 committee members abstained. It is not clear who abstained or why. Anthony Martin, director of CLP, cast the lone vote against the proposed plan.[128]

The final product published by the Citizens Study Committee was a typewritten, plastic-spine-bound report titled *Public Libraries in Allegheny County... A Plan of Action.*[129] The publication consisted of three sections: the action plan, the consultant report, and results of the data collection. Each of these sections were separate, stand-alone documents that were placed one after another to form a final report. Each section was authored by different people. The action plan was authored by the Citizens Study Committee; the consultant report was authored by Ed and May Dowlin, and the data was presented by the three research assistants. The final publication of the project was not a coherent document, and it lacked consistency across its sections.

The first section was the action plan that had previously circulated in draft form in July. The intended audience of the action plan was broad but clear: it addressed stakeholders and decision-makers across the county, including librarians, library users, trustees, citizens, and government officials. The finalized action plan was 4 pages long. The goal and mission of the final action plan were identical to what had appeared in the earlier draft, but the objectives and action steps of the final plan had been slightly modified. The plan recommended the creation of a city-county library board, an executive director position funded by the county, and the creation of an advisory council. These groups would lead countywide library service initiatives, including a computerized countywide circulation and interlibrary loan system.[130] Unlike the Blasingame report, the new action plan did not include any mechanism, such as a funding formula, for determining funding allocations to local libraries. Also, unlike Blasingame, the action plan proposed a governance structure independent of CLP.

The next section, the consultants' report, appeared after the action plan. The report noted several problems and opportunities for county libraries, including population decline, economic uncertainty, inflation, and shifting demographics. These were among the most pressing topics noted nationally. The report claimed that few improvements in library services had been made from 1970 to 1978, and there were even indications of a decline in library service due to how an increasing number of people lived in service areas not meeting state standards. Areas for improvement included ILL, centralized purchasing, collective cataloging and processing, and professional training.[131] The consultants seemed to doubt that librarians in the county were prepared to participate in a regional system, yet they did not provide a basis for these claims.[132] The consultants placed blame on the county commissioners for not implementing policy for the county and not holding CLP accountable for the funds it already received.[133] Like the preceding action plan, the consultants' report also recommended the creation of a new county board, but there was not an exact correspondence between the two sets of recommendations.[134] Plus, additional suggestions in the consultants' report were vague. The inconsistent presentation of recommendations made the consultants' recommendations unclear and confusing. Beyond an estimate for the proposed executive director position, the consultants' report did not provide a budget.

The third and final section of the citizens study was the appendices. The section updated the statistics from Blasingame, provided comparative data on libraries, presented data from the survey and site visits, included the materials availability analysis, survey instruments, community profile base data, and a list of libraries surveyed, and it reported on the public hearings. It was found that 41 autonomous libraries existed within the county at the time. The report noted that the recommendations of the Blasingame study had not been implemented. While CLP offered bookmobile services through its contract with the county, and it served as district library center under the state plan, the smaller, independently operated libraries would continue to face overwhelming challenges without a broader, countywide system that facilitated joint action. The county population continued to decline, especially in the cities, and while the number of people in the county served by libraries had increased since 1970, so too had the number of people living in areas not meeting library state standards.[135]

Public hearings were held at 5 libraries in mid-September 1978 to discuss the action plan. Members of the Citizens Committee attended one or more of the meetings with the Dowlins to solicit feedback on the plan. At these hearings, there was confusion about the order and sequence of the study, about why hearings were held once the study was close to completion instead of at the beginning where they could have informed the plan from the start. Attendees also commented that the "'plan' does not do what mission statement says it should," which was to "create a workable, affordable pattern of public library service in Allegheny County leading toward reduction of inequities in services and improvement of the services which currently exist."[136] This criticism was accurate—the study had not conducted a needs analysis to identify what the inequities and needs across the county were.

The committee met with the commissioners to discuss the plan, and the plan was distributed to librarians and library trustees across the county, soliciting feedback.[137] There were mixed responses to the plan by librarians and trustees around the county. Some librarians and trustees generally supported the plan.[138] Some trustees were concerned with how the proposed county board might affect local control of libraries.[139] Establishing a separate advisory council alongside the district advisory committee was said to be redundant by some.[140] The lack of sufficient research about the existing structure within the county as well as other workable plans was perhaps the most notable criticism.[141] Other respondents expressed support for continued county funding for CLP.[142] Some saw the proposed county board as unnecessary and wasteful, especially in light of existing district support through CLP.[143] Some respondents noted funding concerns.[144] But there seemed to be consensus on the plan's high-level principles, such as equitable access, increased financial support, cooperation, and public awareness.[145] In the end, the plan and recommendations in the citizens study restated the fallback position recommended earlier by Blasingame which was to "incorporate an independent County-chartered county library system governed by a board of directors broadly representative of various interest groups in the County."[146] This fallback position was suggested both as an alternative to establishing an office of county librarian within the county government and in the event that CLP could not be restructured to become a city-county library. This fallback position was ultimately rejected by

Blasingame because it was thought that a CLP-led county library system would be better integrated than a government-administered one. The advisory council mentioned in the new action plan, which was a new idea, was perhaps added to remedy any disconnect between the citizen-led county board and the professional expertise among county librarians.

The new action plan and the consultant report flatly rejected the recommendation from Blasingame that a county system form around CLP. This rejection is especially striking given how the board of CLP was reconstituted in January 1978 to include representation from the county. Two years earlier, in 1976, administrators at CLP had even planned to rename the director position the Director of the Carnegie Library of Pittsburgh and Allegheny County. Moreover, they planned to create a new administrative position that would oversee countywide services.[147] In the consultants' view, however, these changes were insufficient. According to the consultants, the county commissioners could no longer assign responsibility and authority for county library development to CLP:

> To an extent, the county/CLP contract amounts to a "de facto" assignment of this responsibility. The County experience has been that this assignment entails acceptance of the urban library's service philosophies and levels of cost. The CLP has been unable to demonstrate the skills of planning and reporting which incorporate the values and concerns of suburban librarians. Considerable credit is due [to] the Carnegie Library for altering its library board to include County representation, but these new representatives are agreed that the change is one of form rather than function. This course of action is considered to be a formal recognition of the de facto situation, with the likely outcome to be preservation of the status quo. That the State and County government felt this study project necessary raises serious doubts about the desirability of this alternative. The actions of a large majority of the Citizen's Study Committee serve to confirm this conclusion.[148]

The consultants "particularly advised against assigning responsibility for [countywide] activities to the Carnegie Library of Pittsburgh" due to the library's "aura of authority."[149] Evidently, the consultants believed that if CLP led the countywide system initiative, then the relationships that would be established among the county libraries would primarily benefit CLP, not the other county libraries. The consultants feared that, due to an existing power imbalance, there would not be reciprocal and equitable exchange of power if CLP took the lead. Some members of the Citizens Committee sharply disagreed with this analysis by the consultants.[150] Indeed, there was not a full explanation for why the committee and the consultants reversed the recommendation of Blasingame.

The Citizens Study Committee chair, Keith Gilbert, and other members of the committee sided with the Dowlins in recommending against forging a county library system around CLP, stating that the county did not get its money's worth from contracting with them for county services. Other local libraries, they believed, should receive a share of that funding instead.[151] Again, it is not clear how these assertions were justified.[152]

Besides its explicit rebuke of the recommendation of Blasingame, the action plan also proposed two other new ideas. The first was the advisory council composed of county librarians. It was recommended that this council work with the proposed county board and executive director. Blasingame did not propose anything similar. The second new idea was a shared, computerized, countywide catalog and circulation system. Such a system had been discussed at the state level by the Governor's Advisory Committee and at the district level by a committee at CLP,[153] but the citizens study was the first instance such a system was proposed for the county. This recommendation foreshadowed future developments.

Its new ideas and proposals notwithstanding, the citizens study was not effective at initiating change. Its proposals were overshadowed by the consultants' general ignorance of public library systems, the lack of documentation in the consultant report to support its claims in general and particularly those against CLP, and the consultants' strange psychological framework. In substance, the plan did not describe in sufficient detail how the proposed system would be formed, governed, or financed, what types of services it would offer, how it would benefit local libraries, how it would fill in service gaps across the county. The plan especially did not fully address the proposed system's potential overlap with the state roles of CLP. These were all points that should have been thoroughly addressed in the plan. In addition, instead of fostering feelings of reciprocity and exchange among county libraries, the plan sowed divisions. It portrayed CLP as a rival whose interests conflicted with other libraries in the county. These feelings of distrust toward CLP that were promoted by the citizens study would be difficult to overcome in the future.

There is no record of how the county commissioners responded to the plan. No evidence exists that the commissioners implemented any part of it. No independent county board was formed in Allegheny County at this time. The position of County Librarian that was established by the LSCA grant disappeared when grant funding concluded. Immediately following the plan's unveiling in October 1978, it was swiftly proposed that CLP be charged with the development of its own plan for a county library system.[154] It is not clear what, if any, planning occurred after this proposal.

In 1979, after the citizens study, the commissioners continued to fund county services as they had. They did this by continuing to contract with CLP. Though CLP did reconstitute its board in 1978, a county library system led by the library was never formed. Neither was an independent county board formed, nor a county library office within the county government. None of the studies conducted in Allegheny County during this period—whether those by Blasingame, Allegheny County Government Study Commission, or the citizens study—effected a substantive change in county library services. Despite the diverse proposals offered from 1970 to 1979, and despite the significant economic and demographic upheaval during this time, the existing county arrangement continued largely unchanged.

Period 4 is significant because it describes a decision cycle where there were multiple opportunities for substantive change, but no change occurred. This was despite other major social and economic changes that were happening throughout the country. In the end, the public library infrastructure in greater Pittsburgh failed to address during this time the very pressing demographic and economic issues it faced and were clearly discussed and identified in the early 1970s. For instance, the infrastructure did not solve the issues raised by urbanization, specifically the racial disparities in living standards that resulted from the Great Migration and the shrinking tax base in the City of Pittsburgh caused by both deindustrialization and demographic shifts. The result of continuity in Period 4 is surprising given all the reasons noted above. The cause of the continuity will be important to identify to fully understand how and why public library infrastructures develop as they do. As shown by Releaser Theory, Period 4 resulted in continuity due to low responsiveness. But the reasons for low responsiveness deserve additional scrutiny, and these reasons are best identified by comparing what happened in Period 4 to what happened in other periods with similar conditions but a different outcome.

In terms of the tessellation in Period 4, this period's decision cycle displays a jostling of control for the location of the infrastructure within the CLP and county circuits. At this time, CLP operated the de facto regional public library infrastructure in greater Pittsburgh. Under contract with the county commissioners, CLP carried out county policy, but it was also left to CLP to decide how county funding would be spent. CLP also functioned as the district library. In Period 4, CLP's status as the agency responsible for the infrastructure was challenged. It was proposed that the infrastructure be wrested from CLP and brought under control of a separate body. This would have restructured the inner periphery of the county circuit and relocated the public library infrastructure elsewhere within the county's inner periphery. Because the effort in Period 4 failed, the inner periphery of the county remained unchanged.

1 Carnegie Library of Pittsburgh, Annual Report, 1970, 20, CLP-PENN.

2 Carnegie Library of Pittsburgh, Annual Report, 1970, 32.

3 Carnegie Library of Pittsburgh, Statistics for Allegheny County Libraries, 1970, District Services Collection, Series 2, Box 4, Folder 5, CLP-OLIVER.

4 Carnegie Library of Pittsburgh, Annual Report, 1970, 16.

5 Natalie Wiest, Lee Lourea, and Brigitte L. Kenney, Inventory of Pennsylvania Interlibrary Cooperatives and Information Networks, Drexel University Graduate School of Library Science, May 1972, SLPARSC; Alexander Strasser, *Bibliographic Access in Pennsylvania* (Pittsburgh: Pittsburgh Regional Library Center, January 1979), 21; Keith Doms, "Pittsburgh Regional Library Center," *PLA Bulletin* 23, no. 3 (February 1968); Marjorie Karlson, "A New National Interlibrary Loan Code," *PLA Bulletin* 23, no. 6 (November 1968).

6 Ruth Corrigan and Stephen Folts, "Programs, Resources, Librarians, Communications," *PLA Bulletin* 32, no. 6 (November 1977): 125; "News: Carnegie Library Changes to LC," *PLA Bulletin* 28, no. 3 (May 1973).

7 Carnegie Library of Pittsburgh, Annual Report, 1970, 24.

8 Daniel Bell, "The Post-Industrial Society," in *Technology and Social Change*, ed. Eli Ginzberg (New York: Columbia University Press, 1964).

9 Daniel Bell, *The Coming of Post-Industrial Society: A Venture in Social Forecasting* (New York: Basic Books, 1973), 134.

10 Roy Lubove, *Twentieth-Century Pittsburgh: Government, Business, and Environmental Change* (Pittsburgh: Wiley, 1969); Roy Lubove, *Twentieth-Century Pittsburgh: The Post-Steel Era*, vol. 2 (Pittsburgh: University of Pittsburgh Press, 1996).

11 Lubove, *Twentieth-Century Pittsburgh: Government, Business, and Environmental Change*, 142 – 76.

12 Lubove, *Twentieth-Century Pittsburgh: The Post-Steel Era*, 2, 3.

13 Lubove, *Twentieth-Century Pittsburgh: The Post-Steel Era*, 2, 24; Lubove, *Twentieth-Century Pittsburgh: The Post-Steel Era*, 2, 29.

14 US Bureau of Labor Statistics, (2020), https://www.bls.gov/data/; Pennsylvania Department of Labor & Industry, *Center for Workforce Information & Analysis* (2020), https://paworkstats.geosolinc.com; Pennsylvania Department of Labor & Industry, Labor Market Letter, 1971 – 1976, Bloomsburg University of Pennsylvania; US Bureau of the Census, *Population of the United States: Trends and Prospects: 1950 – 1990* (Washington, DC, 1974); Federal Reserve Bank of St. Louis, *Unemployment Rate: Aged 15 – 64: All Persons for the United States* (2020, January 19), https://fred.stlouisfed.org/series/LRUN64TTUSA156N.

15 Ralph Upshaw Blasingame, Ernest R. DeProspo, and Robert L. Goldberg, Organization for Public Library Services in an Urban Field: Plan for Establishment of a County Public Library Service System for Allegheny County, Pennsylvania, 1970, 14, District Services Collection, Series 2, Box 2, Folder 1, CLP-OLIVER.

16 John E. Bebout and David Popenoe, "America as an Urban Society: A Review of Some Trends, Problems and Responses," in *Research on Library Service in Metropolitan Areas: Report of a Rutgers Seminar, 1964/1965*, ed. Ralph Upshaw Blasingame (New Bruswick: Graduate School of Library Service, Rutgers, the State University, 1967), 6.

17 Blasingame, DeProspo, and Goldberg, Organization for Public Library Services in an Urban Field, 1970, 14.

18 United States Census Bureau, *The Great Migration, 1910 to 1970* (13 September 2012), https://www.census.gov/dataviz/visualizations/020/.

19 State Library of Pennsylvania, *A Five Year Plan: Pennsylvania Library Development, 1971 – 1976* (Harrisburg: State Library of Pennsylvania, 1 September 1972), 50.

20 Pittsburgh Board of Public Education, *The Quest for Racial Equality in the Pittsburgh Public Schools: Problems, Principles, Practices, Plans*, Pittsburgh Board of Public Education (Pittsburgh, 1965).

21 Lubove, *Twentieth-Century Pittsburgh: Government, Business, and Environmental Change*, 142–76.
22 Pittsburgh Board of Public Education, *The Quest for Racial Equality in the Pittsburgh Public Schools* (1965), 10.
23 Lubove, *Twentieth-Century Pittsburgh: Government, Business, and Environmental Change*, 131.
24 Henry Cummings. Campbell, *Public Libraries in the Urban Metropolitan Setting*, The Management of Change: Studies in the Evolution of Library Systems, (Hamden: Linnet Books, 1973), 259.
25 Government Studies and Systems, *Evaluation of the Effectiveness of Federal Funding of Public Libraries* (Washington, DC: US GPO, 1976), 13–16; Pennsylvania Library Master Plan Committee, *Working Report of the Pennsyvania Library Master Plan Committee* (Harrisburg: Bureau of Library Development, September 1973), 7.
26 Government Studies and Systems, *Alternatives for Financing the Public Library* (Washington, DC: US GPO, 1974); Government Studies and Systems, *Evaluation of the Effectiveness of Federal Funding of Public Libraries*; Government Studies and Systems, *Improving State Aid to Public Libraries* (Washington, DC: National Commission on Libraries and Information Science, 1977); Rodney P. Lane, *Basic Issues in the Governmental Financing of Public Library Services* (New York: Teachers College, Columbia University, 1973); Rodney P. Lane, *The Role of the State in the Development of Public Library Services* (Washington, DC: Division of Library Programs, 1974).
27 Bebout and Popenoe, "America as an Urban Society," 14–15.
28 Ralph Upshaw Blasingame, "Chapter II: Libraries in an Urbanized Society," in *Research on Library Service in Metropolitan Areas: Report of a Rutgers Seminar, 1964/65*, ed. Ralph Upshaw Blasingame (New Bruswick: Graduate School of Library Service, Rutgers, the State University, 1967), 22–23.
29 John C. Frantz, "The Changing Environment and Changing Institution: The Urban Library," *Library Trends* 20, no. 2 (1971): 372.
30 Ralph Upshaw Blasingame, "The Future of the Urban Main Library: I," *Library Trends* 20, no. 4 (1972): 772; Ralph Upshaw Blasingame, "Equalization of Opportunity," *Library Journal* 90, no. 9 (1965): 2072–73, 75.
31 Lowell A. Martin, "The Future of the Urban Main Library: II," *Library Trends* 20, no. 4 (1972): 786.
32 Lowell A. Martin, *Library Response to Urban Change: A Study of the Chicago Public Library* (Chicago: American Library Association, 1969).
33 Bernard Berelson, *The Library's Public: A Report of the Public Library Inquiry* (New York: Columbia University Press, 1949).
34 Georgianne Petrosky, "Project Outreach," *PLA Bulletin* 23, no. 5 (August 1968).
35 Carnegie Library of Pittsburgh, Annual Report, 1970, 19.
36 Blasingame, DeProspo, and Goldberg, Organization for Public Library Services in an Urban Field, 1970, 8.
37 Ralph Upshaw Blasingame, "The Public Library as an Urban Phenomenon" (Doctor of Library Service Columbia University, 1973), ii.
38 US Office of Education Library Services Branch, *Library Research in Progress* (1960, April). Blasingame, "The Public Library as an Urban Phenomenon."; Blasingame, DeProspo, and Goldberg, Organization for Public Library Services in an Urban Field, 1970; Ralph Upshaw Blasingame and Ernest R. DeProspo, "Effectiveness in Cooperation and Consolidation in Public Libraries," in *Advances in Librarianship* (1970); Andrea Dragon, "In Memoriam: Ernest R. Deprospo, 1937–1983," *Journal of Library Administration* 5, no. 1 (1984).

39 Blasingame, "The Public Library as an Urban Phenomenon," 14; John Friedmann and John Miller, "The Urban Field," *Journal of the American Institute of Planners* 31, no. 4 (1965): 313 – 14.
40 Blasingame, "The Public Library as an Urban Phenomenon," 6 – 7.
41 Blasingame, "The Public Library as an Urban Phenomenon," 19 – 27.
42 Paul Wasserman, *The New Librarianship: A Challenge for Change* (New York: R. R. Bowker, 1972), 17 – 18.
43 Ralph Upshaw Blasingame, "Libraries in a Changing Society," *Library Journal* 97, no. 9 (1972): 1669.
44 Ralph W. Conant, "Sociological and Institutional Changes in American Life: Their Implications for the Library," *ALA Bulletin* 61, no. 5 (1967).
45 Conant, "Sociological and Institutional Changes in American Life: Their Implications for the Library," 533.
46 Campbell, *Public Libraries in the Urban Metropolitan Setting*, 263 – 64.
47 Blasingame, DeProspo, and Goldberg, Organization for Public Library Services in an Urban Field, 1970, 22.
48 Blasingame, DeProspo, and Goldberg, Organization for Public Library Services in an Urban Field, 1970, 25.
49 Blasingame, DeProspo, and Goldberg, Organization for Public Library Services in an Urban Field, 1970, 8.
50 Blasingame, DeProspo, and Goldberg, Organization for Public Library Services in an Urban Field, 1970, 48.
51 Blasingame, DeProspo, and Goldberg, Organization for Public Library Services in an Urban Field, 1970, 55.
52 Blasingame, DeProspo, and Goldberg, Organization for Public Library Services in an Urban Field, 1970, 60 – 61.
53 Blasingame, DeProspo, and Goldberg, Organization for Public Library Services in an Urban Field, 1970, 62 – 63.
54 Blasingame, DeProspo, and Goldberg, Organization for Public Library Services in an Urban Field, 1970, 68.
55 Blasingame, DeProspo, and Goldberg, Organization for Public Library Services in an Urban Field, 1970, 66.
56 Blasingame, DeProspo, and Goldberg, Organization for Public Library Services in an Urban Field, 1970, 72.
57 Blasingame, DeProspo, and Goldberg, Organization for Public Library Services in an Urban Field, 1970, 82.
58 Walter T. Brahm and Mildred W. Sandoe, *Allegheny County Library Survey* (1950); Metropolitan Study Commission of Allegheny County, *An Urban Home Rule Charter for Allegheny County: A Report of the Metropolitan Study Commission of Allegheny County* (Pittsburgh, 1955).
59 Carnegie Library of Pittsburgh, Annual Report, 1970, 19.
60 Carnegie Library of Pittsburgh, List of District Meeting, 16 October 1970, District Services Collection, Series 2, Box 2, Folder 4, CLP-OLIVER; Donald C. Potter, Letter from Donald C. Potter to Library Trustees and Librarians of Allegheny County, 19 October 1970, District Services Collection, Series 2, Box 2, Folder 4, CLP-OLIVER.
61 Elsie W. Milton, Letter from Elsie W. Milton to Donald C. Potter, 19 November 1970, District Services Collection, Series 2, Box 4, Folder 8, CLP-OLIVER; Donald C. Potter, Letter from Donald C. Potter to CLP Admins: Meeting with Mt. Lebanon Board, 28 November 1970, District Services Collection, Series 2, Box 4, Folder 8, CLP-OLIVER; Donald C. Potter, Letter from Donald C. Potter to CLP Admins:

Meeting with Usc Board, 5 January 1971, District Services Collection, Series 2, Box 4, Folder 8, CLP-OLIVER; Donald C. Potter, Letter from Donald C. Potter to CLP Admins: Meeting with Dormont, 10 March 1971, District Services Collection, Series 2, Box 4, Folder 8, CLP-OLIVER; Donald C. Potter, Letter from Donald C. Potter to CLP Admins: Meeting with Northland, 11 May 1971, District Services Collection, Series 2, Box 4, Folder 8, CLP-OLIVER.

62 Carnegie Library of Pittsburgh, Allegheny County Public Library Service: A Discussion of the Blasingame Study, 26 July 1971, District Services Collection, Series 2, Box 5, Folder 3, CLP-OLIVER.

63 Donald C. Potter, Survey of Allegheny County Public Libraries, 25 January 1972, District Services Collection, Series 2, Box 5, Folder 8, CLP-OLIVER.

64 Donald C. Potter, Letter from Donald C. Potter to Joseph B. Rounds, 9 March 1971, District Services Collection, Series 2, Box 2, Folder 4, CLP-OLIVER; Donald C. Potter, Letter from Donald C. Potter to Alice F. Reilly, 9 March 1971, District Services Collection, Series 2, Box 2, Folder 4, CLP-OLIVER.

65 Leonard Staisey (Democrat, 1968 – 1976), Tom Foerster (Democrat, 1968 – 1996), and William R. Hunt (Republican, 1968 – 1976, 1980 – 1984). Peter A. Caputo, Letter from Peter A. Caputo to Donald C. Potter, 13 August 1975, District Services Collection, Series 2, Box 5, Folder 8, CLP-OLIVER; Carnegie Library of Pittsburgh, Allegheny County Commissioners, Typescript, Vertical File, CLP-PENN.

66 Donald C. Potter, Letter from Donald C. Potter to Bertha N. Miller, 1 October 1971, District Services Collection, Series 2, Box 5, Folder 8, CLP-OLIVER; Donald C. Potter, Proposed County Grants to Independent Libraries, 27 September 1971, District Services Collection, Series 2, Box 5, Folder 6, CLP-OLIVER.

67 Donald C. Potter, Letter from Donald C. Potter to Ms. Toni Grant, 5 May 1975, District Services Collection, Series 2, Box 5, Folder 8, CLP-OLIVER.

68 Florence Karnofsky, Minutes of Library Meeting, 28 June 1955, Movements Toward Library Services in Allegheny County, 1935 – 1961: Documents, with Connecting Narrative, compiled and annotated by Ralph Munn, Director's Office Collection, Box 21, Folder 4, CLP-OLIVER.

69 Dorothea M. Strang, Citizens Committee for County Library Service, Minutes of Meeting, 19 April 1960, Movements Toward Library Services in Allegheny County, 1935 – 1961: Documents, with Connecting Narrative, compiled and annotated by Ralph Munn, Director's Office Collection, Box 21, Folder 4, CLP-OLIVER.

70 Donald C. Potter, Letter from Donald C. Potter to Dorothea M. Strang, 26 February 1971, District Services Collection, Series 2, Box 2, Folder 4, CLP-OLIVER.

71 Dorothea M. Strang, Toward a County Library System, 5 January 1971, 5, District Services Collection, Series 2, Box 2, Folder 2, CLP-OLIVER; Potter, Letter from Donald C. Potter to Dorothea M. Strang, 26 February 1971.

72 State Library of Pennsylvania, A Five-Year Plan: Pennsylvania Library Development, 1971 – 1976, Library Services and Construction Act, 1971, District Services Collection, Series 2, Box 9, Manual 2, CLP-OLIVER.

73 State Library of Pennsylvania, A Five-Year Plan, 1971, 30 – 31.

74 State Library of Pennsylvania, A Five-Year Plan, 1971, 36.

75 Pennsylvania Library Master Plan Committee, *Working Report of the Pennsylvania Library Master Plan Committee*, Bureau of Library Development, State Library of Pennsylvania (Harrisburg, 1973).

76 Government Study Commission, Final Report of the Allegheny County Government Study Commission, Its Findings and Recommendations, Fiscal Report and Proposed Home Rule Charter for Allegheny County, 15 March 1974, District Services Collection, Series 2, Box 4, Folder 2, CLP-OLIVER.

77 Government Study Commission, Final Report of the Allegheny County Government Study Commission, 15 March 1974, ii.

78 Government Study Commission, Final Report of the Allegheny County Government Study Commission, 15 March 1974, iv.
79 Beatrice Hirschl, Home Rule Study Report – County Library System, 1974, District Services Collection, Series 2, Box 5, Folder 5, CLP-OLIVER.
80 Allegheny County Government Study Commission, Report from Hearing before Allegheny County Gov't. Study Commission on County Library System, 23 May 1973, District Services Collection, Series 2, Box 5, Folder 5, CLP-OLIVER.
81 Pittsburgh Post-Gazette, "A County Library System?," *Pittsburgh Post-Gazette*, 10 August 1973.
82 Government Study Commission, Final Report of the Allegheny County Government Study Commission, 15 March 1974, 19.
83 James O'Toole, "Allegheny County Has a Long History of Commissioners," *Pittsburgh Post-Gazettte*, 2 January 2000.
84 Joseph H. Myers, Allegheny County Plan for Countywide Libary System, 4 February 1974, District Services Collection, Series 2, Box 5, Folder 6, CLP-OLIVER.
85 State Library of Pennsylvania, County Library Service: Outline of Presentation at the Trustees Meeting, Erie, April 21, 1975, 21 April 1975, District Services Collection, Series 2, Box 4, Folder 6, CLP-OLIVER.
86 Donald C. Potter, County Commissioners, Strang, State Library Meeting, 28 January 1975, District Services Collection, Series 2, Box 5, Folder 8, CLP-OLIVER.
87 Thomas J. Foerster, Letter from Thomas J. Foerster to M. J. Johns, 26 March 1976, District Services Collection, Series 2, Box 5, Folder 8, CLP-OLIVER; Carnegie Library of Pittsburgh, Allegheny County Commissioners.
88 Ernest E. Doerschuk, "View from the State Library," *PLA Bulletin* 28, no. 5 (September 1973): 208.
89 Foerster, Letter from Thomas J. Foerster to M. J. Johns, 26 March 1976; Frank J. Lucchino, A Quiet Crisis: Libraries in Allegheny County, March 1991, 13, District Services Collection, Series 2, Box 3, Manual 4, CLP-OLVER.
90 James M. Walton, Letter from James M. Walton to Trustees, Staff, and Friends of Carnegie Library of Pittsburgh, 11 March 1976, District Services Collection, Series 2, Box 5, Folder 8, CLP-OLIVER.
91 Walton, Letter from James M. Walton to Trustees, Staff, and Friends of Carnegie Library of Pittsburgh, 11 March 1976; Joseph F. Falgione, Priorities, Carnegie Library of Pittsburgh, 3 March 1976, District Services Collection, Series 2, Box 5, Folder 8, CLP-OLIVER; "Potpourri: $$ Cutback in Allegheny County," *PLA Bulletin* 31, no. 3 (May 1976).
92 Joseph F. Falgione, Revised Draft for a Possible County Library Program, 25 June 1976, District Services Collection, Series 2, Box 5, Folder 5, CLP-OLIVER; Joseph F. Falgione, Revised Draft of the Proposal for County-Wide Library Service for Allegheny County, 6 July 1976, District Services Collection, Series 2, Box 5, Folder 5, CLP-OLIVER.
93 Falgione, Revised Draft of the Proposal for County-Wide Library Service for Allegheny County, 6 July 1976.
94 Falgione, Revised Draft for a Possible County Library Program, 25 June 1976, 2.
95 Carnegie Library of Pittsburgh, Comments on the Action Plan for Public Libraries in Allegheny County, 31 July 1978, District Services Collection, Series 2, Box 5, Folder 1, CLP-OLIVER.
96 State Library of Pennsylvania, Various Formulas by Which Pennsylvania County Libraries Distribute County and State Money among Libraries, 1976, District Services Collection, Series 2, Box 4, Folder 6, CLP-OLIVER.
97 Falgione, Revised Draft of the Proposal for County-Wide Library Service for Allegheny County, 6 July 1976.

98 Joseph H. Myers, Letter from Joseph H. Myers to Dorothea M. Strang, 10 March 1976, District Services Collection, Series 2, Box 4, Folder 6, CLP-OLIVER; Joseph H. Myers, 1973 Public Library Income Per Capita County Population, March 1976, District Services Collection, Series 2, Box 4, Folder 6, CLP-OLIVER; Dorothea M. Strang, Letter from Dorothea M. Strang to Joseph H. Myers, 18 December 1976, District Services Collection, Series 2, Box 4, Folder 6, CLP-OLIVER.

99 Patricia M. Broderick, Letter from Patricia M. Broderick to Della E. Britton, 29 July 1976, District Services Collection, Series 2, Box 5, Folder 3, CLP-OLIVER.

100 Broderick, Letter from Patricia M. Broderick to Della E. Britton, 29 July 1976, 1–2.

101 Donald C. Potter, Letter from Donald C. Potter to Anthony A. Martin and Joseph F. Falgione, 30 July 1976, District Services Collection, Series 2, Box 5, Folder 3, CLP-OLIVER; Ross J. Nese, Letter from Ross J. Nese to Donald C. Potter, 24 August 1976, District Services Collection, Series 2, Box 5, Folder 3, CLP-OLIVER; Ross J. Nese, Letter from Ross J. Nese to Patricia Broderick, 25 August 1976, District Services Collection, Series 2, Box 4, Folder 5, CLP-OLIVER; Board of Commissioners of Allegheny County, Project Application, LSCA Title I: Citizens' Commission for Library Service, 26 August 1976, 1, District Services Collection, Series 2, Box 4, Folder 5, CLP-OLIVER.

102 Board of Commissioners of Allegheny County, Project Application, LSCA Title I: Citizens' Commission for Library Service, 26 August 1976, 5.

103 Ernest E. Doerschuk, Letter from Ernest E. Doerschuk to Jim Flaherty, 10 December 1976, District Services Collection, Series 2, Box 4, Folder 6, CLP-OLIVER; Donald C. Potter, Letter from Donald C. Potter to Gordon Mulleneaux, 24 January 1977, District Services Collection, Series 2, Box 5, Folder 1, CLP-OLIVER; Bureau of Library Development State Library of Pennsylvania, Allegheny County LSCA Contract Process, Dates, Etc., 26 January 1977, District Services Collection, Series 2, Box 5, Folder 1, CLP-OLIVER; Carnegie Library of Pittsburgh, Library Statistics, 1977, District Services Collection, Series 2, Box 5, Folder 1, CLP-OLIVER; State Library of Pennsylvania Bureau of Library Development, Types of County Library Systems in Pennsylvania, 4 January 1977, District Services Collection, Series 2, Box 4, Folder 6, CLP-OLIVER.

104 Brahm and Sandoe, *Allegheny County Library Survey* (1950); Blasingame, DeProspo, and Goldberg, Organization for Public Library Services in an Urban Field, 1970.

105 Lois K. Albrecht, Letter from Lois K. Albrecht to William Rumpf, 8 April 1977, District Services Collection, Series 2, Box 5, Folder 1, CLP-OLIVER.

106 Lawrence D. Mann, "Studies in Community Decision-Making," in *Readings in Community Organization Practice*, ed. Ralph M. Kramer and Harry Specht (Englewood Cliffs: Prentice Hall, 1969); Albrecht, Letter from Lois K. Albrecht to William Rumpf, 8 April 1977; State Library of Pennsylvania Bureau of Library Development, Allegheny County Activity Timetable, 3 February 1977, District Services Collection, Series 2, Box 5, Folder 1, CLP-OLIVER; Citizens Study Committee on Libraries, Jottings, 15 March 1978, District Services Collection, Series 2, Box 5, Folder 1, CLP-OLIVER.

107 Dorothea M. Strang, Letter from Dorothea M. Strang to League of Women Voters, Library Resource Committee Members, 8 August 1977, District Services Collection, Series 2, Box 5, Folder 1, CLP-OLIVER.

108 Lois K. Albrecht, Letter from Lois K. Albrecht to Mr. & Mrs. C. Edwin Dowlin, 15 November 1977, District Services Collection, Series 2, Box 5, Folder 1, CLP-OLIVER.

109 Grace I. Wozniak, Letter from Grace I. Wozniak to Martha Reimers, 25 January 1978, District Services Collection, Series 2, Box 5, Folder 1, CLP-OLIVER.

110 Martha M. Reimers, Letter from Martha M. Reimers to Carol Walton, 30 January 1978, District Services Collection, Series 2, Box 5, Folder 1, CLP-OLIVER.

111 Albrecht, Letter from Lois K. Albrecht to William Rumpf, 9 June 1977; Citizens Study Committee on Libraries, Public Libraries in Allegheny County…A Plan of Action, September 1978, District Services Collection, Series 2, Box 1, Manual 10, CLP-OLIVER.
112 Citizens Study Committee, Jottings, 6 February 1978, District Services Collection, Series 2, Box 5, Folder 1, CLP-OLIVER; William Rumpf, Letter from William Rumpf to Study Committee, 1977, May 26, District Services Collection, Series 2, Box 5, Folder 1, CLP-OLIVER; Lois K. Albrecht, Letter from Lois K. Albrecht to William Rumpf, 1977, August 19, District Services Collection, Series 2, Box 5, Folder 1, CLP-OLIVER; Lois K. Albrecht, Letter from Lois K. Albrecht to Robert N. Peirce, Jr., 1978, January 10, District Services Collection, Series 2, Box 5, Folder 1, CLP-OLIVER; Albrecht, Letter from Lois K. Albrecht to William Rumpf, 8 April 1977.
113 Strang, Letter from Dorothea M. Strang to League of Women Voters, Library Resource Committee Members, 8 August 1977; Citizens Study Committee, Jottings, 6 February 1978; Association of College and Research Libraries, "People," *College & Research Libraries News* 47, no. 2 (1986); Albrecht, Letter from Lois K. Albrecht to William Rumpf, 8 April 1977; Albrecht, Letter from Lois K. Albrecht to Mr. & Mrs. C. Edwin Dowlin, 15 November 1977; Albrecht, Letter from Lois K. Albrecht to William Rumpf, 8 April 1977; Lois K. Albrecht, Letter from Lois K. Albrecht to Mr. C. Edwin Dowlin, 29 November 1977, District Services Collection, Series 2, Box 5, Folder 1, CLP-OLIVER.
114 Kenneth Everett Dowlin, *Noisy River: The Saga of Captain Paul Dowlin* (Bloomington: AuthorHouse, 2013), 425.
115 Brook E. Sheldon, "Application of an Operational Model for Library Project Evaluation" (PhD University of Pittsburgh, 1977); Charles Edwin Dowlin, "A New Approach to the Evaluation of Public Libraries: A Pilot Study" (PhD University of Pittsburgh, 1980).
116 Charles Edwin Dowlin, Letter from Ed Dowlin to Tony Martin: Citizens Study Committee on Libraries, 24 February 1978, 2, District Services Collection, Series 2, Box 5, Folder 1, CLP-OLIVER.
117 Desmond L. Cook, *Educational Project Management* (Columbus: Merrill, 1971).
118 Dowlin, Letter from Ed Dowlin to Tony Martin: Citizens Study Committee on Libraries, 24 February 1978, 5–6; Citizens Study Committee on Libraries, Discussion Topics, March 1978, District Services Collection, Series 2, Box 5, Folder 1, CLP-OLIVER.
119 Citizens Study Committee, Info Packet for County Library Study, 1977, District Services Collection, Series 2, Box 5, Folder 1, CLP-OLIVER; Citizens Study Committee on Libraries, Jottings, 15 March 1978; Citizens Study Committee on Libraries, Public Libraries in Allegheny County, September 1978, 1; State Library of Pennsylvania Bureau of Library Development, Minimum Standards and Guidelines for Pennsylvania Public Library Systems Receiving State Aid, June 1978, District Services Collection, Series 2, Box 5, Folder 9, CLP-OLIVER.
120 Keith L. Gilbert, Letter from Keith Gilbert to Committee Members, 19 July 1978, District Services Collection, Series 2, Box 5, Folder 1, CLP-OLIVER.
121 Charles Edwin Dowlin, Letter from C. Edwin Dowlin to the State Library of Pennsylvania, 15 May 1978, District Services Collection, Series 2, Box 5, Folder 1, CLP-OLIVER; Citizens Study Committee on Libraries, Jottings, 15 March 1978; Citizens Study Committee on Libraries, Jottings, 15 March 1978; Ernest E. Doerschuk, State Librarian's Report December 1977-February 1978, 6 March 1978, Department of Education, Commonwealth Libraries, Minutes of the Governor's Advisory Council on Library Development 1962–2009 collection, Carton 1, Folder 2, PASA.
122 Gilbert, Letter from Keith Gilbert to Committee Members, 19 July 1978.
123 Donald C. Potter, Comments on the Action Plan, 26 July 1978, District Services Collection, Series 2, Box 5, Folder 9, CLP-OLIVER.
124 Carnegie Library of Pittsburgh, Comments on the Action Plan for Public Libraries in Allegheny County, 31 July 1978.

125 Lois K. Albrecht, Letter from Lois K. Albrecht to Keith Gilbert, 27 July 1978, District Services Collection, Series 2, Box 5, Folder 1, CLP-OLIVER.
126 Citizens Study Committee on Libraries, You Are Invited to a Discussion of the Services and Problems of Public Libraries in Allegheny County, September 1978, District Services Collection, Series 2, Box 5, Folder 5, CLP-OLIVER.
127 Citizens Study Committee on Libraries, Public Libraries in Allegheny County, September 1978.
128 Matthew Kennedy, "Study of County Libraries Criticizes Lack of Progress," *Pittsburgh Press*, 15 January 1979.
129 Citizens Study Committee on Libraries, Public Libraries in Allegheny County, September 1978.
130 Citizens Study Committee on Libraries, Public Libraries in Allegheny County, September 1978.
131 Citizens Study Committee on Libraries, Public Libraries in Allegheny County, September 1978, 14 – 15.
132 Citizens Study Committee on Libraries, Public Libraries in Allegheny County, September 1978, 18.
133 Citizens Study Committee on Libraries, Public Libraries in Allegheny County, September 1978.
134 Citizens Study Committee on Libraries, Public Libraries in Allegheny County, September 1978, 3 – 4.
135 Citizens Study Committee on Libraries, Public Libraries in Allegheny County, September 1978, Appendix B.
136 Donald C. Potter, Notes on Community Library Meetings, September 1978, District Services Collection, Series 2, Box 5, Folder 1, CLP-OLIVER.
137 Keith L. Gilbert, Study Commission Letter on County Commissioner Meeting, 17 November 1978, Director's Office Collection, Box 49, Folder 11, CLP-OLIVER; Ada Anderson, Letter from Ada Anderson to Keith Gilbert, 7 December 1978, District Services Collection, Series 2, Box 5, Folder 1, CLP-OLIVER.
138 David B. Dodds, Letter from David B. Dodds to Keith Gilbert, 23 January 1979, District Services Collection, Series 2, Box 5, Folder 1, CLP-OLIVER; Monica Stoicovy, Letter from Monica R. Stoicovy to Keith Gilbert, 29 January 1979, District Services Collection, Series 2, Box 5, Folder 1, CLP-OLIVER.
139 Truman A. Staggers, Letter from Truman A. Staggers to Keith Gilbert, 1979, District Services Collection, Series 2, Box 5, Folder 1, CLP-OLIVER; Harry Saperstein, Letter from Harry Saperstein to Keith Gilbert, 22 January 1979, District Services Collection, Series 2, Box 5, Folder 1, CLP-OLIVER.
140 Deborah Schlesinger, Letter from Deborah Schlesinger to Keith Gilbert, 3 January 1979, District Services Collection, Series 2, Box 5, Folder 1, CLP-OLIVER; Joyce E. Schmidt, Letter from Joyce E. Schmidt to Keith Gilbert, 31 January 1979, District Services Collection, Series 2, Box 5, Folder 1, CLP-OLIVER.
141 Schlesinger, Letter from Deborah Schlesinger to Keith Gilbert, 3 January 1979.
142 John E. Henderson, Letter from John E. Henderson to Keith Gilbert, 4 January 1979, District Services Collection, Series 2, Box 5, Folder 1, CLP-OLIVER.
143 Pleasant Hills Library Board, Letter from Pleasant Hills Library Board to Keith Gilbert, 21 January 1979, District Services Collection, Series 2, Box 5, Folder 1, CLP-OLIVER.
144 Debra Conn, Letter from Debra Conn to Keith Gilbert, 23 January 1979, District Services Collection, Series 2, Box 5, Folder 1, CLP-OLIVER.
145 Ada Anderson, Letter from Ada Anderson to Anthony Martin, 27 February 1979, District Services Collection, Series 2, Box 5, Folder 1, CLP-OLIVER; Keith L. Gilbert, Letter from Keith L. Gilbert to Committee Members, 22 February 1979, District Services Collection, Series 2, Box 5, Folder 1, CLP-OLIVER.

146 Blasingame, DeProspo, and Goldberg, Organization for Public Library Services in an Urban Field, 1970, 67.
147 Joseph F. Falgione, Letter from Joseph Falgione to Anthony A. Martin, 6 July 1976, District Services Collection, Series 2, Box 5, Folder 5, CLP-OLIVER.
148 Citizens Study Committee on Libraries, Public Libraries in Allegheny County, September 1978, 31.
149 Citizens Study Committee on Libraries, Public Libraries in Allegheny County, September 1978, 26.
150 Citizens Study Committee on Libraries, Public Libraries in Allegheny County, September 1978, 26 – 27.
151 Kennedy, "Study of County Libraries Criticizes Lack of Progress."
152 Jean Y. Mutschler, Letter from Jean Y. Mutschler to Keith Gilbert, 31 January 1979, District Services Collection, Series 2, Box 5, Folder 1, CLP-OLIVER.
153 Carnegie Library of Pittsburgh, Comments on the Action Plan for Public Libraries in Allegheny County, 31 July 1978.
154 Carnegie Library of Pittsburgh, Toward an Allegheny County Library System, 4 October 1978, District Services Collection, Series 2, Box 5, Folder 9, CLP-OLIVER.

Period 5: Fiscal Crises and Digital Visions, 1980–1993

In 1980, the Pittsburgh region continued to decline economically. The years 1979 to 1983 marked a recession era in the U.S. and especially greater Pittsburgh.[1] The recession was precipitated by the oil crises of 1978 and 1979, but a much more palpable problem in Pittsburgh was the fallout of the steel industry.[2] In the 1970s and 1980s, the populations, wages, and taxes in the region fell, infrastructure declined, and schools in the area closed.

Due in part to the recession, the public library infrastructure in greater Pittsburgh in the 1980s remained largely unchanged from what it had been in the late 1960s. CLP had 21 total branches in 1989.[3] It played the same roles as it had in 1969 under the state plan: local library for the city, county library for Allegheny, district library center, regional resource center, and library for the blind and physically handicapped.[4] There were discussions at CLP to incorporate the Steel Valley libraries, those libraries located in towns along the Monongahela, Ohio, and Allegheny Rivers, though these mergers were never implemented.[5] In 1989, CLP served 40 libraries in Allegheny County as the district library center.[6] The county agreement from 1956 was still in effect: residents from any municipality in the county had borrowing, ILL, and references privileges at CLP. Moreover, CLP and its bookmobiles were still funded in part by the county.[7] However, there was not yet a countywide library card. The same county residents from outside the city who could access CLP services for free could not expect these same services from a library in a neighboring town. The countywide library card, which was considered in the 1978 the Citizens Study Committee report, had not been implemented.[8]

By 1980, independent public libraries in Allegheny County shared some services and collaborated in a limited number of ways. Despite efforts to strengthen the countywide system from 1970 to 1979, as noted in Period 4, and despite the pressures imposed by the nationwide economic recession, county libraries had not developed their regional services beyond those already provided through district services, which took effect in 1963. In 1980, the primary means of resource-sharing was ILL. Within the county, this was facilitated primarily through CLP. At this time, libraries could not view the global holdings for all libraries in the county, they could only view the holdings of CLP, making it necessary to order ILLs only from them.[9] ILL traffic in the county was minimal, comprising only one-fourth of 1 percent of all circulation.[10] There had not been significant advances made toward collective cataloging or processing of materials for countywide libraries. CLP pulled records from OCLC internally, but these records were not shared broadly among other county libraries.[11] In terms of joint purchasing, CLP's district center

https://doi.org/10.1515/9783111013404-007

facilitated a blanket contract with a publishing company that allowed county libraries to purchase materials at a discount, which some did.[12] Thus, CLP was still at the center of most cooperative practices within the county.

There were other limited cooperative efforts within the county that did not involve CLP. South Hills libraries, for example, cooperated for joint purchasing projects by forming South Hills Library Association (SHLA), an association of trustees, and Library Association of South Hills (LASH), an association of librarians.[13] LASH created a list of periodicals that could be shared amongst the libraries. They also had coordinated collection development projects.[14] There existed other regional and specialty groups operating within the county, such as among East Hills libraries, North Hills libraries, and among traveling children's librarians.[15] These collaborative efforts were exceptional.

Despite the otherwise relative inactivity of the public library infrastructure in the 1980s, and despite the dire economic circumstances, there was one major background change underway nationwide, and it was technological. Libraries across the country had begun the early stages of automation. Automation referred to the utilization of computers and telecommunications equipment to carry out library tasks. Bibliographic records, such as library catalogs, began to be created, stored, pooled, transmitted, and shared electronically using computers and phone lines. Card catalogs were still in full use at this time, but these paper machines were supplemented by electronic ones. The computerization underway nationwide had implications for the public library infrastructure in greater Pittsburgh.

It is necessary to step back and view the big picture of automation across the U.S. to understand how automation affected infrastructure in Pittsburgh. The automation movement in the library world was not merely technological: it had a political, economic, and ideological history, and it was supported by national-level events and policies. In the second half of the twentieth century, library leadership at the federal level in the U.S. shifted. The Library Programs Office in the U.S. Department of Education continued to administer federal grants-in-aid funding to state library agencies. This was from LSCA. From 1957 to 1997, during the time when LSA and LSCA were in place, over 3 billion dollars were allocated by the U.S. Congress for use by public libraries. Yet, despite their management of the funding, the Library Service Division never achieved the leadership role that was once envisioned for it. The office served an administrative function, not a coordinating and planning one.[16] This meant that there was a leadership void within the federal government. Prior to 1956 and leading up to the passage of LSA, the American Library Association (ALA) was in a position to fill this void. It played a singular and central leadership role in directing public library development nationally. After 1956, however, ALA faded into the background as a policy director. It was joined on the national stage by other organizations: the Council on Library Resources (CLR), the Library of Congress (LC), and the National Commission on Libraries and Information Science (NCLIS). The automation movement was largely spurred on by these organizations, not ALA or the Library Programs Office.

ALA's policy documents demonstrate ALA's shift away from national planning. This is seen, for example, in ALA's revised public library standards from 1956, which was an attempt to establish a new direction for public libraries, one that was in touch with the times, but one that resulted in failure.[17] The standards did not present a clear vision for public libraries moving forward.[18] The same was true for its standards from 1979 and 1980. [19] These standards abandoned attempts to establish a singular, coherent vision for public library service nationwide, and they instead focused on the development of libraries' individual and community-based planning documents.[20] Later manuals and accreditation standards focused attention away from national planning. [21] ALA's ILL code, which had governed ILL practices across the country since 1919, was considered obsolete by the early 1970s.[22] The result of these failures was that ALA transitioned from a professional lobbying group, not a leadership organization.[23] With few exceptions, ALA had not kept pace with technological trends.[24]

Due in part to these shortcomings, ALA was no longer the single, dominant force on the national stage post-1956. The first other major entity that appeared was the Council on Library Resources (CLR). CLR was established in 1956 at a time when library use was increasing as were library costs. Computer technologies were on the horizon, but it was not yet clear what roles they would play in libraries. CLR's founders were optimistic that computers could benefit libraries especially as cost-saving devices. [25] ALA was excluded from participation in CLR's formative meetings because it was thought that ALA leaders were not visionary enough.[26] CLR was described by some as a catalyst for library development because the creation of a national library network from 1956 to 1996 was guided in part by CLR's funding decisions.[27]

Another significant organization that appeared on the national scene after 1956 was the National Commission on Libraries and Information Science (NCLIS). In 1969, after a series of hearings and studies on a scale not unlike the Public Library Inquiry two decades earlier, NCLIS was signed into law in 1970 as an independent federal agency within the executive branch. [28] The mission of NCLIS was to study and propose national library policy and to orient federal, state, local, and private agencies toward that vision.[29] In light of its charge, NCLIS sponsored influential studies related to the planning and funding of a national library network.[30]

NCLIS was charged with national library planning, but CLR already had its hands in research projects related to the development of a national library system.

Activities of the two organizations therefore overlapped and intersected. Both organizations sought to build a nationwide library network, even if their visions for this network differed. NCLIS envisioned a centralized system with LC as its locus but whose components included existing regional and commercial entities.[31] NCLIS emphasized the centralized development of national standards to guide the creation of a national network.[32] In contrast to NCLIS' vision, CLR's position on a nationwide library network stressed a bottom-up confederation of interlinked library networks rather than a completely centralized one administered by LC.[33] Both CLR and NCLIS called for a national library network as a way to reduce library costs. This could be done by sharing bibliographic resources—catalogs, indexes, and holdings lists—so that libraries would not duplicate their efforts in creating them. The projects funded by CLR and NCLIS were oriented toward facilitating the development of this nationwide bibliographic network. In the 1970s, however, it was not quite clear what shape this emerging nationwide network would take.

Even if the exact nature of the emerging nationwide system was not yet known, there emerged a consensus in the 1960s and 1970s on new terminology to describe it. Policy documents leading up to 1956 had emphasized library systems.[34] By contrast, post-1956, the terminology shifted to networks. This terminology was visible, for example, in the amendments to LSCA in 1966, ALA's standards for state library agencies, and the 1970 legislation that established NCLIS.[35] This terminological shift emphasized how libraries should not just cooperate and share resources, but they should do it using advanced electronic equipment, in other words, computers.[36] It was evident that CLR, NCLIS, LC, and even ALA believed that solutions for problems in libraries were technological in nature.

This process of incorporating digital electronic computers into library operations, and connecting them using new telecommunications technologies, was known as automation. Automation was the defining theme of U.S. library service in the second half of the twentieth century. It was not just a set of practices, but an ideology, one that supported the adoption of electronic machines to perform library functions.[37] In libraries across the U.S., several library functions came to be increasingly automated from the 1960s onward. These functions included circulation, cataloging, acquisitions, information retrieval, and ILL. At first, automation referred to the adoption by libraries of machines of any kind, but as time went on, automation came to refer more exclusively to the use of digital, electronic computers.[38]

Computers and library networks developed contemporaneously and in related ways. Changes in library routines and library organization, especially network formation, were determined largely by the computing options available at that moment. This included the change from analog telephone wires to modems and digital communications.[39] The development of computers, telecommunications, and storage equipment partly explains how, why, and when library networks developed the way that they did. The earliest computers were mainframes. Mainframes were operated in a star structure, with the central computer at the center and command terminals on the periphery.[40] This organization became the basis for online search service networks.[41] Examples of online search services that were used by libraries in the 1970s and 1980s included the National Library of Medicine (NLM), the New York Times (NYT), System Development Corporation (SDC), Lockheed DIALOG, and Bibliographic Retrieval Services (BRS).[42] These services relied on the existing national telephone network to connect libraries to their databases. Mainframes were too expensive for libraries to build and maintain on site. They therefore paid for access to them. For libraries, the purpose of an online search service was to identify journal articles that were relevant to some question or topic. Physical items could then be located and obtained by the library by consulting local library catalogs and holdings lists. Libraries connected to online search services using remote terminals connected to telephone wires.[43]

Mainframe computers also formed the basis of the bibliographic utility networks that emerged in the 1970s, 1980s, and early 1990s. Like the online search services used by libraries at this time, bibliographic utilities used mainframe computers to maintain centralized databases of bibliographic records. Just as with online search services, libraries accessed the centralized databases of bibliographic utilities using remote, online, and dedicated terminals. Unlike online search services, however, bibliographic utilities were not used for finding articles, they were used for cataloging purposes. Bibliographic utilities maintained pools of catalog records. Libraries searched these databases to find and print out the records they needed to add them to their card catalogs. The card catalog, with its author, title, and subject indexes, with each record consisting of a paper index card, was still the dominant cataloging tool used in libraries. Bibliographic utilities meant that catalogers at local libraries could create fewer new and original cards for new items added to the library collection. If a record for an item did not yet exist in the utility's database, the cataloger could create and add one there for others to use. Bibliographic utilities thus facilitated the construction of cooperative online cataloging networks.

By 1980, there were four major bibliographic utilities in North America.[44] The oldest and most used utility was Ohio College Library Center (OCLC), later renamed to Online Computer Library Center.[45] OCLC was the service most used by libraries in greater Pittsburgh.[46] The Machine-Readable Cataloging (MARC) standard developed by LC served as the formative basis for bibliographic utilities. [47] MARC tapes distributed by LC were essential to building the utilities' centralized databases of bibliographic records.[48] Distribution of MARC records grew to overlap with LC's card distribution service, which began in 1902.[49] By 1980, LC was no longer the principal source of printed catalog cards. The primary distributors of catalog records were bibliographic utilities.[50]

Regional and state service networks were an integral part of the larger, centralized OCLC network that emerged across the U.S. These networks provided regional services to member libraries, and they worked as OCLC service brokers.[51] Pennsylvania had two regional networks: PRLC and PALINET. PRLC, or Pittsburgh Regional Library Center, was one of the first out-of-state networks to connect to OCLC, and it supplied OCLC services to libraries throughout the greater Pittsburgh region.[52] PALINET provided similar services to libraries in the greater Philadelphia region.[53]

Contrary to the NCLIS planning documents, a national library network centrally administered by LC did not emerge by 1980. It was said that LC should lead but not manage information services.[54] A distinctly American network emerged nationwide instead, one managed by the private sector and supported by federal government investment. OCLC, together with the other utilities, became the de facto nationwide bibliographic network in the U.S. State and region networks acted as nodes within this larger network.

In the mid-1980s, however, as computer technology evolved, the nature of the nationwide library network also evolved beyond the centralized networks that defined it previously. Online search services such as DIALOG, bibliographic utilities such as OCLC, and regional service networks such as PRLC were facilitated by mainframe computers. This stage of computerization promoted a single, centralized network structure. As mainframe computers gradually gave way to minicomputers and microcomputers, however, libraries incorporated these devices and used them to perform networking functions locally. Centralized networks were no longer the only automation option. Microcomputers, which proliferated in the 1980s, came to be called personal computers or desktop computers.[55] Microcomputers looked like terminals and were the same size, but they possessed a microprocessor, and microprocessors gave microcomputers their own computing power, allowing them to perform their own operations and run their own programs.[56] Coinciding with the miniaturization and decreasing costs of computing equipment, commercial vendors began to market turnkey systems to libraries in the 1980s, allowing them to run commercial software locally on their own devices.[57] Turnkey systems were software packages designed to perform library tasks such as acquisition, circulation, and cataloging. Eventually, turnkey systems gave way to integrated library systems (ILSs), where each module in the ILS represented a library function (e.g., circulation, patron accounts, cataloging), and the modules communicated with one another within the single software application. Functions came to be automated using both the larger, centralized networks such as OCLC as well as smaller, locally administered ones.[58]

State policy in Pennsylvania reflected these national trends. Public libraries across the state in the 1980s utilized national-scale online search services and bibliographic utilities, but there was also a movement at the state level to establish a state bibliographic network and other local bibliographic networks.[59] In the 1960s, the state library promoted the formation of federated library systems, but in the 1970s and 1980s, the state encouraged the formation of electronic networks.[60] Early efforts toward this end in Pennsylvania included a centralized cataloging and classification center and a book delivery service for ILLs.[61] In the early 1970s, teletypewriters (TWX) became a successful and popular tool for handling ILLs.[62] These and other aspects of network development were facilitated by the state through LSCA grants.[63] Public libraries were connected to OCLC through regional brokers, such as PRLC and PALINET.

Momentum for the development of an enhanced statewide resource-sharing network in Pennsylvania came about in 1983 with the creation of a new comprehensive plan.[64] The result was ACCESS PENNSYLVANIA, a proposed agenda for enhanced public library services. The agenda included the creation of a statewide library card system, a reimbursement plan for libraries for the ILLs they shared, revisions to the state aid formula to better support distressed communities, and information literacy instruction in schools.[65] Enhancement of the ILL system within the state required the creation of a statewide database of library holdings. [66] The statewide library card program was introduced in 1985, which was intended to reimburse local public libraries for lending materials to patrons from outside of their service area.[67] In 1985, an ILL compensation plan was also implemented under the ACESS PENNSYLVANIA initiative.[68] To carry out enhanced ILLs throughout the state, a state union catalog of bibliographic records from libraries throughout the state was created and distributed on CD-ROM discs.[69]

Automation unfolded in greater Pittsburgh as it did at the national and state levels. Like other public libraries across the country and state, CLP used the national, centralized, and electronic networks. These included both online search services and bibliographic utilities. All members of the staff were trained to perform searches and demonstrations using DIALOG and ORBIT databases.[70] CLP also continued to create and pull bibliographic records from OCLC throughout the 1980s.[71] CLP was one of the few libraries in Pennsylvania with access to OCLC holdings. Using OCLC, it provided other district library centers in Pennsylvania the holdings information of materials in libraries across the U.S.[72] CLP was one of the few libraries in the U.S. that was authorized by OCLC to make bibliographic changes to the international database.[73] At the same time, CLP coordinated closely with the state to develop a statewide, computerized circulation system for the AC-

CESS PENNSYLVANIA project.[74] The statewide library card program of the ACCESS agenda began in 1985, and Allegheny County joined in 1988.[75] Other libraries in Allegheny County were resistant to the statewide library card system, but CLP became a full participant, using the holdings data on CD-ROM discs to facilitate ILLs.[76]

While CLP participated in the larger state and national networks, it also created its own local one which was funded in part through state LSCA money.[77] Terminals and modems connected CLP's branch libraries to the central library.[78] Expansion of the automated system continued into 1988. The system was named CAROLINE, a portmanteau of *Carnegie online.*[79] In 1989, CAROLINE, the online public access catalog, was unveiled in the main branch.[80] All CLP branches were added to circulation system. There were 12 terminals, all items were barcoded, and book slips were no longer used. The paper-based card catalog, composed of index cards, was still maintained, but it was clear that the card catalog would soon be superseded by an entirely electronic database.

At the same time an electronic network was emerging within Allegheny County, the central node in that network, CLP, faced potentially crippling budget shortfalls.[81] Other libraries in the county also faced financial constraints that might force them to close.[82] By 1992, the fiscal crisis stretched across Pennsylvania, characterized by reduced or flat support from local and state governments and increased costs for the statewide library card system.[83] In discussions that occurred from 1988 through 1990, CLP revisited the idea of a federated county system. The particulars of the plan differed from previous proposals.[84] The final plan proposed a 4-tier countywide system that included reading centers, Level One Library Agencies, Level Two Library Agencies, and an In-Depth Reference and Research Library. Each tier incorporated existing public libraries and their facilities, but the roles of existing libraries were redefined with the system with a view to reducing service gaps.[85] The new system would redistribute services and collections within the county, it would require upgrades in existing facilities, and it would standardize salaries across the county.[86] This system would have solved CLP's funding problems by inheriting other libraries' funding streams, but it would have meant a loss of autonomy by the other libraries. There was some support for the new proposed system from other third-sector organizations, such as the League of Women Voters, but independent libraries and their municipalities in the county were divided on the proposal.[87] The system was not implemented.

Following CLP's proposal, in June 1990, a separate but related proposal for a federated library system appeared in a report authored by Frank Lucchino, the County Controller for Allegheny County. The report, *Looking Beyond Our Past*, included several proposals related to the government organization and funding within Allegheny County. [88] The purpose of the report was to promote county government as the most suitable entity for guiding the development of regional infrastructure.[89] Lucchino's report also proposed a federated public library system, but in contrast to CLP's plan, Lucchino's plan charged the county with its coordination, not CLP.[90] Lucchino's plan resembled those from 1974 and 1978. The release of *Looking Beyond Our Past* marked Lucchino's entrance into library politics in Allegheny County. He was a library outsider, but he would soon become a lightning rod for libraries, building support for what would become major infrastructural change.[91] Lucchino chose to focus on public libraries for a combination of reasons, some political.[92] But primarily, Lucchino sought to inspire public libraries in greater Pittsburgh to form a regional system because it would showcase how regional cooperation could work effectively.[93]

The spring of 1991 marked a major turning point in the development of public library infrastructure in Allegheny County. During this time, Lucchino continued to concentrate on the funding and administration of libraries. The funding issue was particularly acute given the anticipated cut in state funding for public libraries. In January 1991, the governor announced an immediate cut of $2 million in state aid to public libraries and a cut in the appropriation for ACCESS PENNSYLVANIA, the statewide library card and interlibrary delivery system.[94] Public libraries in Allegheny County were under increased pressure. By 1991, CLP's total budget spent on the purchase of materials had fallen below 12 percent, which was the standard required by the state to receive state aid. CLP's Oakland facility was the only branch open 65 hours a week, another requirement for state aid. CLP still received state aid, but it was issued several warnings from the state regarding their state funding.[95] In the spring of 1991, Lucchino met with independent public libraries in the county, excluding CLP. Based in part on these discussions, he and his staff issued a second report that focused exclusively on public libraries in Allegheny County. The title of the report was *A Quiet Crisis*. Released in 1991, it garnered immediate attention because, unlike previous reports and studies about public libraries in the county, it was written by an active elected official within county government.

A Quiet Crisis was released in March of 1991. In anticipation of its publication, Lucchino met with public librarians and trustees throughout the county in advance to prepare them for its release and to form a public library coalition. In one of his meetings with public libraries in the county, Lucchino urged them to capitalize on the timing of the report's release:

> Let me tell you where we are going. In about one month we will issue a report on libraries in Allegheny County, making statements about it and it may cause a flurry in the papers and some activity. If you do nothing for months, there is no synergy. Our effort will be ignored, unless you folks follow up quickly. If you have an idea, you have a chance to succeed. If you just ask for money, you will fail.[96]

Throughout his meetings with libraries, Lucchino stressed the need for a countywide coalition of public libraries. [97] Lucchino saw a federated system as the only plausible option for public libraries in the county to overcome their funding woes.[98]

As Lucchino met with the county libraries, some resisted including CLP in the discussions.[99] One library director indicated that some libraries felt resentful that CLP proposed a countywide library system in 1990, with CLP at its center, without first consulting the other libraries. Some libraries felt bullied by CLP.[100] While some county libraries looked to CLP for leadership, in part because it served as district library, libraries nevertheless described an "us-versus-them" situation where independent county libraires competed with CLP for county funding.[101] Lucchino stressed that CLP had to be approached if there was ever to be a countywide coalition. County government would not fund a coalition of libraries if there was controversy.[102]

One important component of forming a new library coalition, according to Lucchino, was building public support. This meant drawing the attention of government officials and the general public. Librarians were too quiet about their funding situation, hence Lucchino's choice in title, *The Quiet Crisis.* Lucchino urged local county libraries to contact their officials and initiate political action.[103] In addition to contacting local government officials, Lucchino also urged local libraries to wage a media campaign through newspaper editorials and television ads to mobilize citizens throughout the county and to convey a strong message to the county commissioners.[104] The reason a federated countywide system was needed, Lucchino argued, was that it would provide public library infrastructure across the county in a more equitable way and with fewer service gaps.[105] Lucchino recommended several steps toward the formation of a federated library system.[106] Public libraries should unite to lobby government for equitable funding. Unity within the group was essential for this to succeed.[107] A department of library services headed by a director should be created within county government. The staff of this department would administer the county's contribution to library services and coordinate them, guided by an advisory board.[108] Libraries would be funded using a needs-based formula, and county funding would match a portion of local government funding while also accounting for per capita income of the service area.[109] Importantly, Lucchino believed a computer link should be created between CLP and the other libraries within the county as a way to incorporate them all into a single countywide system. In effect, Lucchino proposed that the libraries form a regional computerized network.[110]

There was an immediate response to Lucchino's meetings on the part of the libraries in the county.[111] Some libraries were skeptical about the formation of a federated system, but a small group of libraries, including representatives from

CLP, met to discuss it. [112] As a result of these meetings in February and March 1991, a new countywide library association was formed.[113] The association was called CLASP, or County Library Association Serving the People. Representatives agreed to seek more funding from municipalities and the state and to press for county aid.[114] A CLASP executive committee was formed.[115] The formation of CLASP anticipated the release of Lucchino's report, *A Quiet Crisis*. The report was released two days after CLASP was formed.[116] CLASP continued to meet in early 1991 to discuss the proposed state and federal cut to public library aid and to approve their bylaws.[117] Soon, 32 county libraries became members of CLASP.[118]

In contrast to the enthusiastic response by libraries to Lucchino's plan, the reaction by county government officials was hostile. One of the three county commissioners rejected Lucchino's plan outright for lack of money.[119] Public opinion toward the plan was also lukewarm at best. An editorial in the Pittsburgh Press agreed with Lucchino's summation of the problems with library service in Allegheny County, but it rejected the idea of a new county office.[120] In the end, the commissioners did not allocate additional money to county libraries, and they were unwilling to fund a new county department. But, as a compromise, rather than fund and staff a new county library department, the county commissioners instead agreed to appoint a non-funded library commission to explore the remaining steps in Lucchino's proposed plan. It would be left to this new commission to facilitate the development of a federated library system, to secure funding for a county library director, and to carry out the remaining steps outlined in *A Quiet Crisis*.[121] The new commission was called The Commission on the Future of Libraries in Allegheny County (CFLAC). Commission appointees were sworn in by the Allegheny County Board of Commissioners in June 1991.[122] It held its first meeting on June 20, 1991. CFLAC decided to pursue funding from the Buhl Foundation for the county library director position, which was successful. [123] A county library director was hired in January 1992. [124] The commission also decided to visit the various libraries within the county and to solicit input from library groups and municipalities.[125] A representative from the State Library of Pennsylvania served on the Commission, raising its profile.[126] Through the creation of CFLAC, regional planning and coordination of the public library infrastructure in greater Pittsburgh was brought under the aegis of county government.

CFLAC was formed in the wake of the release of *A Quiet Crisis*, and the group was guided by the goals set forth there. Lucchino served as chair. Central to CFLAC's vision for the regional public library infrastructure was the establishment of a federated library system, but to do this, CFLAC had to develop a means to bring the libraries together in a functional way beyond the loose advocacy association established by CLASP. All public libraries faced similar economic challenges from potential federal, state, and local funding shortfalls, and libraries sought a collaborative means to share resources, but there were few cooperative efforts already underway.[127] Libraries were in a position to automate, and they desired to do so, but they required an additional stimulus.[128] As proposed in *A Quiet Crisis*, the way to bring the county libraries together was through a shared electronic catalog.[129]

Towards the close of 1991, CFLAC began to investigate the gradual incorporation of county libraries into this shared computer network. County planning documents from September 1991 emphasized an electronic network of libraries.[130] CLP would serve as the center of this network since they already had one connecting its branches to CAROLINE. By the end of 1991, CLP had already announced its plans to expand its computer network to include the county libraries.[131] CLP needed to upgrade its online catalog anyway, and at this moment, other libraries throughout the county were interested in automating their systems as well.[132] It would be a benefit for everyone to create a countywide network. [133]

Upgrades to CLP's system, and the equipment placed at other county libraries, were funded through public- and private-sector grants, including those from the county and state.[134] The campaign to connect county libraries to CLP's catalog was dubbed Project Link-Up.[135] All participants in Project Link-Up had to be members of CLASP. [136] The goal of this initiative was to connect the catalogs of all the county libraries, including CLP. [137] By early 1992, limited dial-in access to CLP's database was made possible by placing computer terminals in the other libraries in the county.[138] Funding in the first phase allowed 15 libraries to connect to CLP. [139] The first node was unveiled at Wilkinsburg Public Library. Libraries could access the collection's books and materials, and they could use information there to aid in cataloging their own collections. They could also access the collections of other member libraries whose holdings were similarly computerized.[140] Some librarians had access to email and electronic bulletin boards for the first time.[141] By the end of 1992, CLP agreed to computer placement of reserves by local libraries, and the turnaround time for ILLs was reduced by at least a week.[142] Supplying fax machines to county libraries was an additional component to Project Link-Up. By March 1993, most fax machines and computers had been delivered to libraries.[143] Phase 1 of Project Link-Up was completed by December 1992 when the first computers were delivered to libraries. In 1993, 11 additional county libraries were connected through Project Link-Up, thus completing Phase 2. In Phase 3, there were further discussions about an overall network system plan and the means to finance it.[144]

CLASP libraries coordinated with CFLAC and CLP on Project Link-Up, but beyond the automation efforts led by CFLAC, the CLASP libraries and the new county library director struggled to identify additional collaborative efforts.[145] Project Link-Up did not immediately result in a federated library system, the central recommendation of *A Quiet Crisis* and a central goal for CFLAC. By early 1993, momentum in CLASP and CFLAC had already faded, and there was no appetite for a federated system at the time.[146] But the development of an automated library network was an essential first step toward a new organizational structure.[147] Due to the formation of CLASP and CFLAC, and the success of Project Link-Up, the public library infrastructure in greater Pittsburgh did undergo substantive change in Period 5. A CLASP newsletter was started in May 1992, and in June CLASP's membership included 25 libraries and 5 individuals.[148] By November 1992, these numbers rose to 32 members and 9 individuals.[149] By the end of 1992, CLP and other libraries in the county had connected to the Internet, and the CLP began the formation of a regional FREENET, a collection of online resources hosted by CLP and accessed through the Internet.[150] Later telecommunications legislation would allow libraries greater access to the Internet through reduced telephone rates.[151]

During Period 5, the decision cycle was initiated by two issues: poor funding and changing technology. CLP and the other independent county libraries were spurred on to address these issues by a new catalyst in the arena: Frank Lucchino. Lucchino's reports and discussions carried the decision cycle forward by building legitimacy for a new approach to resource sharing among libraries. Two new decision-making centers were established during this time: CLASP and CFLAC. CLASP was a third-sector civil society group, an advocacy organization for public libraries. CFLAC became a planning and administration wing for the public library infrastructure within the county government. Working together, CLP, CLASP, and CFLAC sought to address the funding and service gap problems identified in *A Quiet Crisis.* Due to the technological circumstances of the moment, the solution to bring libraries together and solve these problems was the creation of a shared computerized network. In real terms, this was Project Link-Up, which represents a formative stage in the development of a computerized regional network.

Period 5 bolsters Releaser Theory. Contrary to what might have been thought, due to the unforgiving economic circumstances of the time, substantive change in the system would not have been expected unless it included major service cuts and closings. Other public libraries throughout the state in fact faced these circumstances. But the public library infrastructure in greater Pittsburgh bucked this trend. This was due to a major outside issue—technological change, i. e. automation—as well as the coalescence of new library groups, led by Lucchino, an active government official. The buildup of consensus about resource sharing among the public libraries led to the formation of CLASP and CFLAC. Planning and coordination of the public library infrastructure came under the auspices of the county government, at least to an extent and for the moment. The vision of a regional computerized network, made possible by the technologies of the day and supported by private and public investment, led to Project Link-Up, which is what served as connective tissue for a growing public library infrastructure in greater Pittsburgh.

Period 5 witnessed the groundwork and genesis for a new circuit, a county library association, or CLASP. There was also the formation of a new agency in the inner periphery of the county circuit, CFLAC, which while not funded by the county operated under the aegis of county government. The formation of the CLASP circuit created the potential for a new home for the regional public library infrastructure. In previous periods, the infrastructure was controlled by the county and CLP. Now, with the emergence of CLASP and Project Link-Up, it was possible for the infrastructure to become self-governing. But only the formative steps for creating this new circuit were implemented in Period 5. The public library infrastructure remained tied to CLP and the county.

1 Roy Lubove, *Twentieth-Century Pittsburgh: The Post-Steel Era*, vol. 2 (Pittsburgh: University of Pittsburgh Press, 1996), 25.

2 Lubove, *Twentieth-Century Pittsburgh: The Post-Steel Era*, 2, 7.

3 Carnegie Library of Pittsburgh, The Carnegie Library of Pittsburgh Branch Service Areas – City of Pittsburgh, February 1989, District Services Collection, Series 2, Box 5, Folder 6, CLP-OLIVER; Presidential Advisory Committee on the Library, The Carnegie: Presidential Advisory Committee for the Library Meeting, 21 February 1989, District Services Collection, Series 2, Box 5, Folder 7, CLP-OLIVER.

4 Carnegie Library of Pittsburgh, Roles of the Carnegie Library of Pittsburgh, 9 January 1989, District Services Collection, Series 2, Box 5, Folder 6, CLP-OLIVER.

5 Albert F. Kamper, Crisis of the Steel Valley Libraries, 1988, District Services Collection, Series 2, Box 5, Folder 6, CLP-OLIVER.

6 Carnegie Library of Pittsburgh, Allegheny County District Library, Library Characteristics, 1990, District Services Collection, Series 2, Box 5, Folder 6, CLP-OLIVER; DeeAnna R. Cavinee, District Library Services Program, 6 February 1989, District Services Collection, Series 2, Box 5, Folder 7, CLP-OLIVER; Carnegie Library of Pittsburgh, Allegheny County Municipalities, 1989, CLP-PENN.

7 Guy A. Tumolo, Letter from Guy A. Tumolo to Robert C. Wilburn, 22 May 1987, District Services Collection, Series 2, Box 4, Folder 4, CLP-OLIVER.

8 Citizens Study Committee on Libraries, Public Libraries in Allegheny County…A Plan of Action, September 1978, District Services Collection, Series 2, Box 1, Manual 10, CLP-OLIVER., Appendix C.

9 Carnegie Library of Pittsburgh, Comments on the Action Plan for Public Libraries in Allegheny County, 31 July 1978, District Services Collection, Series 2, Box 5, Folder 1, CLP-OLIVER. Donald C. Potter, Comments on the Action Plan, 26 July 1978, District Services Collection, Series 2, Box 5, Folder 9, CLP-OLIVER.

10 Potter, Comments on the Action Plan, 26 July 1978.

11 Citizens Study Committee on Libraries, Public Libraries in Allegheny County, September 1978, 15 – 16.

12 Citizens Study Committee on Libraries, Public Libraries in Allegheny County, September 1978, 15.

13 Violet Hutchison, Letter from Violet Hutchison to Keith Gilbert, 9 February 1979, District Services Collection, Series 2, Box 5, Folder 1, CLP-OLIVER; Deborah Schlesinger, Letter from Deborah Schlesinger to Keith Gilbert, 3 January 1979, District Services Collection, Series 2, Box 5, Folder 1, CLP-OLIVER.

14 Librarian 1, Interview Transcript, 7 January 2016, interviewed by Michael M. Widdersheim.

15 Schlesinger, Letter from Deborah Schlesinger to Keith Gilbert, 3 January 1979.

16 Herbert A. Carl, "Division of Library Services and Educational Facilities, U.S. Office of Education," in *The Bowker Annual of Library and Book Trade Information*, ed. Phyllis B. Steckler (New York: Bowker, 1967), 127.

17 Public Library Association, *Minimum Standards for Public Library Systems, 1966* (Chicago: American Library Association, 1967).

18 Lowell A. Martin, "Standards for Public Libraries," *Library Trends* 21, no. 2 (1972): 164, 68. Public Library Association, *Minimum Standards for Public Library Systems, 1966*, 1–7.

19 Redmond Kathleen Molz and Phyllis Dain, *Civic Space/Cyberspace: The American Public Library in the Information Age* (Cambridge: MIT Press, 1999), 24; Public Library Association, *The Public Library Mission Statement and Its Imperatives for Service* (Chicago: American Library Association, 1979); Vernon E. Palmour, Marcia C. Bellassai, and Nancy V. De Wath, *A Planning Process for Public Libraries* (Chicago: American Library Association, 1980).

20 Public Library Association, *The Public Library Mission Statement and Its Imperatives for Service*, v.

21 Douglas L. Zweizig and Eleanor Jo Rodger, *Output Measures for Public Libraries* (Chicago: American Library Association, 1982); Nancy A. Van House et al., *Output Measures for Public Libraries*, 2nd ed. (Chicago: American Library Association, 1987); Charles R. McClure et al., *Planning and Role Setting for Public Libraries: A Manual of Options and Procedures* (Chicago: American Library Association, 1987); American Library Association, *Standards for Library Functions at the State Level* (Chicago: American Library Association, 1963); American Association of State Libraries, *Standards for Library Functions at the State Level* (Chicago: American Library Association, 1970); American Library Association, *Standards for Library Functions at the State Level* (Chicago: American Library Association, 1985); Board of Education for Librarianship, "First Annual Report of the Board of Education for Librarianship," *Bulletin of the American Library Association* 19, no. 4 (1925); Board of Education for Librarianship, "Minimum Requirements for Library Schools," *Bulletin of the American Library Association* 27, no. 13 (1933); Board of Education for Librarianship, "The Librarian," *ALA Bulletin* 42, no. 11 (1948).

22 Margaret W. Ellingson and Susan D. Morris, "Interlibrary Loan: Evolution to Revolution," in *Interlibrary Loan Practices Handbook*, ed. Cherié L. Weible and Karen L. Janke (Chicago: American Library Association, 2011), 5; Vernon E. Palmour, Marcia C. Bellassai, and Nancy K. Roderer, *Resources and Bibliographic Support for a Nationwide Library Program: Final Report to the National Commission for Libraries and Information Science* (Washington, DC: US GPO, 1974), 5.

23 Redmond Kathleen Molz, *National Planning for Library Service, 1935–1975* (Chicago: American Library Association, 1984), 85, 91.

24 Roger H. McDonough and David C. Palmer, "Standards for State Libraries," *Library Trends* 21, no. 2 (1972): 179; W. Lyle Eberhart, "Standards for State Library Agencies," *Library Trends* 27, no. 2 (1978): 211; Stephen R. Salmon, "LITA's First Twenty-Five Years: A Brief History," *Information Technology and Libraries* 12, no. 1 (1977); Charles P. Bourne and Trudi Bellardo Hahn, *A History of Online Information Services, 1963–1976* (Cambridge: MIT Press, 2003), 48–52; Henriette D. Avram and Lenore S. Maruyama, eds., *Toward a National Library and Information Service Network: The Bibliographic Component* (Washington, DC: Library of Congress, 1977), 54; Henriette D. Avram and Beacher Wiggins, "The Linked Systems Project: Introduction and Background," in *The Linked Systems Project: A Networking Tool for Libraries*, ed. Judith G. Fenly and Beacher Wiggins (Dublin: OCLC, 1988), 3–4.

25 Council on Library Resources, *1st Annual Report* (Washington, DC, 1957), 5.

26 Deanna B. Marcum, "Automating the Library: The Council on Library Resources," *IEEE Annals of the History of Computing* 24, no. 3 (2002): 3.

27 Gilbert W. King et al., *Automation and the Library of Congress: A Survey Sponsored by the Council on Library Resources* (Washington, DC: US GPO, 1963); Marcum, "Automating the Library: The Council on Library Resources."; Warren J. Haas, Nancy E. Gwinn, and C. Lee Jones, "Managing the Information Revolution: CLR's Bibliographic Service Development Program," *Library Journal* 104 (1979).

28 Douglas N. Knight and E. Shepley Nourse, eds., *Libraries at Large: Tradition, Innovation, and the National Interest* (New York: Bowker, 1969); Molz, *National Planning for Library Service, 1935–1975*, ix-x; National Commission on Libraries and Information Science, *Annual Report, 1971–1972* (Washington, DC: US GPO, 1973), 5.

29 National Commission on Libraries and Information Science Act, 91–345 (1970).

30 Palmour, Bellassai, and Roderer, *Resources and Bibliographic Support for a Nationwide Library Program: Final Report to the National Commission for Libraries and Information Science*; National Commission on Libraries and Information Science, *Toward a National Program for Library and Information Services: Goals for Action*, National Commission on Libraries and Information Science (Washington, DC: US GPO, 1975); Molz, *National Planning for Library Service, 1935–1975*, x.

31 National Commission on Libraries and Information Science, *Toward a National Program for Library and Information Services: Goals for Action* (1975), xi-xii; Lawrence F. Buckland and William L. Basinski, *The Role of the Library of Congress in the Evolving National Network: Final Report*, National Commission on Libraries and Information Science (Washington, DC: US GPO, 1978).

32 National Commission on Libraries and Information Science, *Toward a National Program for Library and Information Services: Goals for Action* (1975), 50.

33 Council on Library Resources, *14th Annual Report* (Washington, DC, 1970), 17–18.

34 American Library Association, "A National Plan for Libraries," *Bulletin of the American Library Association* 29, no. 2 (1935); American Library Association, "A National Plan for Libraries," *ALA Bulletin* 33, no. 2 (1939); Committee on Post-War Planning, *Post-War Standards for Public Libraries* (Chicago: American Library Association, 1943); Committee on Post-War Planning, *A National Plan for Public Library Service: Prepared for the Committee on Postwar Planning of the American Library Association* (Chicago: American Library Association, 1948); American Library Association, *Public Library Service: A Guide to Evaluation, with Minimum Standards* (Chicago: ALA, 1956).

35 *Library Services and Construction Act Amendments of 1966, 511 20 USC 89 (1966); American Library Association, Standards for Library Functions at the State Level, 19; American Association of State Libraries, Standards for Library Functions at the State Level, 14; National Commission on Libraries and Information Science Act.*

36 Molz, *National Planning for Library Service, 1935–1975*, 109; James G. Williams and Roger Flynn, "Network Topology: Functions of Existing Networks," in *The Structure and Governance of Library Networks: Proceedings of the 1978 Conference in Pittsburgh, Pennsylvania, Co-Sponsored by National Commission on Libraries and Information Science and University of Pittsburgh*, ed. Allen Kent and Thomas J. Galvin (New York: Marcel Dekker, 1979).

37 Frederick G. Kilgour, "Historical Note: A Personalized Prehistory of OCLC," *Journal of the American Society for Information Science* 38, no. 5 (1987): 381.

38 National Commission on Libraries and Information Science, *Toward a National Program for Library and Information Services: Goals for Action* (1975), 32.
39 Network Advisory Committee, *Networks for Networkers II Conference: A Synthesis of Conference Papers and a Summary of Conference Resolutions*, Library of Congress (Washington, DC, 1991), 29.
40 Bourne and Hahn, *A History of Online Information Services, 1963 – 1976*, 11.
41 Charles R. Hildreth, *Library Automation in North America: A Reassessment of the Impact of New Technologies on Networking* (München: Saur, 1987), 2.
42 Hank Epstein, "The Technology of Library and Information Networks," *Journal of the American Society for Information Science* 31, no. 6 (1980).
43 Bourne and Hahn, *A History of Online Information Services, 1963 – 1976*; Epstein, "The Technology of Library and Information Networks," 427.
44 *Barbara M. Robinson, "Cooperation and Competition among Library Networks," Journal of the American Society for Information Science 31, no. 6 (1980); Hildreth, Library Automation in North America: A Reassessment of the Impact of New Technologies on Networking.*
45 David H. Brunell, "The Strategic Alliance between OCLC and Networks: Partnerships That Work," in *OCLC 1967 – 1997: Thirty Years of Furthering Access to the World's Information*, ed. K. Wayne Smith (New York: Haworth, 1998); Arthur D. Little, *A New Governance Structure for OCLC: Principles and Recommendations* (Metuchen: Scarecrow, 1977); OCLC, *Annual Report 1977/1978* (Columbus: OCLC, 1978); OCLC, *Annual Report 1978/1979* (Columbus: OCLC, 1979); OCLC, *Annual Report 1980/1981* (Dublin: OCLC, 1981); Hildreth, *Library Automation in North America: A Reassessment of the Impact of New Technologies on Networking*, 37.
46 Joseph R. Matthews, "The Four Online Bibliographic Utilities: A Comparison," *Library Technology Reports* 15, no. 6 (1979): 758; Buckland and Basinski, *The Role of the Library of Congress in the Evolving National Network: Final Report* (1978), 119 – 20; Robinson, "Cooperation and Competition among Library Networks," 420; Avram and Maruyama, *Toward a National Library and Information Service Network: The Bibliographic Component*; Henriette D. Avram and David C. Hartmann, "Objectives and Accomplishments of the Network Technical Architecture Group," *Program: electronic library and information systems* 13, no. 1 (1979); Avram and Wiggins, "The Linked Systems Project: Introduction and Background."; Network Technical Architecture Group, *Message Delivery System for the National Library and Information Service Network: General Requirements* (Washington, DC: Library of Congress, 1978).
47 Michele Seikel and Thomas Steele, "How MARC Has Changed: The History of the Format and Its Forthcoming Relationship to RDA," *Technical Services Quarterly* 28, no. 3 (2011): 324; Karen M. Spicher, "The Development of the MARC Format," *Cataloging & Classification Quarterly* 21, no. 3 – 4 (1996): 84.
48 John Y. Cole and Jane Aikin, "Machine-Readable Cataloging (MARC)," in *Encyclopedia of the Library of Congress: For Congress, the Nation & the World*, ed. John Y. Cole and Jane Aikin (Lanham: Bernan, 2004), 321; Seikel and Steele, "How MARC Has Changed: The History of the Format and Its Forthcoming Relationship to RDA," 324; King et al., *Automation and the Library of Congress: A Survey Sponsored by the Council on Library Resources*, 1; Spicher, "The Development of the MARC Format," 79.
49 John Y. Cole and Jane Aikin, "Cataloging Distribution Service," in *Encyclopedia of the Library of Congress: For Congress, the Nation, & the World*, ed. John Y. Cole and Jane Aikin (Lanham: Bernan, 2004), 187; Cole and Aikin, "Machine-Readable Cataloging (MARC)," 322; Seikel and Steele, "How MARC Has Changed: The History of the Format and Its Forthcoming Relationship to RDA," 326.

50 Cole and Aikin, "Cataloging Distribution Service," 189; Library of Congress, *Annual Report of the Librarian of Congress for the Fiscal Year Ending September 30, 1979* (Washington, DC: US GPO, 1980), 64; Spicher, "The Development of the MARC Format," 80.
51 *Buckland and Basinski, The Role of the Library of Congress in the Evolving National Network: Final Report (1978); Hildreth, Library Automation in North America: A Reassessment of the Impact of New Technologies on Networking.*
52 Kathleen L. Maciuszko, *OCLC: A Decade of Development, 1967 – 1977* (Littleton: Libraries Unlimited, 1984), 48 – 49; Robinson, "Cooperation and Competition among Library Networks," 419.
53 Buckland and Basinski, *The Role of the Library of Congress in the Evolving National Network: Final Report* (1978), 104 – 05; Robinson, "Cooperation and Competition among Library Networks," 417. Robinson, "Cooperation and Competition among Library Networks," 419.
54 National Commission on Libraries and Information Science, *Public Sector/Private Sector Interaction in Providing Information Services: Report to the NCLIS from the Public Sector/Private Sector Task Force* (Washington, DC: US GPO, 1982).
55 National Research Council, *Evolving the High Performance Computing and Communications Initiative to Support the Nation's Information Infrastructure* (Washington, DC: National Academy Press, 1995), 12.
56 Hildreth, *Library Automation in North America: A Reassessment of the Impact of New Technologies on Networking*, 140 – 41.
57 Hildreth, *Library Automation in North America: A Reassessment of the Impact of New Technologies on Networking*, 9 – 10.
58 Robinson, "Cooperation and Competition among Library Networks," 415 – 16; Molz and Dain, *Civic Space/Cyberspace: The American Public Library in the Information Age*, 124 – 25; Network Advisory Committee, *Networks for Networkers II Conference: A Synthesis of Conference Papers and a Summary of Conference Resolutions* (1991), 28; National Commission on Libraries and Information Science, *The National Information Infrastructure and the Recommendations of the 1991 White House Conference on Library and Information Services* (Silver Spring: Washington Information Services, 1994), 9; Marshall Breeding, "Telecommunications Options Connect OCLC and Libraries to the Future: The Co-Evolution of OCLC Connectivity Options and the Library Computing Environment," in *OCLC 1967 – 1997: Thirty Years of Furthering Access to the World's Information*, ed. K. Wayne Smith (New York: Haworth, 1998), 121; Charles R. McClure, John Carlo Bertot, and Douglas L. Zweizig, *Public Libraries and the Internet: Study Results, Policy Issues, and Recommendations* (Washington, DC: National Commission on Libraries and Information Science, 1994); Charles R. McClure, John Carlo Bertot, and John C. Beachboard, *Internet Costs and Cost Models for Public Libraries* (Washington, DC: US GPO, 1995).
59 State library of Pennsylvania, Comprehensive Plan for Library Service in Pennsylvania, 1983, District Services Collection, Series 2, Box 7, Manual 7. CLP-OLIVER; Charles T. Meadow et al., A Plan for Library Cooperation in Pennsylvania, 1976, 3, District Services Collection, Series 2, Box 7, Manual 6. CLP-OLIVER; Pennsylvania Library Master Plan Committee, *Working Report of the Pennsylvania Library Master Plan Committee*, Bureau of Library Development, State Library of Pennsylvania (Harrisburg, 1973), 1.
60 State library of Pennsylvania, Federated Library Systems, October 1988, ACLA historical file (Folder: Documents). ACLA-OFFICE; State Library of Pennsylvania Bureau of Library Development, Minimum Standards and Guidelines for Pennsylvania Public Library Systems Receiving State Aid, June 1978, District Services Collection, Series 2, Box 5, Folder 9, CLP-OLIVER; Pennsylvania Library Master Plan Committee, *Working Report of the Pennsylvania Library Master Plan Committee* (1973), 19.

61 Sarah K. Vann, *Southeastern Pennsylvania Processing Center Feasibility Study: Final Report* (Harrisburg: Pennsylvania State Library, 1967), 1; Stephen Wood, "Developing Library Cooperation through IDS," *PLA Bulletin* 28, no. 2 (March 1973): 61; Meadow et al., A Plan for Library Cooperation in Pennsylvania, 1976, v; State library of Pennsylvania, Comprehensive Plan for Library Service in Pennsylvania, 1983, 6; State Library of Pennsylvania, *LSCA Long Range Plan, 1982–1987* (Harrisburg: Pennsylvania State Library, 1981), 9–10; Alexander Strasser, *Bibliographic Access in Pennsylvania* (Pittsburgh: Pittsburgh Regional Library Center, January 1979).
62 Harry R. Courtright and Annette McAlister, "Pennsylvania's Interlibrary TWX Demonstration," *PLA Bulletin* 24, no. 4 (May 1969): 145; Pennsylvania Library Master Plan Committee, *Pennsylvania Library Master Plan Committee Report* (Harrisburg: State Library of Pennsylvania, 1974); Natalie Wiest, Lee Lourea, and Brigitte L. Kenney, Inventory of Pennsylvania Interlibrary Cooperatives and Information Networks, Drexel University Graduate School of Library Science, May 1972, SLPARSC; Pennsylvania Library Master Plan Committee, *Working Report of the Pennsyvania Library Master Plan Committee* (Harrisburg: Bureau of Library Development, September 1973); John A. McCrossan, "Cooperative Library Networks–Today and Tomorrow," *PLA Bulletin* 27, no. 5 (September 1972): 252; "News: New Union List," *PLA Bulletin* 29, no. 1 (January 1974); Charles H. Ness, "The President's Page," *PLA Bulletin* 30, no. 2 (March 1975): 31; Marvin W. Mounce, "Library Development in Pennsylvania: A Review," *PLA Bulletin* 27, no. 5 (September 1972): 247; Robert C. Stewart, "Cataloging with a Computer–OCLC Comes to Pennsylvania," *PLA Bulletin* 28, no. 1 (1973).
63 State Library of Pennsylvania, *LSCA Long Range Plan, 1982–1987*, 8; State Library of Pennsylvania, *LSCA Long Range Plan, 1982–1987*, 9.
64 State library of Pennsylvania, Comprehensive Plan for Library Service in Pennsylvania, 1983, Appendix G; Alexander Strasser, *Bibliographic Access in Pennsylvania* (Washington, DC: U.S. Office of Education, 1979).
65 ""Access Pennsylvania" Proposal Endorsed by Governor Thornburgh," *Pennsylvania Library Association Bulletin* (October 1984); State Library of Pennsylvania, *Access Pennsylvania: An Agenda for Knowledge and Information through Libraries* (Harrisburg: State Library of Pennsylvania, September 1984), 1; State Library of Pennsylvania, Access Pennsylvania Scrapbook, 1984–85, Department of Education, Commonwealth Libraries, State Aid to Public Libraries, Box 6, Access Pennsylvania Scrapbook 1984–1986, PASA.
66 State Library of Pennsylvania, *Access Pennsylvania: An Agenda for Knowledge and Information through Libraries*, 13.
67 Pennsylvania Library Association, *The Platform for 21st Century Libraries*, Pennsylvania Library Association (Harrisburg, 1998); J. Matthews and Associates, Recommendations for the Implementation of a Statewide Library Card System: Executive Summary, March 1985, 1, Department of Education, Commonwealth Libraries, State Aid to Public Libraries, Box 6, Access Pennsylvania Scrapbook 1984–1986, PASA; State Library of Pennsylvania, Statewide Library Card System, Proposed Regulations, April 1985, April 1985, Department of Education, Commonwealth Libraries, State Aid to Public Libraries, Box 6, Access Pennsylvania Scrapbook 1984–1986, PASA; Elliot L. Shelkrot, State Librarian's Report to the Advisory Council on Library Development/Federal Advisory Council, 4 February 1986, Department of Education, Commonwealth Libraries, Minutes of the Governor's Advisory Council on Library Development 1962–2009 collection, Carton 1, Folder 2, PASA.
68 King Research, *Interlibrary Loan Compensation Plan for the Commonwealth of Pennsylvania* (Rockville: King Research, 15 October 1985), 1; State Library of Pennsylvania, Interlibrary Loan Compensation Plan, Proposed Guidelines, 19 June 1985, Department of Education, Commonwealth Libraries, State Aid to Public Libraries, Box 6, Access Pennsylvania Scrapbook 1984–1986, PASA; "Pilot

Project Termed a Success," *Pennsylvania Library Association Bulletin* (May/June 1986); "Governor Robert P. Casey Signs FY 1989 – 90 PA Budget Including $5 Million for Library Access," *PLA Bulletin* (September 1989).
69 Hildreth, *Library Automation in North America: A Reassessment of the Impact of New Technologies on Networking*, 81.
70 Carnegie Library of Pittsburgh, Annual Report, 1980, 12, CLP-PENN.
71 Carnegie Library of Pittsburgh, Annual Report, 1980, 12 – 13.
72 Carnegie Library of Pittsburgh, Annual Report, 1980, 13.
73 Carnegie Library of Pittsburgh, 1988 Annual Report, 1988, 2, CLP-PENN.
74 Carnegie Library of Pittsburgh, Comments on the Action Plan for Public Libraries in Allegheny County, 31 July 1978.
75 Patti Murphy, "Library Cooperation Works–for Some," *Pittsburgh Post-Gazette*, 28 April 1988.
76 Carnegie Library of Pittsburgh, 1987 Annual Report, 1987, 5, CLP-PENN. Carnegie Library of Pittsburgh, 1988 Annual Report, 1988, 3; Northland Public Library and Richland Public Library, Implications of the Proposed Pennsylvania Statewide Library Card System, June 1985, Obtained from a library director.
77 State Library of Pennsylvania, Letter from Dave Mitchell to Council Members, 8 September 1986, Department of Education, Commonwealth Libraries, Minutes of the Governor's Advisory Council on Library Development 1962 – 2009 collection, Carton 1, Folder 8, PASA.
78 Carnegie Library of Pittsburgh, 1987 Annual Report, 1987, 5.
79 Carnegie Library of Pittsburgh, 1988 Annual Report, 1988, 2.
80 Carnegie Library of Pittsburgh, 1989 Annual Report, 1989, CLP-PENN; Carnegie Library of Pittsburgh, 1991 Annual Report, 1991, 3, CLP-PENN.
81 Presidential Advisory Committee on the Library, Presidential Advisory Committee on the Library, 23 January 1989, District Services Collection, Series 2, Box 5, Folder 6, CLP-OLIVER; President's Advisory Committee on the Library, Report of the President's Advisory Committee on the Library, April 1990, 1, Director's Office Collection, Box 44, Folder 9, CLP-OLIVER; Presidential Advisory Committee on the Library, The Carnegie: Presidential Advisory Committee for the Library Meeting, 21 February 1989.
82 Governor's Advisory Council on Library Development, Current Articles, 1992, Department of Education, Commonwealth Libraries, Minutes of the Governor's Advisory Council on Library Development 1962 – 2009 collection, Carton 1, Folder 5, PASA.
83 Associated Press, "Libraries across the State Struggling with Funding," *Citizen Voice* (Wilkes Barre) 12 June 1992.
84 DeeAnna R. Cavinee, Letter from DeeAnna Cavinee to Al Kamper, 13 March 1989, District Services Collection, Series 2, Box 5, Folder 6, CLP-OLIVER; Albert F. Kamper, Future Outlets: Response to Dee Cavinee, 17 March 1989, District Services Collection, Series 2, Box 5, Folder 6, CLP-OLIVER; President's Advisory Committee on the Library, Report of the President's Advisory Committee on the Library, April 1990, 1; Carnegie Library of Pittsburgh, Annual Report, 1990, 1, CLP-PENN; Mary Kane, "Rumor of Closing Creates Sounding Board for Library Needs," *Pittsburgh Press* 1990, April 19.
85 President's Advisory Committee on the Library, Report of the President's Advisory Committee on the Library, April 1990, 24.
86 President's Advisory Committee on the Library, Report of the President's Advisory Committee on the Library, April 1990, 25 – 26.
87 Bob Hoover, "Countywide Library System Urged," *Pittsburgh Post-Gazette*, 3 May 1990; "Essential Reading," *Pittsburgh Post-Gazette*, 15 May 1990; Anne Schifferle, Letter from Anne Schifferle to

Albert F. Kamper, 27 April 1990, District Services Collection, Series 2, Box 3, Folder 1, CLP-OLIVER; Bohdan Hodiak, "Libraries' Lending, Use Increase," *Pittsburgh Post-Gazette*, 12 July 1990.
88 Frank J. Lucchino, Looking Beyond Our Past, 12 June 1990, 3, District Services Collection, Series 2, Box 3, Manual 2, CLP-OLIVER.
89 Lucchino, Looking Beyond Our Past, 12 June 1990, 17–34.
90 Lucchino, Looking Beyond Our Past, 12 June 1990, 73.
91 Arthur Curley, "Anatomy of an Advocate," *American Libraries*, 1994, November; Albert F. Kamper, Letter from Al Kamper to Cabinet, 7 February 1991, District Services Collection, Series 2, Box 3, Folder 2, CLP-OLIVER.
92 Lucchino as quoted in Kamper, Letter from Al Kamper to Cabinet, 7 February 1991; Annette E. Kovic, Library Association of the South Hills, Meeting Minutes, 23 January 1991, 2, District Services Collection, Series 2, Box 3, Folder 2, CLP-OLIVER.
93 Curley, "Anatomy of an Advocate."; Kovic, Library Association of the South Hills, Meeting Minutes, 23 January 1991, 3.
94 Dennis P. Leeper, *Pennsylvania Citizens for Better Libraries Newsletter* 14, no. 2 (Spring 1991).
95 Frank J. Lucchino, A Quiet Crisis: Libraries in Allegheny County, March 1991, 38, District Services Collection, Series 2, Box 3, Manual 4, CLP-OLVER.
96 Lucchino as quoted in Kamper, Letter from Al Kamper to Cabinet, 7 February 1991.
97 Kamper, Letter from Al Kamper to Cabinet, 7 February 1991; Lucchino as paraphrased in Albert F. Kamper, Notes on Frank Lucchino Meeting, Northland, February 13, 1991, 18 February 1991, District Services Collection, Series 2, Box 3, Folder 2, CLP-OLIVER.
98 Kamper, Letter from Al Kamper to Cabinet, 7 February 1991.
99 Kovic, Library Association of the South Hills, Meeting Minutes, 23 January 1991, 2.
100 Peter Leonard, Letter from Peter Leonard to County Librarians, Trustees, and Friends, 29 January 1991, District Services Collection, Series 2, Box 3, Folder 2, CLP-OLIVER.
101 Kamper, Letter from Al Kamper to Cabinet, 7 February 1991.
102 Lucchino as quoted in Kamper, Notes on Frank Lucchino Meeting, Northland, February 13, 1991, 18 February 1991.
103 Lucchino as quoted in Kamper, Letter from Al Kamper to Cabinet, 7 February 1991.
104 Lucchino paraphrased in Kamper, Letter from Al Kamper to Cabinet, 7 February 1991; Kovic, Library Association of the South Hills, Meeting Minutes, 23 January 1991, 2.
105 Lucchino, A Quiet Crisis: Libraries in Allegheny County, March 1991, 37.
106 Lucchino, A Quiet Crisis: Libraries in Allegheny County, March 1991, 43–47.
107 Lucchino, A Quiet Crisis: Libraries in Allegheny County, March 1991, 43.
108 Lucchino, A Quiet Crisis: Libraries in Allegheny County, March 1991, 43–44.
109 Lucchino, A Quiet Crisis: Libraries in Allegheny County, March 1991, 44.
110 Lucchino as paraphrased in Kamper, Notes on Frank Lucchino Meeting, Northland, February 13, 1991, 18 February 1991; Lucchino, A Quiet Crisis: Libraries in Allegheny County, March 1991, 47.
111 Kamper, Notes on Frank Lucchino Meeting, Northland, February 13, 1991, 18 February 1991.
112 Megan O'Matz, "Funding Proposed for County Libraries," *Pittsburgh Press*, 13 March 1991.
113 Albert F. Kamper, Letter from Albert F. Kamper to Cabinet, 27 February 1991, 1–2, District Services Collection, Series 2, Box 3, Folder 2, CLP-OLIVER; Albert F. Kamper, Letter from Albert F. Kamper to Main Lib. Dept. Heads/Branch Librarians, 28 February 1991, District Services Collection, Series 2, Box 3, Folder 2, CLP-OLIVER; Peter Leonard, Letter from Peter Leonard to Albert F. Kamper, 11 March 1991, District Services Collection, Series 2, Box 3, Folder 2, CLP-OLIVER.
114 O'Matz, "Funding Proposed for County Libraries."; "Quiet Crisis Getting Noisy: Library Community Fights Back; Bucks National Trends," *PLA Bulletin* (March 1992).

115 Jennifer Fox Rabold, Minutes of the CLASP Executive Committee Meeting, 7 December 1992, ACLA historical file (Folder: CLASP Executive Committee), ACLA-OFFICE; Jennifer Fox Rabold, Minutes of the Executive Meeting of CLASP, 12 May 1992, ACLA historical file (Folder: CLASP Executive Committee), ACLA-OFFICE; Jennifer Fox Rabold, Minutes of the General Meeting of CLASP, 6 November 1991, ACLA historical file (Folder: CLASP), ACLA-OFFICE.

116 O'Matz, "Funding Proposed for County Libraries."; Kamper, Letter from Albert F. Kamper to Cabinet, 27 February 1991.

117 Peter Leonard, C.L.A.S.P. Flyer, 15 April 1991, District Services Collection, Series 2, Box 3, Folder 2, CLP-OLIVER; CLASP, County Library Association Serving the People: By-Laws, 28 May 1991, ACLA historical file (Folder: CLASP), ACLA-OFFICE.

118 Carnegie Library of Pittsburgh, 1992 Annual Report, 1992, 3, CLP-PENN.

119 "Leadership on Libraries," *Pittsburgh Post-Gazette*, 19 March 1991.

120 "Not So Quiet Library Crisis," *Pittsburgh Press*, April 1991.

121 Carnegie Library of Pittsburgh, 1991 Annual Report, 1991, 1.

122 Linda Wilson Fuoco, "Commission Probing Ways to Fund Libraries," *Pittsburgh Post-Gazette*, 10 July 1991; "Quiet Crisis Getting Noisy: Library Community Fights Back; Bucks National Trends."

123 CFLAC, Meeting Minutes: Buhl Grant for Director, Search Committee, 21 August 1991, ACLA historical file (Folder: CFLAC-Minutes), ACLA-OFFICE; "Quiet Crisis Getting Noisy: Library Community Fights Back; Bucks National Trends."; Frank J. Lucchino, Position Announcement for Library Director for Allegheny County, 21 August 1991, ACLA historical file (Folder: CLASP), ACLA-OFFICE; Frank J. Lucchino, Grant Proposal for Allegheny County Library Director, 23 August 1991, 1, ACLA historical file (Folder: CLASP), ACLA-OFFICE.

124 CFLAC, Meeting Minutes: Search Committee, Library Center, Library Tours, 4 December 1991, ACLA historical file (Folder: CFLAC-Minutes), ACLA-OFFICE; DeeAnna R. Cavinee, CFLAC: An Annotated Timeline of Its History and Important Future Dates, January 1992, ACLA historical file (Folder: CFLAC-Minutes), ACLA-OFFICE; Jean Bryant, "County Library Chief Has Volumes of Ideas to Sort," *Pittsburgh Press*, 26 March 1992; CLASP, CLASP Meeting Announcement, 5 February 1992, ACLA historical file (Folder: CLASP), ACLA-OFFICE; CFLAC, Meeting Minutes: CFLAC Roles, Access to CLP CAROLINE, 19 February 1992, ACLA historical file (Folder: CFLAC-Minutes), ACLA-OFFICE.

125 Fuoco, "Commission Probing Ways to Fund Libraries."; CFLAC, Meeting Minutes: CFLAC Roles, Access to CLP CAROLINE, 19 February 1992.

126 CFLAC, Meeting Minutes: Swearing in, Library Visits, 20 June 1991, ACLA historical file (Folder: CFLAC-Minutes), ACLA-OFFICE.

127 CLASP, CLASP Survey on Services and Needs in Allegheny County Libraries, January 1992, ACLA historical file (Folder: CLASP), ACLA-OFFICE.

128 CLASP, CLASP Survey on Services and Needs in Allegheny County Libraries, January 1992.

129 Lucchino, Grant Proposal for Allegheny County Library Director, 23 August 1991, 4.

130 Allegheny County Planning Department, Allegheny County 2001: Our Future, Our Choice, 10 September 1991, District Services Collection, Series 2, Box 3, Folder 1, CLP-OLIVER.

131 CFLAC, Meeting Minutes: Search Committee, Bell of PA Telelcommunications Grant, 1991, ACLA historical file (Folder: CFLAC-Minutes), ACLA-OFFICE; Fox Rabold, Minutes of the General Meeting of CLASP, 6 November 1991.

132 Library Administrator 5, Interview Transcript, 16 July 2015, interviewed by Michael M. Widdersheim.

133 Library Administrator 1, Interview Transcript, 28 July 2015, interviewed by Michael M. Widdersheim.

134 "Quiet Crisis Getting Noisy: Library Community Fights Back; Bucks National Trends."; Fox Rabold, Minutes of the Executive Meeting of CLASP, 12 May 1992; CFLAC, Meeting Minutes: CFLAC Role, Fax Machines, Electronic Link-Up, 20 May 1992, ACLA historical file (Folder: CFLAC-Minutes), ACLA-OFFICE; CFLAC, Meeting Minutes: Project Link-up Participation, CLASP Automation Proposal, 29 July 1992, ACLA historical file (Folder: CFLAC-Minutes), ACLA-OFFICE.

135 CFLAC, Meeting Minutes: CFLAC Roles, Access to CLP CAROLINE, 19 February 1992; CFLAC, Meeting Minutes: Hiring of Director, Link-up Computers at Wilkinsburg, 15 January 1992, ACLA historical file (Folder: CFLAC-Minutes), ACLA-OFFICE.

136 CFLAC, Meeting Minutes: Fax Machines, Project Link-up, CLASP Membership, Reading Centers, 23 June 1992, ACLA historical file (Folder: CFLAC-Minutes), ACLA-OFFICE.

137 CFLAC, Meeting Minutes: Goals for Project Link-up, Multi-Municipal Projects, 18 March 1992, ACLA historical file (Folder: CFLAC-Minutes), ACLA-OFFICE.

138 DeeAnna R. Cavinee, Report to Buhl Foundation: Allegheny County Library Director Grant, 18 December 1992, ACLA historical file (Folder: CLASP), ACLA-OFFICE; CFLAC, Meeting Minutes: Hiring of Director, Link-up Computers at Wilkinsburg, 15 January 1992.

139 Cavinee, Report to Buhl Foundation: Allegheny County Library Director Grant, 18 December 1992, 2; CFLAC, Meeting Minutes: Project Link-up Participation, CLASP Automation Proposal, 29 July 1992.

140 CLASP, The CLASP Report 1(1): Commission on the Future of Libraries Institutes "Project Link-up;" CLASP Plans Enlargement of the Program, September 1992, ACLA historical file (Folder: ACLA-CLASP Newsletters), ACLA-OFFICE.

141 Cavinee, Report to Buhl Foundation: Allegheny County Library Director Grant, 18 December 1992.

142 Cavinee, Report to Buhl Foundation: Allegheny County Library Director Grant, 18 December 1992, 2; Fox Rabold, Minutes of the CLASP Executive Committee Meeting, 7 December 1992; CFLAC, Meeting Minutes: Bookmobile, First Project Link-up Computers, Knowledge Connections, CLASP Newsletter, 20 October 1992, ACLA historical file (Folder: CFLAC-Minutes), ACLA-OFFICE.

143 Jim Lutton, Minutes of the CLASP Executive Committee Meeting, 9 March 1993, ACLA historical file (Folder: CLASP Executive Committee), ACLA-OFFICE; CFLAC, Meeting Minutes: CFLAC Role, Fax Machines, Electronic Link-Up, 20 May 1992; Cavinee, Report to Buhl Foundation: Allegheny County Library Director Grant, 18 December 1992; CFLAC, Press Release: Document Express Links up Libraries, 9 February 1993, ACLA historical file (Folder: CFLAC-Projects), ACLA-OFFICE; CFLAC, Meeting Minutes: CFLAC Roles, Access to CLP CAROLINE, 19 February 1992; CFLAC, Meeting Minutes: Goals for Project Link-up, Multi-Municipal Projects, 18 March 1992.

144 Fox Rabold, Minutes of the CLASP Executive Committee Meeting, 7 December 1992; Lutton, Minutes of the CLASP Executive Committee Meeting, 9 March 1993, 2.

145 CFLAC, Meeting Minutes: Goals for Project Link-up, Multi-Municipal Projects, 18 March 1992; Cavinee as quoted in Bryant, "County Library Chief Has Volumes of Ideas to Sort."; Christine Vorce, "Library Panel Catalogs Shortcomings of County System," *Pittsburgh Press*, 23 April 1992; Cavinee, Report to Buhl Foundation: Allegheny County Library Director Grant, 18 December 1992; Fox Rabold, Minutes of the Executive Meeting of CLASP, 12 May 1992; CFLAC, Meeting Minutes: Project Link-up Participation, CLASP Automation Proposal, 29 July 1992; CLASP, The CLASP Report 1(1): Commission on the Future of Libraries Institutes "Project Linkup;" CLASP Plans Enlargement of the Program, September 1992.

146 CLASP, The CLASP Report 2(1): CLASP Meets to Consider Public Library Project, February 1993, ACLA historical file (Folder: ACLA-CLASP Newsletters), ACLA-OFFICE; CFLAC, Meeting Minutes: SLIS and FreeNet, CLASP Morale, State Library Code, 16 February 1993, ACLA historical file (Folder: CFLAC-Minutes), ACLA-OFFICE.

147 Cavinee, Report to Buhl Foundation: Allegheny County Library Director Grant, 18 December 1992; Jennifer Fox Rabold, Minutes of the CLASP Executive Committee Meeting, 2 June 1993, ACLA historical file (Folder: CLASP Executive Committee), ACLA-OFFICE.

148 Fox Rabold, Minutes of the Executive Meeting of CLASP, 12 May 1992; CFLAC, Meeting Minutes: Fax Machines, Project Link-up, CLASP Membership, Reading Centers, 23 June 1992.

149 CLASP, Celebrate Children's Book Week, 14 November 1992, ACLA historical file (Folder: ACLA-CLASP Newsletters), ACLA-OFFICE; CFLAC, Meeting Minutes: Project Link-up, CLASP Membership, 17 November 1992, ACLA historical file (Folder: CFLAC-Minutes), ACLA-OFFICE; CFLAC, Meeting Minutes: Project Link-up First Phase, Reading Centers, 9 September 1992, ACLA historical file (Folder: CFLAC-Minutes), ACLA-OFFICE.

150 Cavinee, Report to Buhl Foundation: Allegheny County Library Director Grant, 18 December 1992; Carnegie Library of Pittsburgh, 1992 Annual Report, 1992, 2; Carnegie Library of Pittsburgh, 1993 Annual Report, 1993, 2, CLP-PENN; Governor's Advisory Council on Library Development/Federal Advisory Council, Minutes, 13–14 September 1993, Department of Education, Commonwealth Libraries, Minutes of the Governor's Advisory Council on Library Development 1962–2009 collection, Carton 2, Folder 1, PASA.

151 Fox Rabold, Minutes of the CLASP Executive Committee Meeting, 2 June 1993.

Period 6: The Pursuit of RAD, 1993–1994

The early 1990s were difficult economic times for public libraries across the state of Pennsylvania. Libraries closed, reduced staff, decreased hours and services, and held local referenda to make up for operating deficits. Additional state support was not forthcoming. In the years leading up to and including 1993, the statewide library card program that was implemented in the mid-1980s and was intended to reduce costs for libraries was not fully funded. In some cases, this hurt rather than helped local libraries. It meant that libraries were expected to share resources to serve non-local users, but they had to expend local funds to do so.[1] Initiated by the State Library of Pennsylvania in 1992, a reform movement called the Public Library Project was supposed to address weaknesses in the state library law related to library governance and funding.[2] Such revisions might have helped to alleviate libraries' funding challenges. By 1995, however, after years of discussions, major changes to the state funding formula were proposed but never implemented.[3] Libraries were therefore left to devise local solutions to funding problems.

In Allegheny County, the groundwork for new funding and governance opportunities for the public library infrastructure was laid by mid-1993. CLASP, CFLAC, and a County Library Director were established in the wake of Lucchino's *Quiet Crisis*. By summer 1993, CLASP membership rose to 34 libraries.[4] CLASP, CLP, CFLAC, and the County Library Director initiated Project Link-Up, which used county and charitable foundation money to place computer terminals, fax machines, and telecommunication equipment in select public libraries around the county. This allowed them to share documents with one another via fax, connect to CLP's online catalog system, CAROLINE, and connect to the Internet.[5] At the end of December 1993, 23 independent public libraries in the county used their dial-up connections to place ILL requests at CLP.[6] Using grant money, CLASP attempted to extend Project Link-Up to all CLASP member libraries and all Allegheny County libraries.[7] Public libraries in the greater Pittsburgh region were therefore drawn together administratively, financially, and technologically, making their fates increasingly intertwined. CLP continued to serve as the county and district library, providing bookmobile, reference, and lending services to county residents. Yet, these substantive developments did not fully address underlying funding problems. CLASP was an informal membership organization without decision-making power, and neither CFLAC nor the County Library Director position were funded by the county. It was not clear how the emerging electronic network would be maintained over time. By the end of Period 5, there was limited interest among county libraries for further collaborative efforts. Attendance at CLASP meetings in early 1993 dwindled.[8] By summer 1993, neither CLASP, CFLAC, nor the County

https://doi.org/10.1515/9783111013404-008

Library Director had found an effective, uniting vision for the future. Adequate funding remained a pressing issue.[9] Development of the public library infrastructure seemed to be at an impasse.

As the previous decision cycle ended, however, a new one began in mid-1993. In September of 1993, building on the success of Project Link-Up, CLP launched planning for a more comprehensive electronic network.[10] The planning process included representation from CLASP, CFLAC, and CLP.[11] This was at a time when local agencies across the country began to form local networks to connect to the Internet.[12] Goals for the proposed electronic network in Allegheny County included access to information databases, bibliographic information, Internet connection, enhanced delivery of materials, and improved automation systems.[13] The imagined electronic network would encompass a wide range of services and functions on the library side, including the automated systems and local area networks at independent libraries, the shared integrated library system (ILS) that included modules for circulation, cataloging, and patron accounts, and the computer equipment already installed under the auspices of Project Link-Up. The network would also include those connections by county residents using their own computers and telecommunication equipment at places like homes, schools, and offices.[14] The new proposed electronic network would build and maintain all these connections in a coordinated way for the entire public library infrastructure across the county.

Part of the challenge of this grand effort was to convince all the independent libraries to use a single, shared ILS. For those libraries still using card catalogs and other manual cataloging and circulation process, they would need to automate to become part of the system.[15] This involved assigning machine-readable barcodes to items and maintaining electronic circulation and catalog records rather than carrying out circulation and cataloging using paper records. By January 1994, there were 36 libraries who were members of CLASP, and there were 46 independent libraries in Allegheny County.[16] All of these libraries were potential nodes in the electronic network. On December 4, 1993, 120 people representing 37 public libraries attended an open meeting where the electronic network plan was revealed. There was general support for the plan across the county. That said, there remained unanswered questions and skepticism about the project.[17] One question was how the multi-million-dollar project would be adequately funded. Another question was how the network would be governed.

With general support from libraries across the county, planning for the network continued through 1994.[18] As the planning progressed, the network project came to be known as the Electronic Information Network.[19] It was billed as a cooperative effort between CFLAC, Allegheny County, CLASP, and CLP. The network would be centrally staffed, maintain a union catalog composed of MARC records, develop common materials and patron databases, accept library cards from other libraries, and share materials between libraries.[20] In terms of network governance, it was decided that the physical location of the hardware and staff for the network would be housed at CLP. CLP would therefore administer the network, and the independent libraries across the county would contract for network services.[21]

The vision proposed by the Electronic Information Network was unprecedented in scope and in the level of collective effort required to carry it out. Public libraries in the greater Pittsburgh region had never coordinated a joint effort of this complexity and scale. The plan included all public libraries in the county, an estimated 1,400 users. Capital costs were projected at $10 million.[22] Ongoing maintenance costs were estimated to be $1 to $2 million per year. [23] By mid-1994, 42 libraries submitted letters of participation in the new network.[24] Funding for the network was the central concern. CFLAC, CLP, and CLASP launched a capital campaign to raise funds for the network. The capital fundraising target was $9.5 million, with the remaining half million contributed by participating libraries.[25] Library leaders met with representatives from several Pittsburgh foundations, including the Buhl Foundation, Hillman Foundation, R.K. Mellon Foundation, Heinz Endowment, Pittsburgh Foundation, and McCune Foundation.[26] All 6 foundations expressed sincere interest in funding the first two phases of the project at approximately $6 million.[27] Funding of the Electronic Information Network began with a check from Hillman Foundation to CLP for $100,000.[28] Allegheny County also agreed to fund the project for $2 to $2.5 million.[29] These successes in the capital campaign meant the Electronic Information Network could transition from the planning stage to the implementation stage. But as the capital campaign continued, foundations expressed concern about ongoing maintenance costs.[30] CFLAC, local politicians, and local libraries contacted Pennsylvania's Public Utility Commission to press for legislation for reduced telecommunications rates for public libraries. [31] A two-tier rate system would help to defray some of the costs of the proposed network. Nevertheless, an additional solution had to be found that would secure ongoing funding for the ambitious new plan.

Around the time that planning and fundraising for the electronic network was underway, a potential solution emerged that could address its long-term funding dilemma. This solution came in the form of a tax reform initiative within Allegheny County. The tax reform proposed a Regional Asset District (RAD). RAD was envisioned as a special governmental district with the power to levy taxes. The proposed tax reform had significant potential implications for the regional public library infrastructure in greater Pittsburgh because the district was designed to distribute a portion of its tax revenue to cultural assets. This tax reform initiative was backed by city and county officials and orchestrated by the Allegheny Conference on Community Development (ACCD) and the Pennsylvania Economy League, two influential organizations in Pittsburgh. It was up to library leaders to ensure that they qualified as a regional cultural asset and had a hand in this new pot of money.

The initial planning for RAD began in 1990 with an investigation by the Pennsylvania Economy League about how to fund the Pittsburgh Zoo, Phipps Conservatory, and the National Aviary, all located in Pittsburgh and funded by the city.[32] Pennsylvania Economy League was the author of the RAD concept. As had been the case decades earlier, the population in the greater Pittsburgh region continued to migrate out of city centers, forming a more dispersed urban field.[33] With the shifting tax base, the City of Pittsburgh could no longer adequately fund its recreational and cultural facilities. A regional asset district was thus proposed to fund and manage them.[34] This new district would be funded by a change in sales tax.[35] The original RAD proposal covered multiple counties, not just Allegheny County.[36] In the end, however, the new district applied only to Allegheny County. It would raise the existing sales tax in Allegheny County from 6 percent to 7 percent. Revenue from the additional 1 percent sales tax was projected to be $106 million. This sum would be divided in half. One half would fund regional assets, and the other half would be directed to local municipalities to allow for a shift in the tax burden. Cultural assets funded by the RAD plan would therefore apply for a portion of that one-half of the revenue from the 1 percent sales tax. The regional asset district would be a special municipal authority governed by a board appointed by the mayor of Pittsburgh and the county commissioners. This board would have the authority to allocate the sales tax money.[37] The proposal had clear implications for public libraries because they could apply for and potentially receive a portion of the sales tax revenue.

The proposed change in sales tax within Allegheny County required new legislation at the state level. In November and December 1993, the RAD plan fast-tracked the state legislature in Harrisburg with the hope that RAD would begin distributing funds in 1995.[38] Pro-RAD legislators in Harrisburg were optimistic because the state legislature had already passed similar legislation allowing for a 1 percent sales tax increase in Philadelphia two years earlier.[39] The RAD bill came to a vote only one week after it was introduced. [40] This was due to the lobbying effort orchestrated by the Pennsylvania Economy League and the ACCD. RAD was presented as a plan that would reduce or eliminate certain other taxes, such as personal property taxes and property assessments of certain low- and moderate-income senior citizens.[41] The RAD proposal became law in December 1993, creating a special unit of local government designed to finance civic, recreational, library, sports, and other regional assets.[42] After becoming state law, RAD was approved within Allegheny County by the county commissioners in 1994 as a sales tax ordinance. [43] In general, the proposal was positively portrayed in the media, and it had popular support within the county.[44] The 1 percent sales tax increase took effect on July 1, 1994.[45] Following its implementation came the process of selecting RAD's governing board. The 8-member RAD board would be composed in part by appointees from the commissioners and the mayor of Pittsburgh.[46] A 27-member advisory board that reflected the general composition of the community was also required by the new statute.[47]

Beginning in April 1993, and during the design and passage of the RAD proposal, legislators and librarians within Allegheny County had an understanding that, if passed, some amount of RAD funding would be allocated to the public library infrastructure.[48] In the statute, CLP was included on the list of assets, together with the outer independent libraries in the county, the Carnegie Institute, Pittsburgh Zoo, Phipps Conservatory, and the National Aviary.[49] But the proposal was vague. It listed public libraries, in general, as a regional asset, not just CLP. Its mention of "libraries" was open-ended, and the question about which libraries would be funded was left unaddressed. [50] In the summer of 1993, the Economy League, CLP, the County Library Director, and CLASP discussed how to craft a compelling argument for why RAD funding should apply to all the independent public libraries in the county.[51] Central to this argument was the increasingly digital, electronic, computerized, and networked technologies that were needed to connect libraries together. In other words, the prospect of the Electronic Information Network that was under development at the time was used to justify funding to libraries.[52] CLASP reached out to ACCD and the Pennsylvania Economy League in order to learn how its member libraries might be funded.[53] Such funding would certainly support the Electronic Information Network, but it could also potentially support additional county services.

Throughout 1993 and early 1994, there continued to be uncertainty about RAD's implications for libraries. [54] Certainly, CLP would receive funding through RAD. It expected to receive $13,500,000.[55] But the other 40 or so independent public libraries around the county were unsure. Not only was it unclear which libraries might be funded and to what degree, it was also unclear how the distribution of RAD money to libraries might affect the funding they already received locally. Some were concerned that any RAD funding libraries received would replace their municipality support. It was important for libraries to characterize RAD funding as a supplement, not a replacement.[56] It was also not clear how RAD funding would affect library governance.[57] Would local libraries lose autonomy and control of their library operations if they agreed to be funded by RAD? What kinds of strings were attached to the funding?

In light of these fears and concerns, CLASP proposed a clause to the RAD legislation that would make it clear that all public libraries in the county are a regional cultural asset. This would ensure all libraries' eligibility. A second clause would make it clear that RAD funding was a supplement, not a substitute. This would ease any fears that local funding would dry up if libraries succeeded in securing RAD funding. All CLASP members, including trustees, patrons, and friends, were urged to contact state legislators and senators to find out what exactly the RAD bill said and to express concern that RAD funding might in fact unintentionally negatively impact public libraries rather than help them.[58] It was also proposed that municipalities sign a contract agreeing that RAD funding would not supplant local funding.[59] Library directors met with their municipal councils to convince them of the importance of local funding.[60]

The independent public libraries across Allegheny County continued to prepare their arguments for RAD funding. Whether CLASP libraries would receive RAD funding was a decision left to the RAD board. Public hearings for potential regional assets before the RAD board were scheduled for September 1994. CLASP began to prepare for these hearings. This required strategic thinking. First, CLASP had to approach the RAD board with a single voice and as a single entity. The group had to demonstrate that it was a single, regional asset, not a fractured assembly of disparate libraries.[61] The RAD legislation that was passed in December 1993 stated explicitly that non-regional assets would not be funded. These included "any library which is not a regional library resource center, a district library center or which is not part of a library system serving multiple municipalities."[62] It was not obvious to some that CLASP member libraries met these conditions.[63] Each CLASP member library, alone, would likely not qualify as a regional asset. By March 1994, all state-aided libraries in Allegheny County were members of CLASP,[64] and most state-aided libraries participated in ACCESS PA, the statewide ILL system. This showed that the libraries applying for RAD funding met minimum state standards and were in good standing. But it was unclear whether this would qualify a public library as a regional asset. After the RAD board was formed in 1994, CFLAC and CLASP developed criteria for deciding which public libraries would qualify for RAD money.[65] It was decided that in order to receive RAD funding libraries had to be state-aided, participants in the new automation network, members of CLASP, and participants in ACCESS PENNSYLVANIA.[66]

Part of the challenge in developing CLASP's message was that there were few regional initiatives that CLASP libraries had accomplished by this point. CLASP was still a fledgling organization, formed only in 1991. Previous or ongoing regional initiatives included a computerized magazine index project organized by the Library Association of South Hills, a cooperative video loan project between Sewickley, Shaler, and Northland, and a joint cataloging project organized by South Hills libraries.[67] CLASP could report on these and other activities of its member libraries, which would show that libraries had a strong user base across the county. But, while some of the activities might be considered regional initiatives, none of them were led by CLASP or involved all libraries in the county.[68] CLASP's rationale for RAD funding depended in large part on the prospect of regional activities. The most conspicuous of these was the Electronic Information Network. The RAD board held preliminary hearings on March 18, 1994. At these preliminary hearings, representatives from potential assets presented their case for RAD funding. Both CLP and CLASP sent representatives.[69] The Electronic Information Network was central to the rationale by libraries for their receiving RAD funding.[70] This was because the network, though not yet implemented, was one of the few regional projects whose participants included public libraries from across the county. The proposed project demonstrated that CLASP libraries and CLP were not at odds—they could work together, clarify shared goals, and coordinate the development of a complex and expensive project on a regional scale. Still, the proposed automation network was only a promise, not a reality.

In addition to presenting an image of a single regional asset, CLASP and the independent local libraries also revisited their organizational and governance structure in pursuit of RAD funding. Initially, CLASP members considered the creation of a federation, but federation proved to be a divisive issue within the membership.[71] Instead, CLASP members characterized the group as a consortium.[72] It was decided that the CLASP consortium must evolve into a 501(c)3 non-profit corporation with revised bylaws and a revised governance structure.[73] This was to ensure better oversight and accountability. CLASP's new official bylaws were adopted June 15, 1994.[74] While filing for incorporation status as a non-profit organization, CLASP discovered that the CLASP acronym was already in use.[75] At the July 28, 1994, meeting, a new name for the organization was proposed: ACLA, or Allegheny County Library Association.[76] ACLA's recomposed board included representatives from the libraries located in the North, South, East, West, and Central regions of Greater Pittsburgh.[77] This division was purely geographical, created by placing two rulers on a map in the shape an X, dividing the county along the Allegheny, Monongahela, and Ohio rivers.[78] The consortium board would interact with the RAD board on behalf of all member libraries, distribute RAD money to the libraries, and ensure that member libraries conformed to specified criteria.[79] The new consortium board was conceived as a pass-through organization for RAD money. In August 1994, there were 47 recorded public libraries in Allegheny County, and 43 of these were ACLA members.[80] Membership included all the state-funded libraries in the county.[81]

As part of ACLA's preparation for the RAD hearings in September 1994, the group scrambled to create a long-term plan. Such a plan would have to include a vision statement and identify strategic initiatives for the next 5 to 10 years. Unlike CLP, which was an established library that regularly conducted strategic planning, ACLA did not have sufficient time to develop a comprehensive strategic plan before the hearings. The timeline for a strategic planning process was about a year and would cost about $200,000 to $250,000 for the entire process.[82] Rather than carry out this process, ACLA instead promised to undertake a strategic planning process should they receive RAD funding. The RAD board required that long-term plans include assessments, targets, goals, and objectives.[83] ACLA established general goals and a high-level mission, but it did not yet have specific objectives or targets, and it had not defined a way to assess its own performance as an asset. By mid-August, ACLA had yet to cement its fiscal management strategies.[84] It did not have an executive director or paid administrative staff. ACLA stated that, once funded, it intended to apply to the State Library of Pennsylvania for federated system status.[85] Recognition as a library system in the eyes of the state would qualify the organization for state funding. Federated system status would require that ACLA apply state-mandated standards to its own practices. ACLA would also have to enforce state standards among its member libraries. ACLA planned to develop a countywide plan, encourage cooperation among libraries, promote awareness of public libraries, and work to secure additional government and private funds.[86]

Central components of ACLA's and CLP's presentations to the RAD board were their budgets. Prospective regional assets had to request a specific amount from RAD and explain how that funding would be used if received.[87] Both ACLA and CLP included in their budget proposals funding for the Electronic Information Network.[88] ACLA had the additional task of developing a distribution formula outlining how RAD money would be allocated to ACLA member libraries. The formula had to allocate funds in a transparent, responsible, and equitable way.[89] ACLA considered several categories of funding: supplemental, equalization, and incentive.[90] Drafts of a distribution formula were considered in the months leading up to the hearings, but a finalized formula was not in place by the time ACLA presented to the RAD board in September 1994.[91] The amount of ACLA's request to RAD also changed over time, between $6.5 million and $9.5 million, suggesting that there was no clear indication of how much money the regional public library infrastructure actually needed overall or how that money would be spent.[92] ACLA used budget reports from all libraries for recent years to bolster their budget request.[93] Of course, these budgets did not establish a true baseline because there had not yet been regional initiatives in effect. ACLA's final ask to the RAD board was $8 million. This included funding for support of basic services countywide, funding for cooperative efforts, and funding for regional initiatives.[94] ACLA's anticipated budget request from RAD for subsequent years increased slightly by about 2 percent. ACLA also anticipated that it would begin to receive increases in local, state, and private funding.[95] ACLA promised to actively pursue other funding sources over and above RAD funding.

In addition to its planning and budgeting efforts, ACLA also focused on building its public relations. ACLA received letters of endorsement for RAD funding from CFLAC, state senators and representatives, local municipalities, and local libraries.[96] Representatives from ACLA met with the RAD board in a number of informational meetings.[97] In a final push for public support, the ACLA public relations committee asked library directors to gather signatures for a petition. The petition was to support ACLA's campaign to receive RAD funds for all public libraries in the county. The goal of the petition was 130,000 signatures, or 10 percent of the population.[98] The committee organized a petition-signing rally in Market Square of downtown Pittsburgh several days before ACLA's scheduled RAD hearing.[99] Librarians and library supporters were encouraged to get a crowd, have supplies ready, bring pens, clipboards, handouts, petitions, and two or three card tables to create a visual event and create good photo opportunities.[100] The rally received coverage from the local newspaper and local TV news station.[101] In the end, about 20,000 library users signed the petition supporting regional funding and cooperation.[102]

As ACLA worked to develop a new plan and bolster its image, CLP worked in the background to tie up its own loose ends in preparation for RAD funding. In early 1994, CLP served as district library center for the Pittsburgh Library District. At that time, the Pittsburgh district included not only all of Allegheny County but also parts of Westmoreland County and parts of Washington County. CLP worked with the state library and the Governor's Advisory Council to realign the Pittsburgh district to coincide with Allegheny County. Realigning the district would mean that the proposed Electronic Information Network, as well as any funds CLP received from RAD, would apply only within the confines of Allegheny County and nowhere else. Under the proposed redistricting, the Monessen District, a library district bordering Pittsburgh, would include all of Westmoreland County and all of Fayette County, and the Washington District, another neighboring district, would include all of Washington County and all of Greene County.[103] Though the redistricting simplified the emerging network and potential RAD funding for CLP, it meant a potential decrease in service level for those libraries placed outside the Pittsburgh district. District services provided by CLP to these libraries included ILL, delivery service, children's services, public relations and printing, reference services, film center, and consulting services.[104] Libraries formerly in the Pittsburgh district had to receive these services elsewhere. The redistricting was ultimately successful, but it was controversial for those libraries removed from the Pittsburgh District.[105] The redistricting ensured that CLP's service area coincided with the Electronic Information Network and the area designated as eligible for RAD funding.

ACLA's oral presentation before the RAD board was held on September 21, 1994. Its written proposal was submitted on September 15.[106] Leading up to the hearing, the uncertainty of ACLA's success was captured by ACLA's consultant, Bill Dodge:

> It appears that we are facing the incredibly challenging task of producing a proposal;
> - for an *unproven idea* (that libraries collectively are a critical regional asset in Allegheny County)
> - by an *untested* organization (the Allegheny County Library Association)
> - to an *unknown organization* (the ARAD Board)
> - that will be endorsed by a an [sic] *unlikely collaboration* (library trustees, municipal and county government officials and the ACLA Board) and
> - result in *unheard of resources* ($10 million—more or less—in new funding for library services).
>
> And, of course, do all of this within *unrealistic deadlines* (by the middle of September!)
> I believe that we are up to the task.[107]

Public hearings in front of the RAD board began with presentations by statutory assets. Statutory assets were those assets that received funding from the city or county in 1992 and were guaranteed to receive some degree of funding through RAD. They included the Pittsburgh Zoo, Three Rivers Stadium, CLP, the Carnegie Museums and Science Center, the National Aviary, Phipps Conservatory, four large city parks, and nine county parks. Non-statutory assets presented next. Non-statutory assets could seek money, but funding was not guaranteed.[108] ACLA was grouped with the non-statutory assets. All assets, whether statutory or non-statutory, competed for a share of the revenue from the additional 1 percent sales tax that went into effect July 1, 1994. By September, RAD already began receiving that tax revenue. The first payment was $2.1 million.[109] Four of the eight statutory assets presented at the first public hearing on September 10.[110] Together, CLP, the Pittsburgh Zoo, Phipps Conservatory, and the National Aviary asked for $33 million, more than half of the $53 million projected to disburse annually.[111] RAD funding requests from other statutory assets continued to be tight, with requests from parks and the stadium.[112] ACLA's presentation urged RAD board members to consider ACLA a regional asset.[113] Following the hearing, ACLA followed up with the RAD board by providing further details about local government funding to state-aided libraries.[114] Some libraries had banked on receiving RAD funding. Castle Shannon, for instance, began saving for an expansion project.[115]

October 3, 1994 was the decision day for the RAD board.[116] The results of the funding distribution would be applied to assets for the 1995 calendar year. In the end, neither CLP nor ACLA received the full amounts they requested.[117] In fact, no asset did. But Three Rivers Stadium Authority received the fullest share of their request at 85 percent. CLP received $11.9 million, about half of the amount it requested. ACLA received $5 million. This amount was short of the $8 million that

ACLA requested, but RAD funding in any amount was a great success for ACLA. This meant ACLA had persuaded the RAD board that the independent public libraries in Allegheny County functioned as a single, regional asset. ACLA members were overjoyed with the results of the RAD funding.[118]

Following the success of the RAD initiative, at the close of 1994, ACLA worked with RAD board to establish a final distribution formula.[119] In order to receive RAD funding, contracts were drawn between library boards, municipalities, and ACLA.[120] Library leaders also began the search for a new county library director, who had resigned.[121]

In Period 6, a new circuit came into existence: RAD. Then, ACLA became part of the inner periphery of the RAD circuit, receiving funding from the RAD board and in effect carrying out RAD policies. This major change occurred due to the high legitimacy of ACLA's claims and the high responsiveness of the RAD board toward those claims. At first glance, it might have appeared that ACLA came into its own as a self-standing organization, one responsible for leading the regional public library infrastructure in new directions. In reality, it was the RAD board that took over as the lead orchestrator for the public library infrastructure. This may not have been apparent to ACLA member libraries at the time. ACLA member libraries realized that as members of ACLA they had to adhere to ACLA-approved policies. This was acceptable because it was felt that there was a short and direct line of communication between membership and ACLA HQ. Members felt they could influence association policy relatively easily. But accepting ACLA also meant accepting RAD. With the acceptance of RAD funding, the county libraries bound themselves together as a single regional asset obligated to carry out policies not just from ACLA, but also those handed down from the RAD board. This was a significant development in the infrastructure. Previously, it was thought that the infrastructure might fall within the inner periphery of the county or become part of CLP. But with the formation of RAD, the county began to divest itself of the public library infrastructure. This meant that control of CLP and ACLA began to move outside the county circuit. It is notable that ACLA established itself as a separate and distinct organization, one recognized as a non-profit corporation with the State of Pennsylvania. Almost immediately, however, ACLA became nested within the RAD circuit as part of its inner periphery, and decision-making by member libraries actually became farther removed from them than previously thought. The so-called independent county libraries were now "quadruple-nested," meaning that they were each part of the inner peripheries of the local government circuit, state circuit, ACLA circuit, and RAD circuit. Local public libraires had to carry out local policies from within their own local municipal circuits, but they were also responsible for functioning within the inner peripheries of other regional circuits. At the end of Period 6, ACLA the organization was not yet recognized by the state as a public library system, so it was not yet part of the state's inner periphery. This issue would be taken up by ACLA in the following decision cycle.

1 "Library Access (Statewide Library Card Program)," *PLA Bulletin* (April 1993).

2 Governor's Advisory Council on Library Development/Federal Advisory Council, Minutes, 14–15 September 1992, Department of Education, Commonwealth Libraries, Minutes of the Governor's Advisory Council on Library Development 1962–2009 collection, Carton 1, Folder 4, PASA; State Library of Pennsylvania, *Public Library Project: Concepts for Discussion* (Harrisburg: State Library of Pennsylvania, 11 June 1993), 1; "Commonwealth Libraries Report on Activities," *PaLA Bulletin* (Februrary 1994); Victoria Dow, "Caucus Held on the Public Library Project," *PLA Bulletin* (August 1993); CFLAC, Meeting Minutes: Project Link-up, Three Rivers FreeNet, 14 September 1993, ACLA historical file (Folder: CFLAC-Minutes), ACLA-OFFICE; CFLAC, Meeting Agenda and Minutes: RMG Consultants and Project Link-Up, 12 January 1994, ACLA historical file (Folder: CFLAC-Minutes), ACLA-OFFICE; Commonwealth Libraries, Public Library Policies, 8 April 1994, ACLA historical file (Folder: Black Binder), ACLA-OFFICE; "Viewpoints Offered on Public Library Project," *PLA Bulletin* (July 1994); Matthew Flynn, "Draft Legislation for a New Library Code," *PLA Bulletin* (December 1994); Governor's Advisory Council on Library Development and Information Science, Meeting Minutes, 10 March 1995, Department of Education, Commonwealth Libraries, Minutes of the Governor's Advisory Council on Library Development 1962–2009 collection, Carton 2, Folder 6, PASA; Public Library Project, Minutes, 14 May 1992, Department of Education, Commonwealth Libraries, Minutes of the Governor's Advisory Council on Library Development 1962–2009 collection, Carton 1, Folder 4, PASA; Letters from Public Libraries, January 1995, Department of Education, Commonwealth Libraries, Minutes of the Governor's Advisory Council on Library Development 1962–2009 collection, Carton 2, Folder 7, PASA.

3 PLA Board of Directors, "Board Endorses Proposed State Aid Formula," *PLA Bulletin* (October 1995).

4 Robert A. Cohen, Progress Report from CLASP to All Allegheny County Libraries, 21 July 1993, ACLA historical file (Folder: ACLA-CLASP Newsletters), ACLA-OFFICE; Jennifer Fox Rabold, Minutes of the CLASP Executive Committee Meeting, 2 June 1993, ACLA historical file (Folder: CLASP Executive Committee), ACLA-OFFICE; CFLAC, Meeting Minutes: Project Link-up, Three Rivers FreeNet, 14 September 1993.

5 Cohen, Progress Report from CLASP to All Allegheny County Libraries, 21 July 1993; Fox Rabold, Minutes of the CLASP Executive Committee Meeting, 2 June 1993.

6 CFLAC, Meeting Agenda and Minutes: RMG Consultants and Project Link-Up, 12 January 1994.

7 Robert A. Cohen, The CLASP Report 2(3): September Meeting Features CD-ROM Workshop, September 1993, 3, ACLA historical file (Folder: ACLA-CLASP Newsletters), ACLA-OFFICE; Cohen, Progress Report from CLASP to All Allegheny County Libraries, 21 July 1993; Fox Rabold, Minutes of the CLASP Executive Committee Meeting, 2 June 1993.

8 CFLAC, Meeting Minutes: SLIS and FreeNet, CLASP Morale, State Library Code, 16 February 1993, ACLA historical file (Folder: CFLAC-Minutes), ACLA-OFFICE.

9 CFLAC, Meeting Minutes: Project Link-up Governance, Adminstrator Position Extension, 15 June 1993, 2, ACLA historical file (Folder: CFLAC-Minutes), ACLA-OFFICE; Robert A. Cohen, Letters from Robert Cohen to Tom Foerster, Pete Flaherty, and Larry Dunn, 12 August 1993, ACLA historical file (Folder: CFLAC-Projects), ACLA-OFFICE; CFLAC, Meeting Agenda and Minutes: RMG Consultants

and Project Link-Up, 12 January 1994; DeeAnna R. Cavinee, Report to Buhl Foundation: Allegheny County Library Director Grant, 18 December 1992, 7, ACLA historical file (Folder: CLASP), ACLA-OFFICE.

10 CLASP, CLASP Newsletter, September 1993, 8, ACLA historical file (Folder: ACLA-CLASP Newsletters), ACLA-OFFICE; CFLAC, Meeting Minutes: Project Link-up, Three Rivers FreeNet, 14 September 1993, 1–2; CFLAC, Meeting Agenda and Minutes: RMG Consultants and Project Link-Up, 12 January 1994.

11 CLASP, CLASP Newsletter, September 1993.

12 Peter B. King, "A Modem for the Common Man," *Pittsburgh Post-Gazette* 1994, September 11; CFLAC, Meeting Minutes: Project Link-up, Three Rivers FreeNet, 14 September 1993, 2; CFLAC, Meeting Minutes: Three Rivers FreeNet, Project Link-Up, 27 October 1993, ACLA historical file (Folder: CFLAC-Minutes), ACLA-OFFICE; DeeAnna R. Cavinee, Director Goals: Northern Tier Study, EIN Review Group, Knowledge Connection Development, November 1993, ACLA historical file (Folder: CFLAC-Minutes), ACLA-OFFICE.

13 RMG Consultants, Goals of an Allegheny County Public Library Information Strategy, November 1993, 2–6, ACLA historical file (Folder: CFLAC-Minutes), ACLA-OFFICE.

14 RMG Consultants, Goals of an Allegheny County Public Library Information Strategy, November 1993, 2–7.

15 CFLAC, Electronic Information Strategies for Public Libraries in Allegheny County, December 1993, ACLA historical file (Folder: CFLAC-Minutes), ACLA-OFFICE.

16 Jennifer Fox Rabold, Minutes of the Executive Board of CLASP, 11 January 1994, ACLA historical file (Folder: CLASP Executive Committee), ACLA-OFFICE.

17 CFLAC, Meeting Agenda and Minutes: RMG Consultants and Project Link-Up, 12 January 1994.

18 DeeAnna R. Cavinee, CLASP Newsletter: Allegheny County Library Director's Comments, February 1994, ACLA historical file (Folder: ACLA-CLASP Newsletters), ACLA-OFFICE; Fox Rabold, Minutes of the Executive Board of CLASP, 11 January 1994; Jennifer Fox Rabold, Minutes of the Meeting of the Executive Board of CLASP, 14 April 1994, ACLA historical file (Folder: CLASP Executive Committee), ACLA-OFFICE.

19 Fox Rabold, Minutes of the Meeting of the Executive Board of CLASP, 14 April 1994; DeeAnna R. Cavinee, Newsletter of the Electronic Information Network of Public Libraries in Allegheny County (3), July 1994, ACLA historical file (Folder: Newspaper Articles General Library), ACLA-OFFICE.

20 CLASP, Electronic Information Network for Public Libraries in Allegheny County, 28 April 1994, ACLA historical file (Folder: ACLA History to Receive Funding), ACLA-OFFICE; CFLAC, Agenda and Minutes: PA Utility Commission Petition, CLASP Funding, EIN Foundation Presentations, 8 June 1994, ACLA historical file (Folder: CFLAC-Minutes), ACLA-OFFICE.

21 CFLAC, Meeting Agenda and Minutes: RMG Consultants and Project Link-Up, 12 January 1994; Fox Rabold, Minutes of the Executive Board of CLASP, 11 January 1994.

22 Cavinee, Newsletter of the Electronic Information Network of Public Libraries in Allegheny County (3), July 1994.

23 CFLAC, Director's Report: EIN Funding, RAD Response Plan, 5 April 1994, ACLA historical file (Folder: CFLAC-Minutes), ACLA-OFFICE; CLASP, Notes on How CLASP Should Restructure to Receive RAD Funds, April 1994, ACLA historical file (Folder: ACLA History to Receive Funding), ACLA-OFFICE.

24 Cavinee, Newsletter of the Electronic Information Network of Public Libraries in Allegheny County (3), July 1994; DeeAnna R. Cavinee, Letter from Dee Cavinee to CLASP Executive Committee Members: Agreements for EIN and Legal Consortium, 16 May 1994, ACLA historical file (Folder:

ACLA History to Receive Funding), ACLA-OFFICE; Carolyn A. Toth, CLASP List, July 1994, ACLA historical file (Folder: Black Binder), ACLA-OFFICE.

25 Cavinee, Newsletter of the Electronic Information Network of Public Libraries in Allegheny County (3), July 1994.

26 CFLAC, Agenda and Minutes: PA Utility Commission Petition, CLASP Funding, EIN Foundation Presentations, 8 June 1994.

27 CFLAC, Meeting Minutes: EIN Capital Funding, CLASP Plans, 27 July 1994, ACLA historical file (Folder: CFLAC-Minutes), ACLA-OFFICE; CLASP, Minutes of the CLASP General Meeting, 21 July 1994, ACLA historical file (Folder: ACLA History), ACLA-OFFICE.

28 CFLAC, Meeting Minutes: Resignation of DeeAnna, EIN Funds, ACLA Funds, 19 October 1994, ACLA historical file (Folder: CFLAC-Minutes), ACLA-OFFICE.

29 CFLAC, Agenda and Meeting Minutes: EIN Funding, Knowledge Connections, 7 September 1994, ACLA historical file (Folder: CFLAC-Minutes), ACLA-OFFICE.

30 CFLAC, Agenda and Meeting Minutes: EIN Funding, Knowledge Connections, 7 September 1994.

31 Fox Rabold, Minutes of the CLASP Executive Committee Meeting, 2 June 1993; CFLAC, Agenda and Minutes: PA Utility Commission Petition, CLASP Funding, EIN Foundation Presentations, 8 June 1994; Carolyn A. Toth, Letter from Carolyn Toth to David Rolka: Telephone Rate for Public Libraries, 21 July 1994, ACLA historical file (Folder: CLASP), ACLA-OFFICE; CFLAC, Agenda and Meeting Minutes: EIN Funding, Knowledge Connections, 7 September 1994; CFLAC, Agenda and Meeting Minutes: Special Telecommunications Rates for Linked Libraries, 19 October 1994, ACLA historical file (Folder: CFLAC-Minutes), ACLA-OFFICE.

32 Pennsylvania Economy League, A Case for the Western Pennsylvania Regional Asset District, 11 June 1993, 2–3, ACLA historical file (Folder: ARAD-Pre-Legislation), ACLA-OFFICE.

33 Pennsylvania Economy League, *Regional Asset Nature of the Pittsburgh Zoo, Phipps Conservatory, and the Pittsburgh Aviary*, Pennsylvania Economy League (Pittsburgh, October 1990); Pennsylvania Economy League, A Case for the Western Pennsylvania Regional Asset District, 11 June 1993, 1–2.

34 Pennsylvania Economy League, A Case for the Western Pennsylvania Regional Asset District, 11 June 1993, 2–3; Lee Boyd, CLASP Minutes, 29 June 1993, 1–2, ACLA historical file (Folder: CLASP Executive Committee), ACLA-OFFICE; Cavinee, Report to Buhl Foundation: Allegheny County Library Director Grant, 18 December 1992, 7; Allegheny Conference on Community Development, The Future of Our Region: New Ways of Working Together, New Alliance for Improving Our Quality of Life, 23 November 1992, 9, ACLA historical file (Folder: Legislative Information on ARAD Tax), ACLA-OFFICE; Robert B. Croneberger, Letter from Bob Croneberger to DeeAnna Cavinee: CLP Data for PA Economy League, 10 June 1993, ACLA historical file (Folder: ACLA-ARAD Background), ACLA-OFFICE.

35 Boyd, CLASP Minutes, 29 June 1993, 1–2; Linda Wilson Fuoco, "Panel Urges Eventual End to Personal Property Levy," *Pittsburgh Post-Gazette*, 24 March 1993.

36 Pennsylvania Economy League, A Case for the Western Pennsylvania Regional Asset District, 11 June 1993.

37 CLASP, Regional Asset District and Allegheny County Sales Tax: Summary Prepared in Reaction to Nov. 21 News Article, November 1993, ACLA historical file (Folder: Black Binder), ACLA-OFFICE.

38 CLASP, Regional Asset District and Allegheny County Sales Tax: Summary Prepared in Reaction to Nov. 21 News Article, November 1993, 1; Robert A. Cohen, Letter from Robert Cohen to Libraries in Allegheny County, 23 November 1993, ACLA historical file (Folder: ACLA-ARAD Background), ACLA-OFFICE; Jon Schmitz and Frank Reeves, "Push Is on for Early Pa. Sales Tax Increase," *Pittsburgh Post-Gazette*, 3 December 1993, B-1.

39 Schmitz and Reeves, "Push Is on for Early Pa. Sales Tax Increase."
40 Tim Reeves, "Singel Says He'll Ok Bill for County Sales Tax," *Pittsburgh Post-Gazette*, 9 December 1993; Tim Reeves, "Tax Push Campaign a Lesson in Lobbying," *Pittsburgh Post-Gazette*, 12 December 1993.
41 Tim Reeves, "Is Plan to Raise Sales Tax an Asset?," *Pittsburgh Post-Gazette*, 12 December 1993.
42 Steve Massey, "Tax Hike Praised by Economists," *Pittsburgh Post-Gazette*, 16 December 1993; Tim Reeves, "Senate OKs Sales Tax," *Pittsburgh Post-Gazette*, 14 December 1993; Pennsylvania General Assembly, Act 77: House Bill No. 659, 14 December 1993, ACLA historical file (Folder: Black Binder), ACLA-OFFICE; Ira Weiss, Summary of Act 77 of 1993, 1994, ACLA historical file (Folder: ACLA History), ACLA-OFFICE.
43 Weiss, Summary of Act 77 of 1993, 1994, 1; James Turner, Letter from Jim Turner to DeeAnna Cavinee: County Commissioners Meeting on RAD, 28 March 1994, ACLA historical file (Folder: ACLA-ARAD Background). ACLA-OFFICE; John M. R. Bull, "7% Tax Is Moving on Fast Track," *Pittsburgh Post-Gazette*, 25 February 1994.
44 David L. Michelmore, "The 1% Solution," *Pittsburgh Post-Gazette*, 16 January 1994.
45 Allegheny Conference on Community Development, Regional Directions, March 1994, ACLA historical file (Folder: Black Binder), ACLA-OFFICE; Tom Barnes, "County Sales Tax May Be 7% by July," *Pittsburgh Post-Gazette*, 23 February 1994; John M. R. Bull, "Sales Tax Boost Near," *Pittsburgh Post-Gazette*, 30 March 1994.
46 ARAD, 27 Advisory Board Selections, 1994, ACLA historical file (Folder: Black Binder). ACLA-OFFICE.
47 Jon Schmitz, "Positions on Assets Board Are a Hot Item," *Pittsburgh Post-Gazette*, 17 April 1994; Tom Barnes and John M. R. Bull, "6 Named to Board to Divide Regional Assets Tax," *Pittsburgh Post-Gazette* (ACLA historical file (Folder: Black Binder). Allegheny County Library Association Office, Pittsburgh, PA), 22 April 1994; Robert A. Cohen, Letter from Robert Cohen to All Library Directors, Trustees, and Interested Parties, March 1994, ACLA historical file (Folder: ACLA-CLASP Newsletters), ACLA-OFFICE.
48 D. Michael Fisher, Letter from Mike Fisher to Monica Stoicovy, 20 December 1993, ACLA historical file (Folder: Legislative Information on ARAD Tax), ACLA-OFFICE.
49 Paul Leger, Letter from Paul Leger to Bob Croneberger: Background on County Libraries and RAD, 20 December 1993, ACLA historical file (Folder: Black Binder), ACLA-OFFICE.
50 Pennsylvania Economy League, A Case for the Western Pennsylvania Regional Asset District, 11 June 1993, 7.
51 Croneberger, Letter from Bob Croneberger to DeeAnna Cavinee: CLP Data for PA Economy League, 10 June 1993.
52 Croneberger, Letter from Bob Croneberger to DeeAnna Cavinee: CLP Data for PA Economy League, 10 June 1993; CFLAC, Meeting Minutes: Project Link-up Governance, Adminstrator Position Extension, 15 June 1993, 2.
53 Fox Rabold, Minutes of the CLASP Executive Committee Meeting, 2 June 1993; Lee Boyd, Letter from Lee Boyd to DeeAnna Cavinee: CLASP Executive Board Meeting with Susan Parker, 19 June 1993, ACLA historical file (Folder: ACLA-ARAD Background), ACLA-OFFICE; Boyd, CLASP Minutes, 29 June 1993; Albert F. Kamper, Letter from Al Kamper to Susan Parker: CLP and County Library Reports Data, 30 June 1993, ACLA historical file (Folder: ACLA-ARAD Background), ACLA-OFFICE; Robert A. Cohen, Letter from Robert Cohen to Alison Spence: Contact to CLASP from the PA Economy League About RAD, 16 August 1993, ACLA historical file (Folder: ACLA-ARAD Background), ACLA-OFFICE.

54 CLASP, Regional Asset District and Allegheny County Sales Tax: Summary Prepared in Reaction to Nov. 21 News Article, November 1993.
55 Robert B. Croneberger, Message from the Director, March 1994, ACLA historical file (Folder: ACLA-ARAD Background), ACLA-OFFICE.
56 Cohen, Progress Report from CLASP to All Allegheny County Libraries, 21 July 1993, 2; CLASP, CLASP Newsletter, September 1993, 5 – 6; Cohen, The CLASP Report 2(3): September Meeting Features CD-ROM Workshop, September 1993, 3; Albert F. Kamper, Letter from Al Kamper to Susan Parker: Enlisting the Pennsylvania Economy League, 17 August 1993, ACLA historical file (Folder: ACLA-ARAD Background), ACLA-OFFICE.
57 CLASP, Regional Asset District and Allegheny County Sales Tax: Summary Prepared in Reaction to Nov. 21 News Article, November 1993.
58 Cohen, Letter from Robert Cohen to Libraries in Allegheny County, 23 November 1993.
59 Fox Rabold, Minutes of the Executive Board of CLASP, 11 January 1994.
60 William Opalka, "Assets Tax Leaves Library Funding in Limbo," *Pittsburgh Tribune Review*, 6 March 1994; Kendra Duerst, "Pleasant Hills Approves County Sales Tax," 2 March 1994; Jim McMahon, "New Sales Tax Gets Grudging Support," *Pittsburgh Post-Gazette*, 2 March 1994.
61 DeeAnna R. Cavinee, Public Libraries: A Critical Asset for Allegheny County Residents, May 1994, ACLA historical file (Folder: ACLA-ARAD Background), ACLA-OFFICE; William R. Dodge, Letter from Bill Dodge to Carolyn Toth: Public Relations Documents, 4 May 1994, ACLA historical file (Folder: ACLA History to Receive Funding), ACLA-OFFICE; Laura Shelley, Letter from Laura Shelley to All CLASP/RAD Committee Members, June 1994, ACLA historical file (Folder: ACLA-ARAD Background), ACLA-OFFICE; Laura Shelley, Letter from Laura Shelley to the CLASP Committee on RAD, 23 June 1994, ACLA historical file (Folder: ACLA-ARAD Background), ACLA-OFFICE; William R. Dodge, Letter from Bill Dodge to Carolyn Toth: Suggested Revisions on RAD Proposal, 25 May 1994, ACLA historical file (Folder: ACLA History to Receive Funding), ACLA-OFFICE; CLASP, Notes on How CLASP Should Restructure to Receive RAD Funds, April 1994; Laura Shelley, Letter from Laura Shelley to Regional Asset District Committee of CLASP, 31 March 1994, ACLA historical file (Folder: ACLA-ARAD Background), ACLA-OFFICE; CFLAC, Director's Report: EIN Funding, RAD Response Plan, 5 April 1994; CLASP, Summary of Meeting, Regional Asset District Committee, 17 February 1994, ACLA historical file (Folder: ACLA History), ACLA-OFFICE; CFLAC, Director's Report: EIN Funding, RAD Response Plan, 5 April 1994; Turner as quoted in Joe Mandak, *Valley News Dispatch*, 8 March 1994.
62 Pennsylvania General Assembly, Act 77: House Bill No. 659, 14 December 1993., Sec. 3130-B.
63 CFLAC, Agenda and Minutes: PA Utility Commission Petition, CLASP Funding, EIN Foundation Presentations, 8 June 1994.
64 Cohen, Letter from Robert Cohen to All Library Directors, Trustees, and Interested Parties, March 1994.
65 CLASP, Notes on How CLASP Should Restructure to Receive RAD Funds, April 1994.
66 CLASP, Notes on How CLASP Should Restructure to Receive RAD Funds, April 1994.
67 Peter Leonard, Letter from Peter Leonard to CLASP RAD Justification Committee, 13 July 1994, ACLA historical file (Folder: Black Binder), ACLA-OFFICE; CLASP member libraries, RAD Supplemental Funds Survey, July 1994, ACLA historical file (Folder: Black Binder), ACLA-OFFICE; CLASP member libraries, Supplemental Funds Survey, 28 July 1994, ACLA historical file (Folder: ACLA History to Receive Funding), ACLA-OFFICE; Carolyn A. Toth, Letter from Carolyn Toth to Public Library Directors in Allegheny County, 9 July 1994, ACLA historical file (Folder: Black Binder), ACLA-OFFICE.
68 ACLA, Allegheny County Library Association Fact Sheet: Prepared for the Regional Asset District Board, August 1994, ACLA historical file (Folder: ARAD Informational Breakfast), ACLA-OFFICE;

Peter Leonard, Letter from Peter Leonard to County Libraries: Survey on County Library Use, 12 August 1994, ACLA historical file (Folder: ACLA History to Receive Funding), ACLA-OFFICE; Peter Leonard, Letter from Peter Leonard to Marilyn Jenkins, 1994, August 12, ACLA historical file (Folder: Black Binder), ACLA-OFFICE; Peter Leonard, Letter from Peter Leonard to County Libraries, 12 August 1994, ACLA historical file (Folder: Black Binder), ACLA-OFFICE.

69 CFLAC, Director's Report: EIN Funding, RAD Response Plan, 5 April 1994; Robert B. Croneberger, Testimony on Regional Asset District, 18 March 1994, ACLA historical file (Folder: ACLA-ARAD Background), ACLA-OFFICE; Jo Ellen Kenney, Testimony on Regional Asset District, 18 March 1994, ACLA historical file (Folder: ACLA History to Receive Funding), ACLA-OFFICE; Laura Shelley, Testimony of Laura Shelley at Public Hearing on Regional Asset District in Allegheny County, 18 March 1994, ACLA historical file (Folder: ACLA-ARAD Background), ACLA-OFFICE.

70 CFLAC, Agenda and Meeting Minutes: EIN Funding, Knowledge Connections, 7 September 1994.

71 CLASP, Summary of Meeting, Regional Asset District Committee, 17 February 1994.

72 Fox Rabold, Minutes of the Meeting of the Executive Board of CLASP, 14 April 1994; CLASP, Notes on How CLASP Should Restructure to Receive RAD Funds, April 1994.

73 CLASP, Notes on How CLASP Should Restructure to Receive RAD Funds, April 1994.

74 CFLAC, Agenda and Minutes: PA Utility Commission Petition, CLASP Funding, EIN Foundation Presentations, 8 June 1994; CFLAC, Meeting Minutes: EIN Capital Funding, CLASP Plans, 27 July 1994; CLASP, Principles for the Regional Asset District to Use with Libraries, June 1994, ACLA historical file (Folder: ACLA-ARAD Background), ACLA-OFFICE; CLASP, County Library Association Serving the People: Official by-Laws, 15 June 1994, ACLA historical file (Folder: CLASP), ACLA-OFFICE.

75 CLASP, Minutes of the CLASP General Meeting, 21 July 1994; CFLAC, Meeting Minutes: EIN Capital Funding, CLASP Plans, 27 July 1994.

76 CLASP executive committee, CLASP Executive Committee Meeting Agenda, 28 July 1994, ACLA historical file (Folder: Black Binder), ACLA-OFFICE; Carolyn A. Toth, Letter from Carolyn Toth to ACLA Library Directors: CLASP Statement on RAD Funding, 4 August 1994, ACLA historical file (Folder: ACLA-ARAD Background), ACLA-OFFICE.

77 CLASP, Notes on How CLASP Should Restructure to Receive RAD Funds, April 1994; CLASP, Minutes of the CLASP General Meeting, 21 July 1994; ACLA, ACLA Executive Board, August 1994, ACLA historical file (Folder: Black Binder), ACLA-OFFICE.

78 Albert F. Kamper, CLASP County Regions, 21 July 1994, ACLA historical file (Folder: ACLA History), ACLA-OFFICE; ACLA, Regional Meeting Minutes: East, 28 August 2015, ACLA-WIKI.

79 CLASP, Notes on How CLASP Should Restructure to Receive RAD Funds, April 1994; Fox Rabold, Minutes of the Meeting of the Executive Board of CLASP, 14 April 1994; Jennifer Fox Rabold, Bases for RAD Funding, 16 April 1994, ACLA historical file (Folder: CFLAC-Minutes), ACLA-OFFICE; CFLAC, Meeting Minutes: Draft RAD Funding Suggestions, 27 April 1994, ACLA historical file (Folder: CFLAC-Minutes), ACLA-OFFICE.

80 ACLA, Allegheny County Library Association Regions and Region Leaders, August 1994, ACLA historical file (Folder: Black Binder), ACLA-OFFICE.

81 Toth, CLASP List, July 1994; CFLAC, Meeting Minutes: EIN Capital Funding, CLASP Plans, 27 July 1994; CLASP, Principles of CLASP for RAD and Purpose of CLASP, June 1994, ACLA historical file (Folder: ACLA History to Receive Funding), ACLA-OFFICE; CLASP, Letter from CLASP to Library Trustees, 14 May 1994, ACLA historical file (Folder: Black Binder), ACLA-OFFICE.

82 William R. Dodge, A Consultant's Thoughts on the Planning Process, August 1994, ACLA historical file (Folder: ACLA Proposal to ARAD Jo Ellen's Speech), ACLA-OFFICE.

83 Allegheny Regional Asset District, Agenda for ARAD Board Meeting, 8 August 1994, ACLA historical file (Folder: ACLA History to Receive Funding), ACLA-OFFICE.
84 Albert F. Kamper, Letter from Al Kamper to ACLA Executive Board and Finance Subcommittee Members: Strategy for ARAD Funding, 11 August 1994, ACLA historical file (Folder: ACLA-ARAD Background), ACLA-OFFICE.
85 ACLA, ARAD Rationale, August 1994, ACLA historical file (Folder: ACLA History to Receive Funding), ACLA-OFFICE.
86 ACLA, ARAD Rationale, August 1994; Toth, Letter from Carolyn Toth to ACLA Library Directors: CLASP Statement on RAD Funding, 4 August 1994.
87 CLASP, CLASP Committee on RAD Proposal Group, 13 July 1994, ACLA historical file (Folder: ACLA-ARAD Background), ACLA-OFFICE; Peter Leonard, Letter from Peter Leonard to Justification Committee Members, 25 July 1994, ACLA historical file (Folder: ACLA History to Receive Funding), ACLA-OFFICE.
88 CLASP, Notes on ACLA Plans for RAD Funding, June 1994, ACLA historical file (Folder: ACLA History to Receive Funding), ACLA-OFFICE.
89 CLASP, Notes on ACLA Plans for RAD Funding, June 1994; CLASP, Minutes of the CLASP General Meeting, 21 July 1994.
90 Peter Leonard, Letter from Peter Leonard to Marilyn Jenkins: ARAD Proposal Files, 12 August 1994, ACLA historical file (Folder: ACLA History to Receive Funding), ACLA-OFFICE; ACLA, ARAD Rationale, August 1994.
91 Fox Rabold, Minutes of the Meeting of the Executive Board of CLASP, 14 April 1994; CLASP, Summary of CLASP Regional Assets Committee Meeting, 5 April 1994, ACLA historical file (Folder: ACLA-ARAD Background), ACLA-OFFICE; CLASP, Proposal for RAD Funding of Allegheny County Libraries, July 1994, ACLA historical file (Folder: Black Binder), ACLA-OFFICE; Dodge, A Consultant's Thoughts on the Planning Process, August 1994.
92 Leonard, Letter from Peter Leonard to Marilyn Jenkins: ARAD Proposal Files, 12 August 1994; ACLA finance committee, Proposed Formula for Distribution of ARAD Funds, August 1994, ACLA historical file (Folder: ACLA Proposal to ARAD Jo Ellen's Speech), ACLA-OFFICE.
93 Carolyn Marks, Regional Asset District Budget Request Sent to Joellen Kinney, 15 August 1994, ACLA historical file (Folder: Black Binder), ACLA-OFFICE; Albert F. Kamper and Laura Shelley, Letter from Al Kamper and Laura Shelley to All Libraries in Allegheny County: Information Gathering for RAD, 6 September 1994, ACLA historical file (Folder: ACLA Proposal to ARAD Jo Ellen's Speech), ACLA-OFFICE.
94 ACLA, Proposal for ARAD Funding, Allegheny County Library Association, 15 September 1994, ACLA historical file (Folder: ACLA Letters of Support for ARAD Funding 1994), ACLA-OFFICE.
95 ACLA, Proposal for ARAD Funding, Allegheny County Library Association, 15 September 1994, 6.
96 ACLA, In Soliciting ARAD Support, ACLA Has Received the Following Letters of Endorsement to Date, August 1994, ACLA historical file (Folder: Black Binder), ACLA-OFFICE; Toth, Letter from Carolyn Toth to Public Library Directors in Allegheny County, 9 July 1994; D. Michael Fisher, Letter from Mike Fisher to Rick Stafford, 20 July 1994, ACLA historical file (Folder: Legislative Information on ARAD Tax), ACLA-OFFICE; CLASP, CLASP Report, 28 July 1994, ACLA historical file (Folder: Black Binder), ACLA-OFFICE; CFLAC, Agenda and Meeting Minutes: EIN Funding, Knowledge Connections, 7 September 1994; William R. Dodge, Letter from Bill Dodge to Jo Ellen Kenney, Marilyn Jenkins, Laura Shelley, and DeeAnna Cavinee, 25 August 1994, ACLA historical file (Folder: Black Binder), ACLA-OFFICE; Albert F. Kamper, Letter from Al Kamper to RAD Proposal Committee: Bill's Notes on Meeting and Proposal, 13 August 1994, ACLA historical file (Folder: Black Binder), ACLA-OFFICE; Frank J. Lucchino, Letter from Frank Lucchino to Linda Dickerson: Request for Endorsement of

RAD Funding, 14 September 1994, ACLA historical file (Folder: ACLA Letters of Support for ARAD Funding 1994), ACLA-OFFICE.

97 ACLA, Calendar of Regional Asset District Proposal Strategy, August-September, 1994, August-October 1994, ACLA historical file (Folder: Black Binder), ACLA-OFFICE; Carolyn A. Toth, Letter from Carolyn Toth to ACLA Library Directors, 22 August 1994, ACLA historical file (Folder: Black Binder), ACLA-OFFICE; DeeAnna R. Cavinee, Good News, 12 August 1994, ACLA historical file (Folder: ACLA-ARAD Background), ACLA-OFFICE; ACLA, Presentation for RAD Board: Agenda, Outline, and Preliminary Draft of Content, 30 August 1994, ACLA historical file (Folder: ARAD Informational Breakfast), ACLA-OFFICE; Marilyn A. Jenkins, ARAD Presentation, 30 August 1994, ACLA historical file (Folder: ARAD Informational Breakfast), ACLA-OFFICE.

98 Kathryn L. Firestone, Letter from Kathy Firestone to ACLA Libraries: Petition Rally, September 1994, ACLA historical file (Folder: Pre-RAD), ACLA-OFFICE.

99 Firestone, Letter from Kathy Firestone to ACLA Libraries: Petition Rally, September 1994.

100 CFLAC, Agenda and Meeting Minutes: EIN Funding, Knowledge Connections, 7 September 1994.

101 Firestone, Letter from Kathy Firestone to ACLA Libraries: Petition Rally, September 1994.

102 Jo Ellen Kenney, ARAD Board Speech, 21 September 1994, Private collection.

103 Robert B. Croneberger, S. Fred Natale, and Robert T. Horvath, Letter from Croneberger, Natale, and Horvath to Sara Parker, 7 June 1994, District Services Collection, Series 2, Box 11, Folder 2, CLP-OLIVER; Albert F. Kamper, Letter from Al Kamper to CLP, 27 October 1994, District Services Collection, Series 2, Box 11, Folder 2, CLP-OLIVER; Sara Parker, Letter from Sara Parker to Governor's Advisory Council: Boundary Alignments of Three Southwestern Pennsylvania Public Library Districts, 26 September 1994, Department of Education, Commonwealth Libraries, Minutes of the Governor's Advisory Council on Library Development 1962–2009 collection, Carton 2, Folder 4, PASA.

104 Robert B. Croneberger, Letter from Bob Croneberger to Pittsburgh District Libraries, CLP Branch, Division and Department Heads, 28 November 1994, Department of Education, Commonwealth Libraries, Minutes of the Governor's Advisory Council on Library Development 1962–2009 collection, Carton 2, Folder 4, PASA; Sara Parker, Letter from Sara Parker to Governor's Advisory Council and Westmoreland Libraries, 26 September 1994, Department of Education, Commonwealth Libraries, Minutes of the Governor's Advisory Council on Library Development 1962–2009 collection, Carton 2, Folder 4, PASA.

105 Governor's Advisory Council on Library Development, Minutes, 6 December 1994, Department of Education, Commonwealth Libraries, Minutes of the Governor's Advisory Council on Library Development 1962–2009 collection, Carton 2, Folder 4, PASA.

106 ACLA, Proposal for ARAD Funding, Allegheny County Library Association, 15 September 1994.

107 Bill Dodge as cited in Kamper, Letter from Al Kamper to RAD Proposal Committee: Bill's Notes on Meeting and Proposal, 13 August 1994.

108 Tom Barnes, "Asset Tax Board Gathers Data," *Pittsburgh Post-Gazette*, 31 August 1994.

109 Tom Barnes, "First Asset District Check: $2.1 Million," *Pittsburgh Post-Gazette*, 13 September 1994.

110 ARAD, Allegheny Regional Asset District Public Hearing: Agenda, 10 September 1994, ACLA historical file (Folder: ACLA Proposal to ARAD Jo Ellen's Speech), ACLA-OFFICE.

111 Tom Barnes, "Officials Fear Asset Tax Will Not Meet Their Needs," *Pittsburgh Post-Gazette*, 11 September 1994; Tonya Vinas, "Four Agencies Make Pitch for Extra Slice of Assets Tax," *North Hills News Record*, 11 September 1994.

112 Tom Barnes, "Lots of Hands out, Not Enough Funds in Assets District," *Pittsburgh Post-Gazette*, 17 September 1994.

113 Kenney, ARAD Board Speech, 21 September 1994, 3, 8.

114 Marilyn A. Jenkins, Letter from Marilyn Jenkins to Henry Ewalt: ACLA Projected Estimates, 27 September 1994, ACLA historical file (Folder: ACLA-ARAD Background), ACLA-OFFICE.
115 Carolyn A. Toth, Letter from Carolyn Toth to Marilyn Jenkins, 26 September 1994, ACLA historical file (Folder: Black Binder), ACLA-OFFICE.
116 Tom Barnes, "Stadium's the Winner in Sales Tax Handout," *Pittsburgh Post-Gazette*, 4 October 1994.
117 Dan Cook and Jennifer Sabol, "Jackpot for Jocks, Snake-Eyes for Books," *In Pittsburgh*, October 1994.
118 Carolyn A. Toth, Letter from Carolyn Toth to Dee Anna Cavinee: Yea for ACLA, 4 October 1994, ACLA historical file (Folder: ACLA-ARAD Background), ACLA-OFFICE.
119 Monica L. Haynes, "Library Group Awaits Meeting before It Budgets Asset Money," *Pittsburgh Post-Gazette*, 13 October 1994.
120 ACLA board, Letter from ACLA Board to Library Directors, Trustees, and Officials: $5 Million Award, 12 October 1994, ACLA historical file (Folder: ACLA-ARAD Proposal), ACLA-OFFICE.
121 CFLAC, Meeting Minutes: Resignation of DeeAnna, EIN Funds, ACLA Funds, 19 October 1994.

Period 7: System Dreams, 1994 – 1997

In 1994, Allegheny County Library Association (ACLA) successfully lobbied for funding from the newly formed Regional Asset District (RAD). In October 1994, RAD made allocations for the fiscal year 1995.[1] ACLA received $5 million, though there was not yet a distribution formula in place within ACLA for allocating this funding to ACLA member libraries. By October 1994, there were 43 ACLA member libraries, including CLP, though CLP received RAD money separately from ACLA.[2] All of these libraries became funded in part through RAD. It was then left to the ACLA organization to distribute and administer RAD money to its member libraries. Public libraries in the county now had to negotiate their relationships with one another and with this new funding body.

For ACLA libraries, there was a sense of accomplishment, excitement, and relief at the close of 1994. Their year-long lobbying effort had been a success. But there was still uncertainty. It was not clear whether ACLA would continue to receive RAD funding beyond 1995.[3] Unlike CLP, ACLA was not initially considered a statutory asset. The organization therefore could not immediately depend on RAD funding moving forward.[4] Moreover, it was still unclear how RAD funding would affect the local funding of ACLA libraries, whether it would supplement their municipal funding or replace it. It was feared that some municipalities would withdraw their funding in light of RAD money.[5] The architects of the RAD concept, ACCD and the Economy League, remained silent on municipal contributions to the public library infrastructure.[6]

Despite the remaining uncertainties about ACLA and RAD, some countywide public library projects moved forward. One countywide project, first administered by CFLAC, was Knowledge Connections. This program placed reading rooms with computer equipment in five public housing projects throughout Allegheny County.[7] The Knowledge Connections gained state attention due to its promotion by CFLAC.[8] The bookmobile, which was now administered in part by ACLA, continued to be streamlined and reworked in light of the Knowledge Connections program and the new resource sharing across the county.[9]

The most expensive and complex project underway at this time was the Electronic Information Network (EIN). By late 1994, EIN began to receive grant money from its capital campaign, enough to initiate the first phase of development.[10] The bulk of the capital expenses for EIN were provided through private foundations, but the county commissioners contributed also.[11] Both ACLA and CLP contributed a portion of their RAD funds to EIN.[12] EIN's board of directors held its first official meeting in December 1994, staff were hired in early 1995, and EIN created a new planning and governance structure.[13] CLP was selected as the designated agent of

https://doi.org/10.1515/9783111013404-009

the system, meaning that it managed EIN's fiscal resources and established the guidelines for the ownership of equipment, liability issues, and operating rights and responsibilities.[14] EIN provided services for both CLP and ACLA libraries, but it was CLP that oversaw the day-to-day operations of EIN.[15] EIN offices were located in CLP's basement.

EIN was an extension of Project Link-Up, which was initiated following *Quiet Crisis*. By the end of 1994, some libraries throughout Allegheny County already had access to CLP's online catalog, CAROLINE. The EIN project greatly expanded online services beyond what Project Link-Up offered.[16] EIN connected 40 public libraries throughout the county and facilitated access to a county-wide union catalog with 1.5 million titles and access to online commercial databases. This was done in part by installing computer workstations throughout the county and by building a private network ring.[17] Libraries across the county continued their own processes of automation in order to join the new network.[18] EIN services were phased in throughout 1995, 1996, and 1997.[19] Through EIN, libraries were able to participate in the shared union catalog, interlibrary loans, the shared patron database, and they were able to connect to the World Wide Web.[20] The first libraries began circulation on the new system in October 1996, and all libraries were online by late spring 1997.[21] There was increased circulation and interlibrary loans between the public libraries, and CLP's district services provided delivery across the county three days a week.[22] All county libraries were connected to EIN by Fall 1997.

The advent of RAD funding presented new challenges and new roles for ACLA. Once it was decided that ACLA would receive RAD funding, the ACLA and RAD boards had to decide how to distribute that money to the 40 county libraries.[23] ACLA was initially conceived as a pass-through organization for RAD money. This meant, in principle, that ACLA HQ received RAD money and merely distributed it according to a formula approved by the RAD board.[24] "ACLA HQ" was the informal term used to distinguish the ACLA board and ACLA administrative staff from ACLA's library membership. In reality, though most RAD money was passed on to member libraries as was envisioned, a portion of RAD money was also retained by ACLA HQ and put toward regional projects, including EIN and the development of a countywide plan.[25] By May 1995, libraries began receiving their share of the $2.7 million distribution in quarterly installments.[26] The other $2.3 million for cooperative projects was used for EIN, the development of a countywide plan, assistance to libraries pursuing state aid, grants for multi-library initiatives and special projects, addressing service in underserved areas, and administrative support.[27] For the year 1995, ACLA's distribution formula was developed in close coordination with RAD, ACCD, and the Pennsylvania Economy League. The league revised the formula and included several factors, including population, local support, and financial distress.[28] Ultimately, the RAD board directed how ACLA distributed RAD money to member libraries and how that money was used.[29]

The first year of RAD funding in 1995 led to several improvements in the public library infrastructure in greater Pittsburgh. These improvements included additional service hours, added professional staff, new collection materials, digital equipment such as CD-ROMs, repairs and renovations, and open access in almost every library for all county residents.[30] Capital expenditures included painting, remodeling, automation equipment, computer equipment, new furniture, reupholstering chairs, shelving, carpeting, ADA compliance renovations, and roof repair.[31] Funding also went toward EIN, which included the purchase of new computers, access to new online databases, and a down payment on the new countywide automation system.[32] In 1995, the first year of RAD funding, exactly $2.7 million was allocated to 38 libraries as ACLA's basic service distribution. Allocations to individual libraries ranged from $19,000 to $140,000.[33] In 1995, CLP received its own separate allocation from RAD totaling $11.9 million.

The initial uncertainties about RAD funding gradually gave way to greater predictability. ACLA became a statutory asset in January 1995. This meant that at least $5 million in annual RAD funding was guaranteed to ACLA until 2005. That is, if RAD agreed to ACLA's distribution plan.[34] RAD renewed the ACLA's term for another five years beginning in 2005, then another term beginning in 2010.[35] In addition to becoming a stable funding source, the RAD money given to the local libraries also came to be considered supplemental money by libraries' municipalities. Contrary to what was feared, local municipalities did not cut back their funding to libraries. Though there was a degree of leveling that occurred, there was no substantial cutback of local funding.[36]

After ACLA began receiving RAD funding, ACLA HQ took on new oversight and administrative duties. Oversight of member libraries by ACLA HQ was a condition for receiving RAD funding. As part of this role, ACLA HQ collected documentation from member libraries, including annual reports, plans for RAD funding, quarterly accounting records, impact reports that tracked each library's use of RAD funds, annual state reports, and certified audits.[37] In cases where a library did not meet the requirements or did not submit the necessary paperwork, ACLA held a library's funding in escrow.[38] Each library signed a Cooperation and Support Agreement (CASA) with ACLA HQ and submitted quarterly reports and regional reimbursement reports.[39]

With its new administrative and oversight roles, ACLA outgrew its status as a volunteer membership organization from its CLASP days. This meant hiring paid staff.[40] ACLA hired a secretary and secured office space.[41] Together with EIN, the ACLA office was located in the basement of CLP's Oakland branch.[42] It later moved to the Allegheny Regional Branch on the North Side.[43] RAD advised ACLA to hire an executive director.[44] When a new County Library Administrator was hired in 1995, that administrator began to manage the operations of ACLA in a proto executive director position.[45]

With the advent of RAD funding and the formation of ACLA, CLP made adjustments to its traditional roles.[46] There were 21 CLP branches in 1995. CLP continued to operate the Allegheny County Bookmobiles, and it continued to serve as the district library and as a regional resource center.[47] But CLP's functions began to change. Once ACLA became a regional asset, ACLA negotiated a new bookmobile contract with CLP.[48] ACLA also began to influence district services. The district services provided by CLP included consulting and coordinated planning between libraries.[49] There began to emerge overlaps and redundancies with the services that CLP provided and the duties that ACLA wanted to take on. These overlaps would become a source of tension and negotiation between ACLA and CLP.

Tensions began to emerge not just between CLP and ACLA, but also among ACLA member libraries. This was because acceptance of RAD funding came with conditions. One of the conditions was that libraries provide free services to any resident in the county. Most county libraries already provided free access to all state residents. This was through Pennsylvania's ACCESS PENNSYLVANIA plan, which provided state money to reimburse libraries for the non-resident access they provided. But six ACLA libraries did not participate in ACCESS PENNSYLVANIA, including Northland Public Library.[50] Two of the largest libraries in the county, Northland and Monroeville, received RAD funding in 1995 but refused to allow open access to county residents.[51] Northland believed that open access to county residents would lead to a loss of local funding.[52] Despite Northland's initial support for ACLA and RAD, Northland remained steadfast in the decision not to provide open access, even if it meant not receiving RAD funding in 1996.[53] By mid-1995, Northland was the only library not to agree to the open access provision.[54] Northland was the second-largest collection in the county after CLP, and not having Northland as a member damaged ACLA's image as a regional infrastructure.

Library leaders tried several punitive measures to convince Northland to participate in ACLA. The EIN board made open access a condition for participation in the network, and the State Library of Pennsylvania even sent a letter to all libraries in Allegheny County stating that open access was a requirement for state funding.[55] But Northland approached the RAD board on its own for separate funding, thus widening the rift between Northland and the ACLA membership.[56] The attempted punitive measures were ineffective. Subsequent revisions to the distribution formula did not help matters. In March 1995, as RAD funding was being distributed to libraries, the cap on the amount of RAD funding libraries received was removed. The ACLA formula factored in local support. This was intended to leverage local funding: the more a local municipality contributed to its library, the larger the share of RAD funding a library would receive. But libraries whose municipalities contributed very little had their RAD funding capped at the amount their municipality provided. This was to prevent municipalities from becoming overly dependent on RAD funding, but the cap was criticized because some municipalities did not have the means to contribute more. The RAD board's decision to remove the cap redirected funding away from the larger and more affluent libraries, such as Northland, and toward smaller and more distressed libraries. Needless to say, the revision did nothing to encourage Northland to join ACLA.[57] It also led to the unintended effect of reducing the overall funding to libraries located in poorer municipalities. This was because as RAD funding increased to those libraries their municipalities saw it as an opportunity to reduce local funding.[58]

Negotiations continued between ACLA HQ, RAD, and Northland through 1995 about the distribution formula for 1996.[59] RAD requested further relief to libraries in distressed communities, but ACLA tried to reward communities who provided local support, such as Northland. Meanwhile, Northland proposed a compromise where it would rejoin ACLA but receive additional RAD funding. Northland did not receive RAD funding in 1996, and it was not a participant in EIN.[60]

ACLA's budget and distribution for 1997 proposed several changes. Instead of the 1 percent limit that RAD set for administrative costs, ACLA HQ requested 3 percent of the annual RAD allocation.[61] The formula was also revised.[62] In contrast to the formulas used in 1995 and 1996, which targeted smaller and distressed libraries, the proposed 1997 formula distributed more money to larger and more affluent libraries. This distribution formula met with resistance from smaller libraries, and it passed the ACLA general membership by only a small margin.[63] In the end, the proposal was rejected by the ACLA board and the RAD board because the formula overly stressed rewarding local support instead of supporting libraries in economically distressed communities.[64] It was revealed that ACLA's proposed revisions of the formula were intended in part to entice Northland to join ACLA.[65] But again, Northland did not join ACLA, and it instead presented a separate proposal for RAD funding.[66] Again, for the second consecutive year, Northland decided not to participate in RAD funding for 1997.[67] ACLA therefore lost Northland's share from its RAD allocation. [68] ACLA HQ disagreed with RAD's proposed revisions, which focused on leveraging local income, but in the end another revised formula that was submitted by ACLA HQ was approved by RAD.[69] This was the one used for 1997.

ACLA continued discussions with RAD about the ACLA budget into 1997. RAD insisted on increased oversight of ACLA's expenditures, especially regarding capital expenditures, administrative costs, and the distribution formula.[70] To develop a distribution formula for 1998, RAD met with local and municipal officials.[71]

Based on these discussions, RAD eliminated the regional grants category of the ACLA budget, moving money from that category into the basic service distribution.[72] EIN continued to be funded by RAD as a separate special grant, but the RAD board urged ACLA and CLP to investigate alternative funding solutions for EIN.[73]

ACLA continued to solicit Northland's participation in the organization for 1998.[74] Some RAD board members urged ACLA to accommodate Northland in any way possible. The ACLA board worked to incorporate these and other ideas.[75] Northland expressed interest in becoming a provisional member.[76] Despite ACLA's efforts at compromise, Northland continued to refuse RAD funding through ACLA. RAD allocations for 1998 were $5,300,000 to ACLA and $13,575,000 to CLP.[77] Northland approached the RAD board with a separate budget request. ACLA was concerned that, if Northland's request were accepted, it would open the door to other member libraries submitting independent requests.[78] But RAD's Library Formula Committee was sympathetic to Northland.[79] Eliminating the regional grants category from the ACLA budget raised the distribution for all ACLA libraries, especially Northland. In addition, RAD's library subcommittee recommended a separate grant for Northland in 1998.[80] ACLA HQ and some ACLA member libraries disagreed with separate funding to Northland.[81] Despite the objections from other libraries in the county, Northland Library participated in RAD funding in 1998.[82] In 1998, Northland received RAD funding through ACLA, and, in addition, Northland also received a special 1-year grant from RAD. With the acceptance of this funding, Northland agreed to free lending service in the county for 1 year. Northland became an ACLA member in 1999.[83]

Development of the electronic network, revisions to the funding formula, and negotiations with libraries were the major challenges facing the public library infrastructure in greater Pittsburgh following the substantive change that occurred in Period 6. These arose from two new entities, ACLA and RAD, making decisions about how RAD funding should be used. ACLA HQ was tasked with distributing and managing $5 million. But ACLA did not yet have a long-term plan for how that funding would be used. ACLA received RAD funding in 1994 on the condition that the organization would develop a long-term plan. In part, the plan had to clarify the responsibilities between ACLA HQ, CLP's district services, and the County Library Administrator.[84] In April 1995, ACLA began a planning process that was completed in May 1996.[85] This was ACLA's first long-term plan.[86]

After studying such things as the population, funding, and staffing related to the public library infrastructure, analysts concluded that ACLA HQ should improve efficiencies through cooperative endeavors like EIN and establish a uniform level of service across the county.[87] In order to do this, it was recommended that ACLA establish a federated library system countywide.[88] In a federated system, individual libraries voluntarily join together to provide enhanced library services. The libraries retain their individual boards and most decision-making power, but a system board determines regional policies and procedures.[89] A federated system, created from ACLA or some successor organization, would become the county-wide agency equipped to coordinate public library service on a regional scale, thereby assuming many of the responsibilities dispersed among various county agencies. Responsibilities would include distributing state aid, distributing RAD funding, administering and funding EIN, delivering bookmobile service, managing the Knowledge Connections program, and overseeing marketing, training, and planning.[90] It was also imagined that this new system would assume all district library responsibilities from CLP.[91] The new system could be formed as part of county government, as part of CLP, or as a separate entity such as a nonprofit agency.[92]

Regardless of its form, the proposed system would resemble ACLA's current status in terms of its funding, but it would differ from ACLA in terms of its governance structure and membership statuses.[93] Unlike ACLA, the system would have three levels of membership: full members, associate members, and provisional members. Full members were state-aided libraries who signed a system agreement. These libraries were voting members. Associate members were non-state-aided libraries with no voting privileges. Provisional members were a temporary option without voting privileges. Like ACLA, the new entity would have a system board, which had decision making power for most issues, but this board would be reconstituted with different representation than the ACLA board. Also, unlike ACLA, the system board would be complemented by a Librarians Advisory Council (LAC), composed of library directors. ACLA HQ, together with the directors of member libraries, acted promptly to design more concrete plans for the federated system proposal.[94] ACLA encouraged discussion and participation in the process of drafting these documents. They were presented for consideration to the general membership at the general meeting in July 1996. There, members had a chance for input into the formation of the federated system.[95] ACLA leaders were clearly in favor of the federated system idea, and they continued to promote it, explaining how the system would be an improvement compared to the current situation.[96]

CLP followed discussions about the proposed federated system from a distance. CLP was concerned about its representation in the system and how much funding it would lose.[97] CLP also had to determine, for the purposes of district services, how to treat the different types of members. For the time being at least, the new system would not affect CLP's role as the state-designated district library center. But the new system would likely receive a portion of the county money that CLP received for the bookmobile and EIN.[98] State aid to CLP would also likely go down with the formation of a new system. Though the funding to CLP for district services would not change, other amounts would.[99] CLP would need to reduce any duplication of services between the new system and district services.

Contrary to claims by ACLA HQ, there was not unanimous support for a federated system among ACLA members. In fact, from the time the federated system was proposed in the consultancy report in early 1996, there was resistance.[100] Criticisms of the proposed system related to the funding tiers and the implementation of system standards.[101] Members were uncomfortable with ACLA HQ's aggressive timeline for implementation, and some members questioned the need for a system at all.[102] Members questioned whether ACLA in its present form could not achieve what the proposed federated system would.[103] There were also concerns about governance, representation, decision-making power, and funding within the new system.[104] It was feared that libraries would lose their influence within the new organization. [105] One of the problems in the proposed member agreement for the system was that the rights and responsibilities of the general membership, the LAC, and the system board were not clearly demarcated.[106]

One prominent detractor of the proposed system was the anonymous author or authors of the "Library Militia of Allegheny County," or LMAC, who faxed counter-information about ACLA, RAD, and EIN to county officials, municipal officials, RAD officials, library directors, librarians, reporters, and others.[107] At first, the publication was issued under the titles *Library Militia News* and *LMAC Reporter.* Early issues had limited circulation to a restricted group. Then, in April 1996, *Library Militia News* began circulating to a wider readership. The circular was designed to disrupt ACLA HQ's push for a federated system.[108] Recipients on LMAC's expanded mailing list included representatives from RAD, CFLAC, the Economy League, and the county commissioners. In October 1996, the title of the publication changed to *Library Reporter.*[109] *Library Reporter* and its predecessors were guerilla publications that broadcast critical views of ACLA, its leadership, the distribution formula, and the proposed federated system. At least 56 issues of *Library Reporter* were published between October 1996 and June 1997. These were in addition to informational booklets and reports.[110]

At times the criticism toward the proposed system seemed overwhelming. ACLA struggled to respond to claims made by the *Library Reporter,* which continued to spread counter-information about ACLA's plans. By April 1996, *Library Reporter* had been in circulation for over a year, and its thirty-seventh issue was published on April 2, 1997. During the first year of circulation of *Library Reporter,* ACLA did not respond directly to its claims. This changed in March 1997 when the claims in *Library Reporter* seemed to gain greater traction, causing public officials and others to contact county library leaders.[111] As a response to the *Library Reporter,* ACLA created the *ACLA Report* to bolster ACLA's credibility and to clear up the perceived confusion created by the *Library Reporter.*[112]

Despite the opposition, the ACLA leadership continued to push forward with federated system plans by tamping down criticisms, developing and revising necessary documents, and forging internal alliances.[113] Facing resistance, ACLA HQ sought approval for the plan from a majority of ACLA members.[114] Though ACLA leadership publicly stated that they hoped to investigate the possibility of forming a federated system, in fact, the decision to form a federated system had already been made by them, and they strove to persuade the membership why it was beneficial. ACLA HQ attempted to build support for the plan through 1996 with the hope to implement the new system in January 1997.[115] Over the course of its preparations, ACLA HQ decided that ACLA itself restructure to become the federated system rather than form some new and separate organization.[116] In order to build additional membership support, ACLA HQ decelerated the restructure process with the aim to implement the new system in 1998.[117] Discussions were held to address the concerns that were raised within the membership.[118] A "system benefit brochure," meant as an informational brochure, was distributed by ACLA leaders to its members. It was met with some resistance because it did not represent membership views.[119] Yet, ACLA leaders remained focused on the federated system goal through 1997.[120]

The summer of 1997 marked a turning point for ACLA. The discussions about the federated system had occurred for over a year, following discussions from the consultants' report in early 1996.[121] Though there was a fair amount of resistance to the federated system idea, by June 1997, ACLA had already received letters of intent to join the federated system from 19 libraries.[122] Budget applications for 1998 RAD funding were due by July 1997.[123] It was in June 1997 that Gilbert Gaul published an exposé on the poor conditions of public libraries in Pennsylvania.[124] This article was circulated widely, and it perhaps served to further galvanize libraries around the cause of increased state recognition through federation. A vote on the formation of the federated system was held on June 17, 1997.[125] The vote was decided by a simple majority. Detractors of the plan orchestrated a final blitz, but the vote ended in a 27–13 decision to form a federated system. [126] ACLA's process, and the resulting vote, were supported by RAD.[127] Following the vote, ACLA contacted the state library to become formally recognized as a federated library system. [128] ACLA began system operations in January 1998. Four major changes occurred within ACLA as a result of system formation: eligibility for state aid to ACLA HQ, board structure and composition, tiered membership, and the establishment of a Librarians Advisory Council (LAC).[129] ACLA informed member libraries of the changes, reminded them of due dates, assuaged outstanding fears and uncertainties, and noted outstanding issues in the new bylaws. These issues included voting procedures for the membership, board selection procedures, and clarification about which bylaws the board could amend without the vote of the membership.[130] Further bylaws amendments would be needed as ACLA transitioned into the new system.[131] In September 1997, ACLA received notification from the state library that it qualified as a federated system and that its bylaws had been approved.[132] ACLA was incorporated and recognized by the Pennsylvania Department of State in December 1997.[133]

Following the vote, ACLA continued to receive more letters of concern from municipalities regarding the new bylaws.[134] Opposition also continued to play out in the newspapers and online listservs.[135] RAD supported ACLA's transition to a federated system, but it urged ACLA HQ to address any outstanding concerns raised within the membership.[136] As of September 1997, ACLA libraries eligible for full status began to sign member library agreements.[137] ACLA clarified that CLP would continue to hold a separate contract with the RAD board, it would continue to serve as the district library, and it would continue to serve as the regional resource center.[138] ACLA leaders promoted their accomplishment, pointing out that a federated public library system had been proposed numerous times throughout the twentieth century.[139]

Period 7 recounts how ACLA became part of the inner periphery of the state circuit. This was achieved by becoming recognized as a federated public library system in the eyes of the state. ACLA leadership successfully overcame high resistance within ACLA membership by promoting the federation issue through communicative power, thus increasing legitimacy within the tessellation. Previously, independent local public libraries that were ACLA members were part of the state circuit because they received state aid in the form of the public library subsidy. They therefore had obligations to carry out state policies, such as maintaining state standards. CLP, as the district library, administered state aid and the state responsibilities of those independent libraries that received it. But ACLA as an organization was not yet part of the state circuit until 1998 when it became recognized by the state as a federated public library system. Now, ACLA was nested within the state and RAD circuits, and it operated as a circuit unto itself as well, with its own decision-making core, inner periphery, and outer periphery. But the complicated internal decision-making structure of ACLA membership and ACLA HQ had yet to be fully worked out. It was not clear who had the authority to make final decisions for ACLA, whether it was the membership, the ACLA board, or the RAD board. The LAC, which was composed of library directors, had only an advisory role. This contestation over the circuitry of the public library infrastructure would remain the decisive issue for future decision cycles.

1 William R. Dodge, Letter from Bill Dodge to DeeAnna Cavinee: Discussions About ARAD Allocations, 17 October 1994, ACLA historical file (Folder: ARAD Newspaper Coverage), ACLA-OFFICE; Geof Becker, "Allegheny County to Cut 5 Mills," *North Hills News Record*, 15 December 1994.
2 ACLA, ACLA Member Libraries, 17 October 1994, 92, ACLA historical file (Folder: Black Binder), ACLA-OFFICE.
3 Carolyn A. Toth, Letter from Carolyn Toth to the Pittsburgh Post-Gazette: On Supplemental Funding, 4 August 1995, ACLA historical file (Folder: ARAD Newspaper Coverage), ACLA-OFFICE; Carolyn A. Toth, "Libraries and RAD," *Pittsburgh Post-Gazette*, 13 August 1995.
4 CFLAC, Meeting Minutes: Resignation of DeeAnna, EIN Funds, ACLA Funds, 19 October 1994, ACLA historical file (Folder: CFLAC-Minutes), ACLA-OFFICE.
5 Jane-Ellen Robinet, "Winning Asset District Funding Could Be Revenue Drag," *Pittsburgh Business Times*, 17–23 October 1994.
6 James Turner, Letter from Jim Turner to Joyce Baskins, 12 December 1994, ACLA historical file (Folder: Black Binder), ACLA-OFFICE.
7 CFLAC, The Knowledge Connection Program, June 1995, ACLA historical file (Folder: 1995 Knowledge Connections), ACLA-OFFICE; Marilyn A. Jenkins, CFLAC Regular Report: County Library Updates, ACLA, Bookmobile, Knowledge Connections, 6 June 1995, ACLA historical file (Folder: CFLAC-Minutes), ACLA-OFFICE; CFLAC, Application for Allegheny County Library Association ARAD Grants, 15 August 1995, ACLA historical file (Folder: KC's-Awards), ACLA-OFFICE; ACLA, Check to Commission for the Future of Libraries, 27 September 1995, ACLA historical file (Folder: 1995 Knowledge Connections), ACLA-OFFICE; Marilyn A. Jenkins, CFLAC Regular Report: Northland Proposal, ACLA Advised by ARAD to Hire Executive Director, EIN Developments, 12 September 1995, ACLA historical file (Folder: CFLAC-Minutes), ACLA-OFFICE.
8 CFLAC, Meeting Minutes: Knowledge Connections, Seniors, Consideration of ACLA Administrator, 12 September 1995, ACLA historical file (Folder: CFLAC-Minutes), ACLA-OFFICE; Gary D. Wolfe, Letter from Gary Wolfe to Frank Lucchino: Appointment to Commissioner for Commonwealth Libraries and Deputy Secretary of Education, 30 August 1995, ACLA historical file (Folder: CFLAC-Minutes), ACLA-OFFICE; CFLAC, Meeting Minutes, 25 October 1995, ACLA historical file (Folder: CFLAC-Minutes-1996), ACLA-OFFICE; William B. Parent, Letter from William Parent to Frank Lucchino: 1996 Innovations in American Government Award, 15 November 1995, ACLA historical file (Folder: KC's-Awards), ACLA-OFFICE; Marilyn A. Jenkins, Letter from Marilyn Jenkins to Jeanne Jvostal: ALCOA School Readiness Grant, 21 November 1995, ACLA historical file (Folder: KC's-Awards), ACLA-OFFICE.
9 Marilyn A. Jenkins, CFLAC, Regular Report: Federated System, Knowledge Connections, Bookmobile, EIN, 8 May 1996, ACLA historical file (Folder: CFLAC-Minutes-1996), ACLA-OFFICE.
10 DeeAnna R. Cavinee, Allegheny County Library Director, Progress Report, CFLAC: EIN Board Appointees, 30 November 1994, ACLA historical file (Folder: CFLAC-Minutes), ACLA-OFFICE; Carnegie Library of Pittsburgh, 1995 Annual Report, 1995, CLP-PENN; CFLAC, Minutes for March 20 Meeting, 20 March 1996, ACLA historical file (Folder: CFLAC-Minutes-1996), ACLA-OFFICE; Marilyn A. Jenkins, Letter from Marilyn Jenkins to Guy Tumolo: County Funds for CLP and EIN, 27 March 1995, ACLA historical file (Folder: eiNetwork Agreements), ACLA-OFFICE.
11 EIN, Funding Report, 26 July 1995, ACLA historical file (Folder: CFLAC-Minutes), ACLA-OFFICE; CFLAC, Meeting Minutes: Contract with County and CLP for EIN, 18 April 1995, ACLA historical file (Folder: CFLAC-Minutes), ACLA-OFFICE; CFLAC, Minutes for March 20 Meeting, 20 March 1996.
12 Marilyn A. Jenkins, Regular Report: CFLAC, 11 September 1996, ACLA historical file (Folder: CFLAC-Minutes-1996), ACLA-OFFICE.

13 Cavinee, Allegheny County Library Director, Progress Report, CFLAC: EIN Board Appointees, 30 November 1994; CFLAC, Meeting Minutes: Knowledge Connections, ACLA Bookkeeper, EIN, 21 February 1995, ACLA historical file (Folder: CFLAC-Minutes), ACLA-OFFICE; Marilyn A. Jenkins, CFLAC Update: RAD, ACLA, EIN, 18 April 1995, ACLA historical file (Folder: CFLAC-Minutes), ACLA-OFFICE; CFLAC, Meeting Minutes: Knowledge Connections, ACLA Bylaw Revisions, EIN Proposals, EIN Project Manager, 6 June 1995, ACLA historical file (Folder: CFLAC-Minutes), ACLA-OFFICE; EIN, EIN: Governance Principles, 17 April 1995, ACLA historical file (Folder: eiNetwork Agreements), ACLA-OFFICE; Judith L. Volkmar, ACLA Board Meeting Minutes, 25 March 1996, Carnegie Library of Pittsburgh, https://web.archive.org/web/*/www.clpgh.org/acla; CFLAC, Meeting Minutes, 30 October 1996, ACLA historical file (Folder: CFLAC-Minutes-1996), ACLA-OFFICE; CFLAC, Meeting Minutes, 25 June 1996, ACLA historical file (Folder: CFLAC-Minutes-1996), ACLA-OFFICE; Judith L. Volkmar, ACLA Board Meeting Minutes, 27 January 1997, Carnegie Library of Pittsburgh, https://web.archive.org/web/*/www.clpgh.org/acla.

14 Marilyn A. Jenkins, CFLAC Regular Report: County Libraries, ACLA Consultants, Knowledge Connections, Bookmobiles, EIN, CCAC Connectivity, 25 October 1995, ACLA historical file (Folder: CFLAC-Minutes), ACLA-OFFICE; Jenkins, CFLAC, Regular Report: Federated System, Knowledge Connections, Bookmobile, EIN, 8 May 1996.

15 Carnegie Library of Pittsburgh, County EIN Audits, 18 November 1997, ACLA historical file (Folder: eiNetwork Audit), ACLA-OFFICE; Lawrence Walsh, "Computer Equipment Is Missing from Library," *Pittsburgh Post-Gazette*, 24 December 1997.

16 Carnegie Library of Pittsburgh, 1996 Annual Report, 1996, CLP-PENN.

17 Karen Commings, "Libraries of the Future … Carnegie Library of Pittsburgh Expands on-Ramps to Its Information Highway," Article, *Computers in Libraries* 15, no. 1 (1995); EIN, Electronic Information Network for Libraries in Allegheny County: Implementation Progress Report Issued July 15, 1995, 15 July 1995, ACLA historical file (Folder: CFLAC-Minutes), ACLA-OFFICE; Jenkins, CFLAC Regular Report: County Library Updates, ACLA, Bookmobile, Knowledge Connections, 6 June 1995; CFLAC, Update Report: Knowledge Connections, EIN, Seniors, ACLA, 4 December 1995, ACLA historical file (Folder: CFLAC-Minutes), ACLA-OFFICE; Jenkins, CFLAC Regular Report: County Libraries, ACLA Consultants, Knowledge Connections, Bookmobiles, EIN, CCAC Connectivity, 25 October 1995; CFLAC, Meeting Minutes, 25 October 1995; EIN, Electronic Information Network for Libraries in Allegheny County: Implementation Progress Report Issued 22 February 1996, 22 February 1996, ACLA historical file (Folder: CFLAC-Minutes-1996), ACLA-OFFICE; Judith L. Volkmar, ACLA Board Meeting Minutes, 15 July 1996, Carnegie Library of Pittsburgh, https://web.archive.org/web/*/www.clpgh.org/acla; CFLAC, Meeting Minutes, 25 June 1996.

18 EIN, Electronic Information Network for Public Libraries in Allegheny County: Six Month Progress Report, 31 July 1995, ACLA historical file (Folder: CFLAC-Minutes), ACLA-OFFICE; EIN, Electronic Information Network for Libraries in Allegheny County: Implementation Progress Report Issued July 15, 1995, 15 July 1995; CFLAC, Minutes for March 20 Meeting, 20 March 1996; Marilyn A. Jenkins, Regular CFLAC Report: Crafton, LA West, Northern Tier, Federation, Knowledge Connections, EIN, New Contract for Marilyn, 25 June 1996, ACLA historical file (Folder: CFLAC-Minutes-1996), ACLA-OFFICE; Volkmar, ACLA Board Meeting Minutes, 25 March 1996; CFLAC, Meeting Minutes, 25 October 1995; CFLAC, Update Report: Knowledge Connections, EIN, Seniors, ACLA, 4 December 1995; EIN, Electronic Information Network for Libraries in Allegheny County: Implementation Progress Report Issued 22 February 1996, 22 February 1996.

19 CFLAC, Agenda and Meeting Minutes: Special Telecommunications Rates for Linked Libraries, 19 October 1994, ACLA historical file (Folder: CFLAC-Minutes), ACLA-OFFICE; Carnegie Library of Pittsburgh, 1995 Annual Report, 1995; John L. Day, Letter from John Day to Frank Lucchino:

CCAC-Public Library Connection, 22 November 1994, ACLA historical file (Folder: CFLAC-Minutes), ACLA-OFFICE; James M. Shaulis, Letter from Jim Shaulis to John Day: CCAC and Project Link-Up, 5 April 1995, ACLA historical file (Folder: CFLAC-Minutes), ACLA-OFFICE; Glenn R. Miller, "Two Year Saga Ends as LSTA Becomes Law," *PaLA Bulletin* (November 1996); Glenn R. Miller, "FCC Endorses Deep Discounts for Libraries," *PaLA Bulletin* (December 1996); Glenn R. Miller, "State PUC Adopts Telecommunications Discounts for Libraries, Schools," *PaLA Bulletin* (June/July 1997); Glenn R. Miller, "Telecomm Discount Applications Mailed to Libraries," *PaLA Bulletin* (January 1998); Barbara W. Cole, Discount Telecommunications for Libraries, 29 May 1997, Department of Education, Commonwealth Libraries, Minutes of the Governor's Advisory Council on Library Development 1962–2009 collection, Carton 2, Folder 5, PASA; CFLAC, EIN Project Organization and Management, May 1995, ACLA historical file (Folder: CFLAC-Minutes), ACLA-OFFICE; EIN, EIN Board Meeting Minutes, 16 September 1996, Carnegie Library of Pittsburgh, https://web.archive.org/web/*/http://www.clpgh.org/ein; CFLAC, Meeting Minutes, 25 June 1996; EIN, Electronic Information Network for Libraries in Allegheny County: Implementation Progress Report Issued 22 February 1996, 22 February 1996; EIN, EIN Board Meeting Minutes, 16 September 1996; EIN, EIN Board Meeting Minutes, 8 April 1996, Carnegie Library of Pittsburgh, https://web.archive.org/web/*/http://www.clpgh.org/ein; Judith L. Volkmar, ACLA Board Meeting Minutes, 28 May 1997, Carnegie Library of Pittsburgh, https://web.archive.org/web/*/www.clpgh.org/acla; EIN, EIN Board Meeting Minutes, 21 October 1996, Carnegie Library of Pittsburgh, https://web.archive.org/web/*/http://www.clpgh.org/ein; Sally J. Coyne, Letter from Sally Coyne to EIN Board of Directors, 22 February 1997, ACLA historical file (Folder: Historical Correspondence). ACLA-OFFICE; EIN, EIN Board Meeting Minutes, 12 December 1996, Carnegie Library of Pittsburgh, https://web.archive.org/web/*/http://www.clpgh.org/ein.

20 Marilyn A. Jenkins, Regular Report, Commission on the Future of Libraries in Allegheny County, 30 October 1996, ACLA historical file (Folder: CFLAC-Minutes-1996), ACLA-OFFICE; Volkmar, ACLA Board Meeting Minutes, 28 May 1997; Carnegie Library of Pittsburgh, 1996 Annual Report, 1996.

21 Volkmar, ACLA Board Meeting Minutes, 27 January 1997; Jenkins, Regular Report: CFLAC, 11 September 1996.

22 CFLAC, Meeting Minutes, 30 October 1996; Coyne, Letter from Sally Coyne to EIN Board of Directors, 22 February 1997; Judith L. Volkmar, ACLA General Membership Meeting Minutes, 29 January 1997, Carnegie Library of Pittsburgh, https://web.archive.org/web/*/www.clpgh.org/acla.

23 ACLA, Proposed Distribution of ARAD Funds, 10 November 1994, ACLA historical file (Folder: CFLAC-Minutes), ACLA-OFFICE; ACLA, ACLA ARAD Draft Distribution, 10 November 1994, ACLA historical file (Folder: Black Binder), ACLA-OFFICE; ACLA, Explanation of Formula Page, December 1994, ACLA historical file (Folder: Black Binder), ACLA-OFFICE.

24 Library Administrator 1, Interview Transcript, 28 July 2015, interviewed by Michael M. Widdersheim.

25 CFLAC, Meeting Minutes, 30 November 1994, ACLA historical file (Folder: CFLAC-Minutes). ACLA-OFFICE.

26 The Times, "Good Returns," *Beaver County Times*, 10 March 1995; Tawanda D. Williams, "Libraries Catalog Their Plans for Money," *Pittsburgh Post-Gazette*, 16 March 1995; CFLAC, Meeting Minutes: Contract with County and CLP for EIN, 18 April 1995.

27 Jo Ellen Kenney, Proposed Distribution of ARAD Funds for 1995, 28 February 1995, ACLA historical file (Folder: Formula 1995), ACLA-OFFICE.

28 ACLA, A Proposed Equalization Distribution Formula for Regional Asset District Funding for Participating Libraries, 6 December 1994, ACLA historical file (Folder: Black Binder), ACLA-OFFICE; Marilyn A. Jenkins, Letter from Marilyn Jenkins to ACLA Ad Hoc Distribution Committee, 7 December 1994, ACLA historical file (Folder: Black Binder), ACLA-OFFICE; Dave Miller, Letter from Dave Miller to Marilyn Jenkins, 6 December 1994, ACLA historical file (Folder: Black Binder), ACLA-OFFICE; Loretta O'Brien, Letter from Loretta O'brien to Marilyn Jenkins and Laura Shelley, 6 December 1994, ACLA historical file (Folder: Black Binder), ACLA-OFFICE.

29 Jo Ellen Kenney, Draft Distribution Formula, 10 November 1994, ACLA historical file (Folder: Formula 1995), ACLA-OFFICE; Jon Schmitz and Tom Barnes, "David Donahoe Named to Head Asset District," *Pittsburgh Post-Gazette*, 15 December 1994; Tom Barnes, "Ex-Revenue Secretary Now Has Job of Doling out Funds," *Pittsburgh Post-Gazette*, 16 December 1994; Turner, Letter from Jim Turner to Joyce Baskins, 12 December 1994; Sandra Fischione Donovan, "One for the Books," *Beaver County Times*, 25 February 1995; David L. Donahoe, Letter from David Donahoe to Carolyn Toth: Revising Minutes, 31 May 1996, ACLA historical file (Folder: ARAD Correspondence), ACLA-OFFICE.

30 ACLA, Allegheny County Library Association: Impact of 1995 ARAD Distribution, 1995, ACLA historical file (Folder: ACLA-ARAD Budget), ACLA-OFFICE.

31 ACLA, Allegheny County Library Association: Impact of 1995 ARAD Distribution, 1995.

32 ACLA, Allegheny County Library Association: Impact of 1995 ARAD Distribution, 1995.

33 ACLA, Allegheny County Library Association: Impact of 1995 ARAD Distribution, 1995; Carole Gilbert Brown, "Libraries Making Wish Lists," *Pittsburgh Post-Gazette*, 2 March 1995; ACLA, ACLibs 2(1), January 1997, ACLA historical file (Folder: ACLA History), ACLA-OFFICE.

34 ACLA, Cooperation and Support Agreement between the Allegheny County Library Association and Member Library, 1 January 2013, ACLA-WIKI.

35 RAD, Request for Comment, July 2013, ACLA-WIKI.

36 Librarian 1, Interview Transcript, 7 January 2016, interviewed by Michael M. Widdersheim.

37 ACLA, Proposed Distribution of ARAD Funds for 1995, December 1994, ACLA historical file (Folder: Black Binder), ACLA-OFFICE; CFLAC, Meeting Minutes: Knowledge Connections, ACLA Bookkeeper, EIN, 21 February 1995; Albert F. Kamper, Letter from Al Kamper to Laura Shelley: Update on Adjustments to Formula Sheet, 30 January 1995, ACLA historical file (Folder: ACLA-ARAD Budget), ACLA-OFFICE.

38 Jenkins, CFLAC Regular Report: Northland Proposal, ACLA Advised by ARAD to Hire Executive Director, EIN Developments, 12 September 1995; Judith L. Volkmar, ACLA Board Meeting Minutes, 28 May 1996, Carnegie Library of Pittsburgh, https://web.archive.org/web/*/www.clpgh.org/acla.

39 ACLA, acLIBS 1(2), June 1996, ACLA historical file (Folder: ACLA History), ACLA-OFFICE.

40 ACLA, Summary and Commentary, Long Term Plan, Allegheny County Library Association (ACLA), March 1996, ACLA historical file (Folder: ACLA History), ACLA-OFFICE.

41 Judith L. Volkmar, ACLA Board Meeting Minutes, 11 January 1996, Carnegie Library of Pittsburgh, https://web.archive.org/web/*/www.clpgh.org/acla; CFLAC, Meeting Minutes: Knowledge Connections, ACLA Bookkeeper, EIN, 21 February 1995.

42 CFLAC, Meeting Minutes: Contract with County and CLP for EIN, 18 April 1995.

43 Volkmar, ACLA Board Meeting Minutes, 11 January 1996.

44 CFLAC, Meeting Minutes, 25 October 1995; CFLAC, Meeting Minutes: Knowledge Connections, Seniors, Consideration of ACLA Administrator, 12 September 1995; Jenkins, CFLAC Regular Report: County Libraries, ACLA Consultants, Knowledge Connections, Bookmobiles, EIN, CCAC Connectivity, 25 October 1995.

45 Cavinee, Allegheny County Library Director, Progress Report, CFLAC: EIN Board Appointees, 30 November 1994; CFLAC, Meeting Agenda, 11 January 1995, ACLA historical file (Folder: CFLAC-Minutes), ACLA-OFFICE; CFLAC, Meeting Minutes: Knowledge Connections, ACLA Bookkeeper, EIN, 21 February 1995; Marilyn A. Jenkins, CFLAC Update: Sto-Rox Opens, Open Access Issues with Northland and Monroeville, 21 February 1995, ACLA historical file (Folder: CFLAC-Minutes), ACLA-OFFICE; Marilyn A. Jenkins, Letter from Marilyn Jenkins to Stephen Gifford: Library in Pine and Hampton, 1 August 1995, ACLA historical file (Folder: Historical Correspondence), ACLA-OFFICE; Frank J. Lucchino, Letter from Frank Lucchino to Robert Schlegel: Northern Tier Library, 16 August 1995, ACLA historical file (Folder: CFLAC-Minutes), ACLA-OFFICE; Frank J. Lucchino, Letter from Frank Lucchino to Herbert Dankmyer: Northern Tier Library, 16 August 1995, ACLA historical file (Folder: CFLAC-Minutes), ACLA-OFFICE; Carolyn A. Toth, Letter from Carolyn Toth to the Commission on the Future of Libraries in Allegheny County: Hiring Marilyn Jenkins, 6 May 1996, ACLA historical file (Folder: Historical Correspondence), ACLA-OFFICE; Frank J. Lucchino, Letter from Frank Lucchino to Carolyn Toth, 24 May 1996, ACLA historical file (Folder: Historical Correspondence), ACLA-OFFICE; Lucchino, Letter from Frank Lucchino to Carolyn Toth, 24 May 1996; Volkmar, ACLA Board Meeting Minutes, 25 March 1996; Judith L. Volkmar, ACLA Board Meeting Minutes, 24 April 1996, Carnegie Library of Pittsburgh, https://web.archive.org/web/*/www.clpgh.org/acla; ACLA, acLIBS 1(2), June 1996; Jenkins, Regular CFLAC Report: Crafton, LA West, Northern Tier, Federation, Knowledge Connections, EIN, New Contract for Marilyn, 25 June 1996.

46 RAD, Allegheny Regional Asset District Public Hearing Agenda, 7 November 1994, ACLA historical file (Folder: ACLA Proposal to ARAD Jo Ellen's Speech), ACLA-OFFICE; Jon Schmitz, "Asset District Adopts $53 Million '95 Budget," *Pittsburgh Post-Gazette*, 2 December 1994.

47 Carnegie Library of Pittsburgh, 1995 Annual Report, 1995.

48 Jenkins, CFLAC Update: Sto-Rox Opens, Open Access Issues with Northland and Monroeville, 21 February 1995; Jenkins, CFLAC Update: RAD, ACLA, EIN, 18 April 1995.

49 Albert F. Kamper, Letter from Al Kamper to Atu Faruquee: DLC Meeting with Kenney, Jenkins, and Croneberger, 12 May 1995, ACLA historical file (Folder: ACLA-ARAD Proposal), ACLA-OFFICE.

50 Emily Tipping, "RAD Board: Libraries Must Open Doors or Skip Funding," *Pittsburgh Tribune-Review*, 12 March 1995.

51 ACLA, Allegheny County Library Association: Impact of 1995 ARAD Distribution, 1995; Jenkins, CFLAC Update: Sto-Rox Opens, Open Access Issues with Northland and Monroeville, 21 February 1995.

52 Tawanda D. Williams, "Ending Nonresident Library Fees Fuels Debate," *Pittsburgh Post-Gazette*, 15 March 1995; William Loeffler, "RAD Money Not a Deal for Library," *North Hills News Record*, 6 March 1995; Laura Shelley, Letter from Laura Shelley to Jo Ellen Kenney: Northland Withdrawal, 25 July 1995, ACLA historical file (Folder: Historical Correspondence), ACLA-OFFICE; Rose Angotti, "Ross Township Library Questions RAD Mandates," *Pittsburgh Tribune-Review*, 26 March 1995.

53 Tipping, "RAD Board: Libraries Must Open Doors or Skip Funding."; Pittsburgh Post-Gazette, "RAD Waste: Why Northland Library Should Treat All Patrons Equally," *Pittsburgh Post-Gazette*, 9 September 1995; CFLAC, Meeting Minutes: Knowledge Connections, ACLA Bookkeeper, EIN, 21 February 1995; Loeffler, "RAD Money Not a Deal for Library."; Eleanor Chute, "Asset Tax Formula for Library Criticized," *Pittsburgh Post-Gazette*, 2 March 1995; Jo Ellen Kenney, Letter from Jo Ellen Kenney to ACLA Board: Bylaws and Regional Use Funds, 10 March 1995, ACLA historical file (Folder: ACLA History to Receive Funding), ACLA-OFFICE; Angotti, "Ross Township Library Questions RAD Mandates."

54 Jenkins, CFLAC Regular Report: County Library Updates, ACLA, Bookmobile, Knowledge Connections, 6 June 1995; CFLAC, Meeting Minutes: Knowledge Connections, ACLA Bylaw Revisions, EIN Proposals, EIN Project Manager, 6 June 1995; Albert F. Kamper, "Why Debate over Funds for Library?," *Pittsburgh Post-Gazette*, 5 April 1995.

55 EIN board, Policy: Electronic Information Network Membership Criteria, 31 July 1995, ACLA historical file (Folder: eiNetwork Agreements), ACLA-OFFICE; CFLAC, Meeting Minutes: Contract with County and CLP for EIN, 18 April 1995.

56 Jo Ellen Kenney, Letter from Jo Ellen Kenney to David Matter: Independent Northland Proposal, 31 July 1995, ACLA historical file (Folder: ARAD Newspaper Coverage), ACLA-OFFICE; Pittsburgh Post-Gazette, "RAD Waste: Why Northland Library Should Treat All Patrons Equally."; Joan B. Homyak, Letter from Joan Homyak to Jo Ellen Kenney: Northland Censure, 4 August 1995, ACLA historical file (Folder: Historical Correspondence), ACLA-OFFICE.

57 Chute, "Asset Tax Formula for Library Criticized."; Jo Ellen Kenney, Letter from Jo Ellen Kenney to ACLA Members: ARAD Approves Revised Formula, 7 March 1995, ACLA historical file (Folder: ACLA History to Receive Funding), ACLA-OFFICE.

58 Tipping, "RAD Board: Libraries Must Open Doors or Skip Funding."

59 CFLAC, Meeting Minutes: Knowledge Connections, ACLA Bylaw Revisions, EIN Proposals, EIN Project Manager, 6 June 1995; Jenkins, CFLAC Regular Report: Northland Proposal, ACLA Advised by ARAD to Hire Executive Director, EIN Developments, 12 September 1995.

60 David L. Donahoe, Letter from David Donahoe to Jo Ellen Kenney: RAD Budget for 1996, 5 December 1995, ACLA historical file (Folder: ACLA-ARAD Budget), ACLA-OFFICE; Jo Ellen Kenney, Letter from Jo Ellen Kenney to Joan Homyak: Northland Rejection, 19 December 1995, ACLA historical file (Folder: Historical Correspondence), ACLA-OFFICE; Marilyn A. Jenkins, CFLAC Regular Report: Northern Tier, South Hills, Sto-Rox, Northland, ACLA, Knowledge Connections, Bookmobile, EIN, 20 March 1996, ACLA historical file (Folder: CFLAC-Minutes-1996), ACLA-OFFICE; CFLAC, Meeting Minutes, 30 October 1996; EIN, EIN Board Meeting Minutes, 16 September 1996.

61 Carolyn A. Toth, Letter from Carolyn Toth to David Donahoe: 1997 Budget Request, 1996, July 18, ACLA historical file (Folder: ARAD Correspondence), ACLA-OFFICE.

62 Volkmar, ACLA Board Meeting Minutes, 15 July 1996.

63 ACLA, Votes on Motions of 7/31/96, 31 July 1996, ACLA historical file (Folder: General Information). ACLA-OFFICE; Christine M. Keith, Letter from Christine Keith to Carolyn Toth: Revised ACLA Budget, 15 August 1996, ACLA historical file (Folder: ACLA-ARAD Budget). ACLA-OFFICE; Keith, Letter from Christine Keith to Carolyn Toth: Revised ACLA Budget, 15 August 1996; Carolyn A. Toth et al., Glenn Walsh Concerns, August to October 1996, ACLA historical file (Folder: Historical Correspondence). ACLA-OFFICE; ACLA, ACLibs 1(3), September 1996, ACLA historical file (Folder: ACLA History), ACLA-OFFICE; B. J. Kochan, ACLA General Membership Meeting Minutes, 13 November 1996, Carnegie Library of Pittsburgh, https://web.archive.org/web/*/www.clpgh.org/acla.

64 Judith L. Volkmar, ACLA Board Meeting Minutes, 26 August 1996, Carnegie Library of Pittsburgh, https://web.archive.org/web/*/www.clpgh.org/acla; Carolyn A. Toth, Letter from Carolyn Toth to Christine Keith: Formula Revisions, 2 September 1996, ACLA historical file (Folder: ACLA-ARAD Budget), ACLA-OFFICE; CFLAC, Meeting Minutes, 30 October 1996; Allegheny Regional Asset District, Report, 1997 Library Committee of the Board, 4 November 1996, RAD files, Contractual Assets–Folders, Library Committee ACLA 1997–2006, ARAD-OFFICE.

65 Marilyn A. Jenkins, Letter from Marilyn Jenkins to David Donahoe: Northland, Formula, Federation, 6 September 1996, ACLA historical file (Folder: ARAD Correspondence), ACLA-OFFICE.

66 Jenkins, Regular Report: CFLAC, 11 September 1996; Judith L. Volkmar, ACLA Board Meeting Minutes, 23 September 1996, Carnegie Library of Pittsburgh, https://web.archive.org/web/*/www.clpgh.org/acla; Judith L. Volkmar, ACLA Board Meeting Minutes, 25 November 1996, Carnegie Library of Pittsburgh, https://web.archive.org/web/*/www.clpgh.org/acla; Marilyn A. Jenkins, Letter from Marilyn Jenkins to Mark Whittaker: North Hills Article, 18 November 1996, ACLA historical file (Folder: Historical Correspondence), ACLA-OFFICE.

67 Marilyn A. Jenkins, Letter from Marilyn Jenkins to David Donahoe: Northland Funds, 2 January 1997, ACLA historical file (Folder: ARAD Correspondence), ACLA-OFFICE; Joan B. Homyak, "About Northland Library," *Pittsburgh Post-Gazette*, 27 April 1997; Northland Public Library, The Impact of Free Lending to Allegheny County, 1996, Obtained from a library director.
68 Jenkins, Letter from Marilyn Jenkins to David Donahoe: Northland Funds, 2 January 1997; Albert F. Kamper, Letter from Al Kamper to David Donahoe: Commonwealth Libraries and Countywide Services, 17 December 1996, ACLA historical file (Folder: ARAD Correspondence), ACLA-OFFICE.
69 Carolyn A. Toth and Marilyn A. Jenkins, Letter from Carolyn Toth and Marilyn Jenkins to ARAD Formula Subcommittee, 1 October 1996, ACLA historical file (Folder: Historical Correspondence), ACLA-OFFICE; Kochan, ACLA General Membership Meeting Minutes, 13 November 1996; Volkmar, ACLA Board Meeting Minutes, 25 November 1996.
70 David L. Donahoe, Letter from David Donahoe to Kathleen Kennedy: ARAD Cooperation Amendment, 21 January 1997, ACLA historical file (Folder: ARAD Correspondence), ACLA-OFFICE; Volkmar, ACLA Board Meeting Minutes, 27 January 1997; Volkmar, ACLA Board Meeting Minutes, 25 November 1996; Marilyn A. Jenkins, Letter from Marilyn Jenkins to David Donahoe: ARAD Amendment Proposal, 6 February 1997, ACLA historical file (Folder: ARAD Correspondence), ACLA-OFFICE; David L. Donahoe, Letter from David Donahoe to Marilyn Jenkins: ACLA Capital Project Requests, 9 April 1997, ACLA historical file (Folder: ARAD Correspondence), ACLA-OFFICE.
71 Carolyn Marks, Letter from Carolyn Marks to Marilyn Jenkins: ARAD Formula Committee, 21 February 1997, ACLA historical file (Folder: ARAD Correspondence), ACLA-OFFICE; Carolyn Marks, Letter from Carolyn Marks to Marilyn Jenkins: ARAD Library Sub-Committee, 2 April 1997, ACLA historical file (Folder: ARAD Correspondence), ACLA-OFFICE; Allegheny Regional Asset District, Comment Summary Report, Library Committee Meeting with Local Municipal Officials, 22 April 1997, RAD files, Contractual Assets–Folders, Library Committee ACLA 1997–2006, ARAD-OFFICE; Carolyn A. Toth, Letter from Carolyn Toth to Gerald Voros, 1 May 1997, ACLA historical file (Folder: ARAD Correspondence), ACLA-OFFICE.
72 ARAD library formula committee, Library Committee Report, 1 October 1997, ACLA historical file (Folder: ARAD Correspondence), ACLA-OFFICE.
73 Allegheny Regional Asset District, Report, 1998 Library Formula Committee of the Board, 1 October 1997, RAD files, Contractual Assets–Folders, Library Committee ACLA 1997–2006, ARAD-OFFICE.
74 Joan B. Homyak, Letter from Marilyn Jenkins to Joan Homyak: Northland Decision, 21 July 1997, ACLA historical file (Folder: Historical Correspondence), ACLA-OFFICE; Joan B. Homyak, Letter from Joan Homyak to Marilyn Jenkins: Deadline Extension, 21 August 1997, ACLA historical file (Folder: Historical Correspondence), ACLA-OFFICE.
75 Marilyn A. Jenkins, Letter from Marilyn Jenkins to Gerald Voros: Recommendations for ACLA, 8 September 1997, ACLA historical file (Folder: ARAD Correspondence), ACLA-OFFICE.
76 Laura Shelley, Letter from Laura Shelley to Marilyn Jenkins, 22 September 1997, ACLA historical file (Folder: Letters of Intent), ACLA-OFFICE.
77 Pittsburgh Post-Gazette, "Allegheny Regional Asset District Allocations," *Pittsburgh Post-Gazette* 1997, October 2.
78 Marilyn A. Jenkins, Letter from Marilyn Jenkins to David Matter: Northland Request, 1997, September 18, ACLA historical file (Folder: ARAD Correspondence), ACLA-OFFICE.
79 ARAD library formula committee, Library Committee Report, 1 October 1997.
80 ARAD library formula committee, Library Committee Report, 1 October 1997.
81 Marilyn A. Jenkins, Letter from Marilyn Jenkins to David Matter: ACLA Adjustments, 6 November 1997, ACLA historical file (Folder: ARAD Correspondence), ACLA-OFFICE; William J. Neely, Letter

from William Neely to ARAD Board: Northland Special Grant, 20 October 1997, ACLA historical file (Folder: ARAD Correspondence), ACLA-OFFICE; Ja Stoddart, Letter from Ja Stoddart to David Matter: Northland Objection, 29 October 1997, ACLA historical file (Folder: ARAD Correspondence), ACLA-OFFICE.

82 Laura Shelley, Letter from Laura Shelley to Marilyn Jenkins: Northland Position, 8 October 1997, ACLA historical file (Folder: Historical Correspondence), ACLA-OFFICE.

83 Northland Public Library, *Northland Public Library Strategic Plan, 2016–2020* (26 January 2016), 23.

84 CFLAC, Meeting Minutes: Knowledge Connections, Seniors, Consideration of ACLA Administrator, 12 September 1995.

85 Jenkins, CFLAC Update: RAD, ACLA, EIN, 18 April 1995; Library Planning Associates, Proposal for Services, 29 May 1995, ACLA historical file (Folder: Background), ACLA-OFFICE; Volkmar, ACLA Board Meeting Minutes, 11 January 1996; Anders C. Dahlgren, Debra Wilcox Johnson, and Robert Bocher, *A Plan for Library Service in Allegheny County,* Library Planning Associates (Streamwood, 1 May 1996); CFLAC, Update Report: Knowledge Connections, EIN, Seniors, ACLA, 4 December 1995; Jenkins, CFLAC Regular Report: County Libraries, ACLA Consultants, Knowledge Connections, Bookmobiles, EIN, CCAC Connectivity, 25 October 1995; David L. Donahoe, Letter from David Donahoe to Carolyn Toth: Extension for Long Term Plan, 19 January 1996, ACLA historical file (Folder: ARAD Correspondence), ACLA-OFFICE; Library Planning Associates, *A Plan for Library Service in Allegheny County,* Library Planning Associates (Streamwood, March 1996); Judith L. Volkmar, ACLA Board Meeting Minutes, 26 February 1996, Carnegie Library of Pittsburgh, https://web.archive.org/web/*/www.clpgh.org/acla; Jenkins, CFLAC Regular Report: Northern Tier, South Hills, Sto-Rox, Northland, ACLA, Knowledge Connections, Bookmobile, EIN, 20 March 1996; Dahlgren, Johnson, and Bocher, *A Plan for Library Service in Allegheny County* (1 May 1996).

86 Volkmar, ACLA Board Meeting Minutes, 28 May 1996; ACLA, acLIBS 1(2), June 1996; ACLA, Allegheny County Library Association: Mission and Purpose, April 1996, ACLA historical file (Folder: ACLA History), ACLA-OFFICE; David L. Donahoe, Letter from David Donahoe to Carolyn Toth: Approval of Long Term Plan, 12 August 1996, ACLA historical file (Folder: ARAD Correspondence), ACLA-OFFICE.

87 Dahlgren, Johnson, and Bocher, *A Plan for Library Service in Allegheny County* (1 May 1996); Library Planning Associates, *A Plan for Library Service in Allegheny County* (March 1996).

88 Library Planning Associates, *A Plan for Library Service in Allegheny County* (March 1996); Commonwealth Libraries, Public Library Policies, 8 April 1994, ACLA historical file (Folder: Black Binder), ACLA-OFFICE.

89 ACLA, acLIBS 1(2), June 1996.

90 Library Planning Associates, *A Plan for Library Service in Allegheny County* (March 1996); Dahlgren, Johnson, and Bocher, *A Plan for Library Service in Allegheny County* (1 May 1996), 92.

91 Dahlgren, Johnson, and Bocher, *A Plan for Library Service in Allegheny County* (1 May 1996), 92.

92 Volkmar, ACLA Board Meeting Minutes, 25 March 1996.

93 Dahlgren, Johnson, and Bocher, *A Plan for Library Service in Allegheny County* (1 May 1996), 94; ACLA, 1996 Board of Directors and Region Reps, January 1996, ACLA historical file (Folder: CFLAC-Minutes-1996), ACLA-OFFICE; Volkmar, ACLA Board Meeting Minutes, 24 April 1996; Library Planning Associates, *A Plan for Library Service in Allegheny County* (March 1996); Marilyn A. Jenkins, Letter from Marilyn Jenkins to ACLA Member Library Directors and Boards, 11 September 1996, 1–2, ACLA historical file (Folder: General Information), ACLA-OFFICE; ACLA, Bylaws Outline, September 1996, ACLA historical file (Folder: Background), ACLA-OFFICE.

94 ACLA, Summary and Commentary, Long Term Plan, Allegheny County Library Association (ACLA), March 1996; ACLA, What Is a "Federated Library System"?, 22 April 1996, ACLA-OFFICE; Albert F. Kamper, PA Federated Public Library Systems: Results of System Survey, 22 April 1996, ACLA historical file (Folder: Background), ACLA-OFFICE; ACLA, acLIBS 1(2), June 1996; CFLAC, Notes from the ACLA Directors' Discussion Re. Federated System Development, 22 April 1996, ACLA historical file (Folder: CFLAC-Minutes-1996), ACLA-OFFICE; Volkmar, ACLA Board Meeting Minutes, 24 April 1996; CFLAC, Meeting Minutes, 30 October 1996; Jenkins, CFLAC, Regular Report: Federated System, Knowledge Connections, Bookmobile, EIN, 8 May 1996; ACLA, Key Questions Regarding Member Library Agreement for Regional Discussion, April 2013, ACLA-WIKI.
95 ACLA, acLIBS 1(2), June 1996.
96 ACLA, acLIBS 1(2), June 1996, 3.
97 Albert F. Kamper, Letter from Al Kamper to Members, Director's Administrative Cabinet, 1997, ACLA historical file (Folder: Background), ACLA-OFFICE.
98 Kamper, Letter from Al Kamper to Members, Director's Administrative Cabinet, 1997.
99 Kamper, Letter from Al Kamper to Members, Director's Administrative Cabinet, 1997.
100 Volkmar, ACLA Board Meeting Minutes, 24 April 1996.
101 Kim Timlin Leonard, "Plan Suggests Allegheny Libraries Form Central System," *Pittsburgh Tribune Review*, 23 June 1996.
102 Volkmar, ACLA Board Meeting Minutes, 26 August 1996; ACLA, Attendance at System Discussion, Green Tree Library, 12 August 1996, ACLA historical file (Folder: General Information), ACLA-OFFICE.
103 Nick S. Fisfis, Letter from Nick Fisfis to Allan Brodsky, 6 March 1997, ACLA historical file (Folder: Correspondence), ACLA-OFFICE; Glenn A. Walsh, Letter from Glenn Walsh to Allegheny County Library Board Presidents, 6 February 1997, ACLA historical file (Folder: Member Library Agreement), ACLA-OFFICE.
104 Fisfis, Letter from Nick Fisfis to Allan Brodsky, 6 March 1997; Judith L. Volkmar, ACLA Special General Membership Meeting Minutes, 18 March 1997, Carnegie Library of Pittsburgh, https://web.archive.org/web/*/www.clpgh.org/acla; Maureen Sheridan, Letter from Maureen Sheridan to ACLA Member Libraries, 8 April 1997, ACLA historical file (Folder: General Information), ACLA-OFFICE; Glenn A. Walsh, Letter from Glenn Walsh to ACLA Members, 9 April 1997, ACLA historical file (Folder: General Information), ACLA-OFFICE.
105 Douglas A. Watkins, Letter from Douglas Watkins to Allan Brodsky, 18 March 1997, ACLA historical file (Folder: General Information), ACLA-OFFICE.
106 ACLA system formation committee, Letter from System Formation Committee to All ACLA Members, 16 January 1997, ACLA historical file (Folder: Committee Meetings), ACLA-OFFICE.
107 CFLAC, Meeting Minutes, 30 October 1996.
108 LMAC, Issue #17, 11 April 1996, Personal copy.
109 Marilyn A. Jenkins, Letter from Marilyn Jenkins to David Donahoe, 7 October 1996, ACLA historical file (Folder: Historical Correspondence), ACLA-OFFICE; Robert B. Croneberger, Letter from Bob Croneberger to David Donohoe: Library Reporter Concerns, 17 October 1996, ACLA historical file (Folder: Historical Correspondence), ACLA-OFFICE; Roxane M. Ostrum, Letter from Roxane Ostrum to BJ Kochan: Avalon and Lou Sander, 24 July 1996, ACLA historical file (Folder: Historical Correspondence), ACLA-OFFICE.
110 LMAC, Useful Library Materials, 20 April 1996, Personal copy; LMAC, ARAD: Library Funding for Basic Services, 27 November 1995, Personal copy.
111 Judith L. Volkmar, ACLA Board Meeting Minutes, 24 March 1997, Carnegie Library of Pittsburgh, https://web.archive.org/web/*/www.clpgh.org/acla.

112 ACLA, The ACLA Report 1(1), 10 April 1997, ACLA historical file (Folder: ACLA History), ACLA-OFFICE; ACLA, The ACLA Report 1(2), 16 April 1997, ACLA historical file (Folder: ACLA History), ACLA-OFFICE; ACLA, The ACLA Report 1(3), 17 April 1997, ACLA historical file (Folder: ACLA History), ACLA-OFFICE; ACLA, The ACLA Report 1(4), 18 April 1997, ACLA historical file (Folder: ACLA History), ACLA-OFFICE; ACLA, The ACLA Report 1(5), 21 April 1997, ACLA historical file (Folder: ACLA History), ACLA-OFFICE; ACLA, The ACLA Report 1(6), 24 April 1997, ACLA historical file (Folder: ACLA History), ACLA-OFFICE; Marilyn A. Jenkins, Letter from Marilyn Jenkins to ACLA Member Libraries, April 1997, ACLA historical file (Folder: General Information), ACLA-OFFICE; Allan J. Brodsky, Letter from Allan Brodsky to Larry Dunn: Annual Retreat, 9 April 1997, ACLA historical file (Folder: Historical Correspondence), ACLA-OFFICE.
113 ACLA system formation committee, System Formation Committee Report, 1 April 1997, ACLA historical file (Folder: Committee Meetings), ACLA-OFFICE; Marks, Letter from Carolyn Marks to Marilyn Jenkins: ARAD Library Sub-Committee, 2 April 1997; Volkmar, ACLA Board Meeting Minutes, 24 March 1997; Marilyn A. Jenkins, Letter from Marilyn Jenkins to ACLA Members, 9 April 1997, ACLA historical file (Folder: General Information), ACLA-OFFICE; ACLA, The ACLA Report 1(4), 18 April 1997; Kim Timlin Leonard, "Libraries Vote on Power Structure," *Pittsburgh Tribune Review*, 27 April 1997; Charles Brown, "Turning a Page for Funding," *Valley News Dispatch*, 20 May 1997; Leonard, "Libraries Vote on Power Structure."; W. David Montz, Letter from W. David Montz to Marilyn Jenkins, 6 May 1997, ACLA historical file (Folder: Historical Correspondence), ACLA-OFFICE; Judith L. Volkmar, ACLA Board Meeting Minutes, 28 April 1997, Carnegie Library of Pittsburgh, https://web.archive.org/web/*/www.clpgh.org/acla; Marcia Taylor, Letter from Marcia Taylor to Wilmer K. Baldwin, 27 May 1997, ACLA historical file (Folder: Bylaws). ACLA-OFFICE; Upper St. Clair Northland, and Baldwin,, Letter from Northland, Upper St. Clair, and Baldwin to ACLA Board, 4 June 1997, ACLA historical file (Folder: Bylaws). ACLA-OFFICE; BJ Mullaney, Letter from BJ Mullaney to Marilyn Jenkins, 5 June 1997, ACLA historical file (Folder: Bylaws). ACLA-OFFICE.
114 Leonard, "Plan Suggests Allegheny Libraries Form Central System."
115 CFLAC, Meeting Minutes, 25 June 1996; Jenkins, Regular CFLAC Report: Crafton, LA West, Northern Tier, Federation, Knowledge Connections, EIN, New Contract for Marilyn, 25 June 1996; Volkmar, ACLA Board Meeting Minutes, 15 July 1996.
116 Kamper, Letter from Al Kamper to Members, Director's Administrative Cabinet, 1997; Garland H. McAdoo, Memorandum to Executive Committee: ACLA and Federated Library System, 26 August 1996, ACLA historical file (Folder: Documents), ACLA-OFFICE; ACLA, ACLibs 1(3), September 1996.
117 ACLA, ACLibs 1(3), September 1996; Jenkins, Letter from Marilyn Jenkins to ACLA Member Library Directors and Boards, 11 September 1996; ACLA libraries, Letters of Intent, September to May 1997, ACLA historical file (Folder: System Letters of Intent), ACLA-OFFICE; Judith L. Volkmar, ACLA General Membership Meeting Minutes, 29 October 1996, Carnegie Library of Pittsburgh, https://web.archive.org/web/*/www.clpgh.org/acla; Carolyn A. Toth, Letter from Carolyn Toth to Fred Baker: Election Bylaws, 5 November 1996, ACLA historical file (Folder: Historical Correspondence), ACLA-OFFICE; Carolyn A. Toth, Letter from Carolyn Toth to Douglas Watkins, 5 November 1996, ACLA historical file (Folder: Historical Correspondence), ACLA-OFFICE; ACLA system formation committee, System Formation Committee Report, 14 January 1997, ACLA historical file (Folder: Committee Meetings), ACLA-OFFICE; ACLA system formation committee, Letter from System Formation Committee to All ACLA Members, 16 January 1997; Kamper, Letter from Al Kamper to Members, Director's Administrative Cabinet, 1997.
118 Volkmar, ACLA Special General Membership Meeting Minutes, 18 March 1997.

119 B. J. Kochan, ACLA Board Meeting Minutes, 24 February 1997, Carnegie Library of Pittsburgh, https://web.archive.org/web/*/www.clpgh.org/acla.
120 Albert F. Kamper, Letter from Al Kamper to Betty Ann O'brien, 26 February 1997, ACLA historical file (Folder: General Information), ACLA-OFFICE.
121 ACLA, History of ACLA Discussion of Federated System, June 1997, ACLA historical file (Folder: Background), ACLA-OFFICE.
122 ACLA, System Formation: Letters of Intent, 4 June 1997, ACLA historical file (Folder: System Letters of Intent), ACLA-OFFICE.
123 "RAD Budget Applications Due July 21," *Pittsburgh Courier*, 11 June 1997.
124 Gilbert M. Gaul and Peter Tobia, "Libraries in Distress: A Reprint of a Series Published in the Philadelphia Inquirer from June 1 through June 4, 1997," *The Philadelphia Inquirer*, June 1997.
125 Kim Timlin Leonard, "Tuesday Vote to Decide Future of Library System," *Pittsburgh Tribune Review*, 16 June 1997.
126 Anne A. Agnew, Letter from Anne Agnew to Marilyn Jenkins: Emsworth Membership, 11 June 1997, ACLA historical file (Folder: Historical Correspondence), ACLA-OFFICE; Glenn A. Walsh, Proposed Bylaw Amendments for Proposed Federated Library System Bylaws, 16 June 1997, ACLA historical file (Folder: Bylaws), ACLA-OFFICE.
127 Kim Timlin Leonard, "Library Delegates Ok Plan for County," *Pittsburgh Tribune Review*, 18 June 1997; Glenn A. Walsh, Letter from Glenn A. Walsh to David M. Matter, 10 July 1997, ACLA historical file (Folder: Bylaws), ACLA-OFFICE; David L. Donahoe, Letter from David Donahoe to Glenn Walsh: ACLA Federation Decision, 31 July 1997, ACLA historical file (Folder: ARAD Correspondence), ACLA-OFFICE.
128 Marilyn A. Jenkins, Letter from Marilyn Jenkins to Gary Wolfe, 14 July 1997, ACLA historical file (Folder: Correspondence), ACLA-OFFICE.
129 ACLA, The ACLA Report 1(11), 18 July 1997, ACLA historical file (Folder: ACLA History), ACLA-OFFICE.
130 ACLA, The ACLA Report 1(11), 18 July 1997.
131 Garland H. Mcadoo, Letter from Garland H. Mcadoo to Marilyn Jenkins, 23 June 1997, ACLA historical file (Folder: Bylaws), ACLA-OFFICE.
132 Marilyn A. Jenkins, Letter from Marilyn Jenkins to Ralph Destefano: Decision to Federate, 8 September 1997, ACLA historical file (Folder: ARAD Correspondence), ACLA-OFFICE.
133 Commonwealth of Pennsylvania, *Second Amendment and Restatement of the Articles of Incorporation of Allegheny County Library Association* (Pennsylvania Department of State, 26 December 1997).
134 Marilyn A. Jenkins, Letter from Marilyn Jenkins to Whitehall Public Library Board of Directors, 11 August 1997, ACLA historical file (Folder: Member Library Agreement), ACLA-OFFICE; Gary J. Klingman, Letter from Gary J. Klingman to Robert Cohen, 21 July 1997, ACLA historical file (Folder: Bylaws), ACLA-OFFICE; Allegheny League of Municipalities, Letter from Allegheny League of Municipalities to Municipal Officials: Federated Library System, 9 July 1997, ACLA historical file (Folder: Historical Correspondence), ACLA-OFFICE; Amie Downs, Letter to Marilyn Jenkins from Amie Downs, Allegheny League of Municipalities, 28 July 1997, ACLA historical file (Folder: Allegheny County League of Municipalities), ACLA-OFFICE; Marilyn A. Jenkins, Letter from Marilyn Jenkins to Amie Downs: Allegheny League of Municipalities Questions, 8 August 1997, ACLA historical file (Folder: Allegheny County League of Municipalities), ACLA-OFFICE.
135 Gordon Ovenshine, "Area Libraries Protest New System," *North Hills News Record*, 30 August 1997; Marilyn A. Jenkins et al., Listserv Discussion on ACLA Bylaws, September 1997, ACLA historical file (Folder: Member Library Agreement), ACLA-OFFICE.

136 David L. Donahoe, Letter from David Donahoe to W. David Montz: Lingering Concerns About ACLA Federation, 2 September 1997, ACLA historical file (Folder: ARAD Correspondence), ACLA-OFFICE.
137 ACLA, ACLA Libraries (42) Eligible for "Full" System Member Status, 1997, September 9, ACLA historical file, ACLA-OFFICE; ACLA, ACLA Libraries (42) Eligible for "Full" System Member Status: Opposition Notes, 9 September 1997, ACLA historical file (Folder: Member Library Agreement), ACLA-OFFICE; Jenkins, Letter from Marilyn Jenkins to Ralph Destefano: Decision to Federate, 8 September 1997.
138 Marilyn A. Jenkins, Letter from Marilyn Jenkins to Al Kamper, 15 September 1997, ACLA historical file (Folder: Member Library Agreement), ACLA-OFFICE; Robert B. Croneberger, Letter from Bob Croneberger to Marilyn Jenkins, 23 September 1997, ACLA historical file (Folder: Member Library Agreement), ACLA-OFFICE; Marilyn A. Jenkins, Letter from Marilyn Jenkins to Bob Croneberger, 29 September 1997, ACLA historical file (Folder: Member Library Agreement), ACLA-OFFICE.
139 ACLA, ACLibs 2(3), October 1997, ACLA historical file (Folder: ACLA History), ACLA-OFFICE.

Period 8: Formula Wars, 1998 – 2011

Allegheny County Library Association (ACLA) became a federated library system in January 1998. This federation represented the whole of the public library infrastructure in greater Pittsburgh. It included both the Carnegie Library of Pittsburgh (CLP) and the other independent public libraries throughout Allegheny County. All the libraries in the new system continued to receive RAD money, though CLP continued to be recognized as a separate regional asset. As a federated system, ACLA became eligible to receive additional funding from the State of Pennsylvania. This funding came in the form of the public library subsidy. ACLA also became eligible to receive grant money from the state. With this additional funding, ACLA HQ had the capacity to grow as an administrative body.

At the same time, in 1998 and into the early 2000s, several developments in the Pittsburgh region threatened to disrupt the current state of public library funding. The first was what came to be known as "Plan B." Plan B was a proposal to use RAD funds to build two sports stadiums, one for the area's professional baseball team, the Pittsburgh Pirates, and one for area's professional American football team, the Pittsburgh Steelers, and to enlarge the Convention Center. The other proposal was the Home Rule Charter in Allegheny County. In the end, Plan B did not significantly affect the RAD funding directed toward public libraries. Public libraries continued to receive the largest share of RAD money. The Home Rule charter, which meant a restructuring of county government, also did not significantly disrupt the public library arrangement. Though CFLAC dissolved with the restructure in 2001, and the County Library Administrator position disappeared, these functions had already transferred to ACLA. In 1999, ACLA hired the County Library Director as its full-time executive director.[1]

From 1998 and into the early 2000s, the public library infrastructure in greater Pittsburgh remained strong. CLP consisted of the main library in Oakland, its 18 additional branches, plus bookmobiles and reading centers.[2] ACLA was the largest federated library system in the state. There were library construction projects underway across the county.[3] EIN, the electronic network system, changed its name to eiNetwork. Cardholder numbers and circulation volume were high, and the libraries continued to add new materials to the shared collection.[4] ACLA increasingly organized coordinated and countywide library projects.[5]

Despite the growth of ACLA membership, and even as ACLA member libraries tried to coordinate regional services, regional cooperative projects could not overcome underlying tensions that were inherent in the organization since its inception. The problems inherent in ACLA all related in some way to existing funding inequities and existing disparities in resources. There were rich and poor libraries,

https://doi.org/10.1515/9783111013404-010

large and small ones, what came to be known as the "haves and have-nots." Some municipalities in the county lacked a public library altogether. This meant that not all municipalities paid for local public library services uniformly. Moreover, residents of non-funding municipalities could still receive public library services from neighboring municipalities at the neighboring municipality's expense. As a condition of receiving RAD money, ACLA libraries had to provide free access to all county residents regardless of whether those residents paid their fair share. As was evidenced by Northland's reluctance to participate in ACLA from 1995 to 1998, this condition of free access presented ACLA members with a freeloader problem, if viewed from one side, and a distributional problem, if viewed from another. Even with a funding formula that attempted to leverage local funding, RAD's condition of free access disincentivized increased municipal support, and it disincentivized participation in ACLA.[6] This meant that libraries that contributed more local resources to the system were not necessarily reimbursed for that contribution. But, seen another way, not all municipalities had the capacity to contribute.

ACLA membership compensated for local funding disparities by redistributing local resources from wealthier libraries.[7] Even so, with such disparate funding across the county, service and collection quality necessarily varied across the county. ACLA attempted early on to maintain a uniform level of service for all member libraries, but ACLA HQ could not devise a way to resolve the disparities.[8] Criticism of the ACLA organization continued. Many of the criticisms that were raised prior to the 1998 vote continued to surface, for instance, that the ACLA distribution formula was unfair to smaller, poorer libraries.[9] Some member libraries felt they were not being served well by ACLA's and RAD's terms.

When it became a federated library system in 1998, ACLA increased its "system funds," or the funds used by ACLA for administrative purposes and for county-wide grants. ACLA administration was limited by RAD in terms of how much RAD money the organization could use for administrative purposes, and the RAD board continued to monitor this spending, but there were no such constraints on how ACLA administration could use other funding sources such as private and state grants.[10] ACLA administration thus began to grow its capacity for carrying out its own projects, outside the reach of what RAD could control. ACLA HQ also began to jostle with CLP for control of service lines, such as Knowledge Connections and the bookmobile.[11]

In light of the new problems and possibilities ACLA faced, in 2000 the organization hired a consultant to conduct an analysis of users and services. The study was called *A New Plan for a New Millennium*, released in April 2001.[12] In light of analysis and observations, the report offered recommendations related to county-wide library development, technology and eiNetwork administration, continuing education, and funding and governance.[13]

The 2001 study highlighted problems with the governance structure of eiNetwork. By this time, eiNetwork had become a comprehensive regional network for the public library infrastructure. Library collections of all members had been weeded and their holdings translated into machine-readable format for inclusion in the regional union catalog. By 2000, there were 4 million items in eiNetwork's union catalog, and there were 1,500 PCs in 70 locations.[14] But the internal governance and funding structure of eiNetwork had changed. Since 1996, RAD funding for eiNetwork had come through ACLA as a direct pass-through.[15] County money also flowed to eiNetwork. In 2000, however, the county commissioners divested themselves of funding eiNetwork, and RAD became the sole funder.[16] Until 2000, CLP was the administrative agent of eiNetwork, meaning that it managed eiNetwork assets, but with the funding change, ACLA became the administrative agent of eiNetwork.[17] EiNetwork staff and facilities were still housed at CLP, and eiNetwork employees were still employees of CLP. The 2001 study recommended that eiNetwork restructure to become a separate non-profit corporation with changes to its governance structure.[18]

In March 2002, the CLP board voted to establish the eiNetwork as an independent 501c(3) corporation.[19] eiNetwork had two members: CLP and ACLA.[20] ACLA supported this change because it provided more direct governing for county libraries about technology initiatives, and CLP supported the change because it established an arm's length separation between CLP and eiNetwork operations.[21] The board changed to a 4–4 structure: the directors of CLP and ACLA served as members of the board, and they each selected 3 additional members for a total of eight.[22] ACLA continued to be the pass-through organization for RAD funding rather than RAD treating eiNetwork as a separate regional asset.[23] CLP modified its arrangement with eiNetwork staff and facilities with the intent to further distance itself from eiNetwork.[24] With the eiNetwork board balanced between CLP and ACLA at 4–4, it often found itself in a deadlock. CLP and ACLA had competing interests, CLP wanted to move at its own pace with IT decisions. ACLA, with over 40 member libraries, had difficulty arriving at consensus.[25] The ineffectiveness of eiNetwork that had been identified in the 2001 study had not yet been resolved.

To address these concerns, and under RAD's direction, eiNetwork undertook its own evaluation process. The reports produced a technology assessment as well as future options.[26] One report was titled *Effectiveness Study of Technological Services to Allegheny County Library Association and the Carnegie Library of Pittsburgh.*[27] The final report was confidential, and it was never released outside select members of the eiNetwork board, the CLP board, and the ACLA board. Nevertheless, it can be gleaned from other sources that eiNetwork considered alternative models related to funding and governance. Four scenarios were considered.[28] Scenario 3 was, at first, the predominant option.[29] It imagined a divided network system, one where CLP and ACLA each made their own technology decisions. Under this scenario, the system no longer operated in a uniform way. CLP had its own technology interests and its own preferred pace for technology adoption.[30] By contrast, ACLA libraries had diverse views, varying levels of sophistication, and a range of budgets.[31] CLP no longer wanted to wait for the other ACLA libraries to come to consensus before implementing new technology. It sought options that would allow it to move at a faster pace, unrestrained by the needs of other libraries in the county and better aligned with its consolidated structure.[32] Under Scenario 3, technology adoption for ACLA libraries would be slower than for CLP, but CLP could operate with greater autonomy. In late 2007, immediately after the report was released, analysis by the eiNetwork board focused exclusively on pursuing Scenario 3.[33] During the discussions about the Scenarios, representatives from RAD were often present, and they were aware of the scenarios, at least superficially. A representative from RAD expressed concern that no perceived differences in service level exist for library users across the county.[34] This meant that if Scenario 3 were implemented, it would have to remain obscured from view.

At least at first, the other scenarios were not thoroughly worked out. Scenario 1 imagined a unified network system but one where funding no longer originated from RAD, at least not completely. Instead, member libraries would be charged for part or all the services they received. Essentially, eiNetwork would become a private utility that served its member libraries in the county.[35] Scenario 4 was never pursued or discussed in detail. It was noted that "Option four is not viable since the six county libraries included in the option are not interested in participating."[36] It is possible that Scenario 4 conceived of several libraries in the county administering the network, drawing funding from a mix of local government, county government, and private sources.

Scenario 2 imagined a unified governance system and a single funding stream from RAD. Under this scenario, CLP would become the administrative host for eiNetwork, thus taking the reins from ACLA. There were certain advantages to Scenario 2. These included financial efficiencies, moving all libraries quickly with new technologies, and retaining a unified system that would more likely meet with RAD approval.[37] The eiNetwork board did not initially consider Scenario 2, but the ACLA board announced its support for this scenario.[38] This recommendation contravened ACLA's own Executive Director and ACLA's own representatives on the eiNetwork board.[39] After learning that the ACLA board supported Scenario 2, the eiNetwork board diverted attention away from Scenario 3.[40]

Details of the four scenarios were not shared with all library directors or the ACLA board.[41] In April, the ACLA board received public comments about the eiNetwork options from member libraries. Though the ACLA board supported Scenario 2, the ACLA membership was not in full support of this recommendation. Based on what little information they were provided, ACLA members expressed concern about the condensed timetable, opacity, and cryptic information provided during the planning process.[42] In response, the ACLA board formed a task force to develop its own solution, one that included "a unified system, cost efficiencies, and a timeline."[43] The long-term approach proposed by the task force was to restructure eiNetwork as an independent vendor. An independent board of directors would be appointed to serve the interests of the new entity exclusively. That board should be self-appointing and self-perpetuating. The new entity could bid on technology needs of libraries and enter into contracts for service, but the entity could provide services to and generate revenue from other types of organizations outside of libraries.[44]

The final decisions regarding eiNetwork's new strategic plan were made in late 2008. These changes were implemented in late 2008 and early 2009.[45] In sharp contrast to the Scenario 3 which was initially envisioned, eiNetwork instead remained a unified system, shared by both CLP and ACLA, and with funding primarily from RAD. Member libraries did not see much change in the services they received. There was little to no increase in costs for individual libraries.[46] Some of the recommendations from the ACLA task force were adopted, but others were not.

Five relatively minor changes to eiNetwork resulted from the discussions. First, related to governance, the eiNetwork board changed from a 4–4 structure to a 3–3 structure with three appointees from ACLA and three from CLP.[47] Second, also related to governance, a Technology Council was formed.[48] The Council included more than 15 representatives from small, medium, and large libraries throughout Allegheny County as well as technology staff from the eiNetwork and a few larger libraries.[49] The third change related to funding. CLP became the fiscal agent of eiNetwork.[50] This meant that RAD money passed through CLP instead of ACLA. The final two changes related to services: eiNetwork withdrew from library operations, meaning that it no longer provided cataloging support or reports to member libraries, and it outsourced the help desk.[51] In June of 2010, after settling on a slightly modified governance and funding structure, eiNetwork moved forward with technology upgrades, including open-source options and the construction of a fiber ring.[52]

Following the study from 2001, and with a review of eiNetwork underway at the same time, ACLA HQ and ACLA member libraries revisited the development of performance standards for the membership. Ostensibly, this was an attempt to improve service delivery quality across the infrastructure.[53] Pennsylvania used minimum standards when qualifying individual libraries for the distribution of state aid, and since its metamorphosis into a state-recognized federated library system in 1998, ACLA HQ also received additional funding from the state for its status as a system that met system standards.[54] ACLA therefore used state standards as a model when developing its own.[55] As was outlined in ACLA's 1996 long-range plan, only libraries that met ACLA's standards should be eligible for system membership and RAD funding. Between 1996 and 2001, however, local standards had not been implemented.[56] What distinguished the standards conversation in ACLA in the early 2000s was the attempt by ACLA HQ to control how state funding flowed to member libraries. There was a push to base both RAD funding and state funding on a single set of performance standards and a single distribution mechanism.

In the early 2000s, at the time ACLA revisited standards, libraries across the state were experiencing great swings in state funding. At first, state funding appeared to be on the rise. In 1997, a series of articles in the *Philadelphia Inquirer* by Gilbert Gaul highlighted the poor funding conditions of public libraries across the state, especially at a time when computers, the Internet, and online services were in demand. In response, the Pennsylvania governor increased state library aid in 1998, and in 1999 the state legislature passed significant increases in state aid funding.[57] This was part of a 10-year plan to increase state aid to public libraries.[58] A report from a task force of the Pennsylvania Library Association, published in 1998 and called *The Platform for Twenty-First Century Libraries*, called for a gradual rise in state aid by 2003.[59] This plan worked out for the first few years, and library appropriations increased.[60] During the governor's tenure, the budget for library services in the state increased from nearly $36 million in fiscal year 1995–1996 to $94 million in fiscal year 2001–2002. The distribution formula for state funding incorporated performance standards. There were several categories of funding to public libraries: 1) county coordination aid, 2) district services aid, 3) regional resource aid, 4) basic aid, 5) incentive aid, and 6) equalization aid. Each year's increase in state funding targeted a single category of the state's distribution formula.[61]

But library funding increases were short-lived. After the governor left office in 2003, and due in part to the economic recession following the September 11 attacks, the state budget for library services dropped.[62] The formula that was used to distribute state money to public libraries across the state became "frozen." This meant that the formula was no longer used for determining what libraries received. Why was the formula frozen? It was because the state's public library subsidy required a minimum funding amount, and after 2003, this minimum amount was no longer present. After running the numbers, the state library determined that applying the formula with less money than required would result in extreme winners and losers. Had the formula been run, some libraries would have received much less money than they had previously. Not all affected libraries agreed to these cuts, and legislators decided not to run the formula.[63] It was instead decided that libraires should take a proportionate cut, referred to as a "pro-rated share" of the cut.[64] After 2003, payments to public libraries across the state were based on what libraries had received in 2002 – 2003. In addition, the state's minimum standards were waived. This was because it would have been unfair to expect libraries to perform at previous levels without at least the same amount of funding. The incentive category of state funding existed in principle, but the state library no longer verified that libraries met those standards. Efforts to revisit the formula and reinstate state standards in later years failed.[65] Problems in state funding continued into the late 2000s. There was some discussion during this time at the state level about how federated library systems, such as ACLA, should distribute state aid, but this issue was not settled.[66]

Within the public library infrastructure in greater Pittsburgh, RAD funding remained steady.[67] Nevertheless, in light of slated funding cuts from the state, CLP planned to close branches.[68] In late 2009, four CLP branches were set to close.[69] In 2011, there was some funding relief from the state because table gaming became allowed at Pennsylvania casinos. Some of that revenue was used to rescue public libraries.[70] The money from table game revenue was used by ACLA libraries to offset cuts in state aid, and CLP's proposed closures did not occur.[71] In 2011, CLP was also successful in its campaign for a dedicated tax income increase that contributed $3.9 million to CLP in 2013.[72]

Despite the effective absence of standards at the state level, the issue of standards remained on the forefront of ACLA's agenda.[73] ACLA formed a task force to review service standards and state and RAD funding tied to them.[74] Intensive discussion of system standards carried on within ACLA for several years, from 2001 to 2006.[75] During this time, new formulas and standards were developed, discussed, revised, and rejected. Tying standards to funding was controversial within the membership, even before the state cuts were implemented. [76] One controversial approach, popularized by ACLA HQ, was referred to as the "Giant Eagle" model. This tiered funding structure was modeled on the local Giant Eagle supermarket chain, which utilized outlets of different sizes—Get Go, Giant Eagle, Market District—to provide different levels of service. The service outlets ranged from gas station convenience stores to luxury grocery markets. It was proposed that the public library infrastructure operate in the same way, with big, medium, and small libraries offering different levels of services.[77] In the end, this tiered model was rejected.

By 2006, the ACLA membership and ACLA board seemed to arrive at a decision point. They voted to adopt standards tied to both state and RAD funding.[78] There were 6 factors in the formula: population, distress, local match, match for other income, hours, and collection use.[79] The adoption of these standards was criticized by some in the membership.[80] Nevertheless, ACLA HQ pushed forward with them and developed plans for withholding funds should libraries not meet the standards.[81]

In late 2006, however, when ACLA presented its revised formula to the RAD board for approval, the RAD board rejected it. As a result, ACLA membership voted to table standards for an additional year.[82] There are several reasons why the standards ultimately failed, including problems with the process used to develop them and consequences of the standards themselves. [83] The main reason for the failure was that the formula incorporated state aid. Based on its own member library agreement, ACLA could not dictate how much state aid members received.[84] Regardless, the bottom line was that the RAD board rejected the formula.[85] Not only that, but the RAD board also required that ACLA members work with consultants to "review the process whereby the formula was adopted and the formula itself."[86] The RAD board was concerned about the contentiousness that arose within ACLA over standards.[87] From its earliest dates, if there was anything that caused controversy within ACLA, it was the RAD distribution formula.

In response to RAD's directive, ACLA retained a consultant to assist with an assessment of the formula and its development process.[88] Meanwhile, CLP, ACLA libraries, and ACLA HQ continued to carry out other regional initiatives. [89]

ACLA's work with the consultant continued from 2007 to 2009.[90] The membership worked with the consultant through a "comprehensive consensus-building process for redesigning the formula for distribution of RAD monies."[91] The goal of the process was to develop a new formula for 2009 – 2010.[92] The formula removed state funding from the equation and allocated RAD funds only.[93] At the same time, the formula continued to incorporate performance metrics tied to funding.[94] What complicated the formula discussion was the global economic recession that occurred from 2007 to 2008. ACLA received almost $1 million less in RAD funding in 2009 than it had in 2008.[95] In the final phase of discussions, it was left to the committee to decide which performance factors to include in the formula and how to weigh them.[96] The committee considered several performance factors, including population, distress, circulation, local income, and public service hours.[97] Three iterations of a 6-factor formula, together with funding calculations for each library, were presented to the general membership to review in the spring and summer of 2009. [98] ACLA member libraries voted to approve a formula option by a vote of 33 out of 42.[99] The accepted formula did not include local income as a performance factor.

Though the formula was approved by the membership, it was not yet clear whether the RAD board would approve it. As might have been predicted, the RAD board rejected ACLA's new proposed formula. The formula was rejected because it failed to encourage local financial contributions from libraries' communities and because it would result in drastic changes in funding levels.[100]

In response to RAD's decision, ACLA formed another team to develop a new formula.[101] In its revision, ACLA included a phase-in process to avoid any sudden funding swings.[102] ACLA's newest formula reincorporated local income as a factor in the calculations and reduced the importance of population and circulation. Of course, the new formula met with resistance from some members, resulting in divisions within the organization.[103] But the RAD board reluctantly agreed to the newest iteration, with the caveat that RAD would continue to scrutinize ACLA's funding formula moving forward.[104] The RAD board directed ACLA to continue to review the formula with a small steering committee composed of representatives from RAD and ACLA.[105] These discussions would continue through 2010.[106] Another slightly modified formula was approved by ACLA and RAD for 2011.[107] There were no substantive changes to the formula, and ACLA as an organization remained deeply divided.[108] Reflecting on the formula development process from 2001 to 2011, one library administrator commented, "fighting over formulas has nearly torn this organization apart over the course of history."[109]

During the formula development process, several other issues were addressed within ACLA. The Librarians Advisory Council (LAC), composed of library directors who served in an advisory capacity to the ACLA board, established a Critical Issues Launch and Oversight Committee (CILOC) to oversee a variety of task forces.[110] These task forces addressed issues such as the structure of ACLA and the development of standards.[111] In the end, proposals from the ACLA restructure task force were not implemented, but the issue of standards recurred.[112] Since 2002–2003, the performance of some of its member libraries had changed. Some libraries had moved up from quality-level to incentive-level funding, and others had moved down from incentive-level funding to quality-level funding.[113] ACLA questioned whether it could instead distribute state money to its member libraries in a way that more accurately reflected their current performance.[114] Despite the suspension of standards at the state, ACLA had continued to monitor whether its member libraries met state standards.[115] It also continued to meet with libraries to develop plans to achieve those standards.[116] But for those libraries who failed to meet state standards, the dilemma facing ACLA was whether it could or should continue to distribute state and RAD funding to them.[117]

The ACLA standards task force that originated from CILOC proposed ACLA service standards in November 2009.[118] There was some pressure from the RAD board for ACLA to development performance standards.[119] The proposed ACLA standards blended the state's standards for quality-level funding and incentive-level funding. [120] By meeting the proposed ACLA standards, a library would qualify for both state aid and RAD funding.[121] ACLA administrators continued to measure the compliance of member libraries against 4 sets of standards: 1) system member standards, 2) quality level standards, 3) incentive level standards, and 4) ACLA proposed standards.[122] In particular, ACLA monitored whether libraries maintained local financial effort and collection expenditures.[123] The point of these standards was to determine whether ACLA member libraries qualified for RAD funding at all.

In June 2010, there was discussion about the new proposed ACLA standards, but there was also resistance to them.[124] The ACLA board postponed a vote on the standards pending additional information and input from the regions.[125] A vote by the General Membership was scheduled for October 2010, but this did not happen. The proposed service standards were set aside "pending outcomes of discussion [sic] at the State level relative to the Library Code and regulations."[126] ACLA moved forward as an organization without immediately implementing service standards.[127]

Period 8 illustrates the nestedness of the ACLA organization and ACLA member libraries within the inner periphery of the RAD circuit. The ACLA member wrestled with underlying funding disparities by trying to develop a new funding formula, and they seemed to arrive at several agreed-upon solutions, but libraries ultimately had to carry out RAD decisions, and during deliberations it was shown that the RAD board was the arbiter of decisions regarding the regional public library infrastructure. This was illustrated by the formula decisions made autonomously by ACLA that were rejected by the RAD board. Moreover, through the formation of and participation in EIN, libraries began to move in unison technologically, and they were too tied together to fragment. With the freezing of state aid, the state circuit became essentially defunct. Allegheny County government further divested itself of the public library infrastructure in greater Pittsburgh. Attempts to implement a new formula, tied to standards and performance metrics, failed due to low legitimacy and high resistance, but the issue of standards was one that could not be forgotten by ACLA HQ and RAD, and it would become the key issue in the next decision cycle.

1 ACLA, A Brief History of the Allegheny County Library Association, 1999, ACLA historical file, ACLA-OFFICE; Tom Barnes, "RAD Members Say They're Open-Minded on Plan B," *Pittsburgh Post-Gazette*, 30 April 1998; Ramesh Santanam, "O'hara Man Appointed to Asset Board," *North Hills News Record*, 8 May 1998; Robert Dvorchak, "RAD Board Played Hand It Was Dealt," *Pittsburgh Post-Gazette*, 12 July 1998; Allegheny County, *Home Rule Charter of Allegheny County* (19 May 1998); Timothy McNulty, "Rapping RAD Is Part of Election Debate Mix," *Pittsburgh Post-Gazette*, 9 April 1998; ACLA, ACLibs 4(2), November 1999, ACLA historical file (Folder: ACLA History), ACLA-OFFICE; Mark Belko, "Lucchino Quitting Politics," *Pittsburgh Post-Gazette*, 26 November 1998; Tom Barnes, "RAD Board to Hear Money Pleas," *Pittsburgh Post-Gazette*, 8 September 1998; Allegheny Regional Asset District, *RAD Fax* (August 1998); David M. Brown, "RAD Releases Early Budget," *Pittsburgh Tribune-Review*, 2 October 1998.

2 Abbie Jones, "Overdue," *Pittsburgh*, January, 2001; CLP, Annual Report, 2000, CLP-PENN.

3 Allegheny County Industrial Development Authority, *Carnegie Library of Pittsburgh, Allegheny County, Pennsylvania, Regional Asset District Revenue Bonds, Series of 2002* (2002); Elizabeth Barczak, "Libraries' High-Tech Offerings Draw in Patrons," *Pittsburgh Tribune-Review*, 18 August 2002.
4 ACLA, acLIBS 3(1), April 1998, ACLA historical file (Folder: ACLA History), ACLA-OFFICE.
5 ACLA, County Wide Library Services Committee: Report to the LAC Executive Committee, 1998, ACLA historical file (Folder: Countywide Services), ACLA-OFFICE; ACLA, County Wide Library Services Committee, Goals for 1998, 28 January 1998, ACLA historical file (Folder: Countywide Services), ACLA-OFFICE; ACLA, County Wide Library Services Committee, 7 April 1998, ACLA historical file (Folder: Countywide Services), ACLA-OFFICE; ACLA, acLIBS 3(1), April 1998; ACLA, County Wide Library Services Committee Continuing Education Survey, September 1998, ACLA historical file (Folder: Countywide Services), ACLA-OFFICE; ACLA, County Wide Library Services Committee, 7 April 1998; ACLA, County Wide Library Services Committee, 7 April 1998; ACLA, County Wide Library Services Committee, 26 April 2000, ACLA historical file (Folder: Countywide Services), ACLA-OFFICE; ACLA, County Wide Library Services Committee, 31 January 2000, ACLA historical file (Folder: Countywide Services), ACLA-OFFICE; ACLA, County Wide Library Services Committee, 2 May 2000, ACLA historical file (Folder: Countywide Services), ACLA-OFFICE; ACLA, County Wide Library Services Committee, 6 June 2000, ACLA historical file (Folder: Countywide Services), ACLA-OFFICE; ACLA, County Wide Library Services Committee, 12 December 2000, ACLA historical file (Folder: Countywide Services), ACLA-OFFICE; Library Administrator 1, Interview Transcript, 28 July 2015, interviewed by Michael M. Widdersheim; Rebecca Riddle, "SHARECO Program Expands for Residents, Businesses," *Daily News*, 17 October 2000; David Titmus, "Area Libraries Share and Shareco Alike," *South Hills Record Star*, 10 August 2000.
6 Jones, "Overdue."
7 Jones, "Overdue."; Marilyn A. Jenkins, Letter from Marilyn Jenkins to Nick Fisfis: Full Member Benefits, 20 October 1998, ACLA historical file (Folder: Historical Correspondence), ACLA-OFFICE.
8 Marilyn A. Jenkins, Commonwealth Libraries LSTA Fund 1998: NetShare Application, 1998, ACLA historical file (Folder: ACLA 1998), ACLA-OFFICE; ACLA, acLIBS 3(1), April 1998.
9 Jonathan D. Silver, "RAD Formula Called Unfair to Poorer Libraries," *Pittsburgh Post-Gazette*, 13 November 1998.
10 Allegheny Regional Asset District, Report, 1999 Library Formula Committee of the Board, 1 October 1998, RAD files, Contractual Assets–Folders, Library Committee ACLA 1997–2006, ARAD-OFFICE; ACLA, ACLibs 4(1), July 1999, ACLA historical file (Folder: ACLA History), ACLA-OFFICE.
11 ACLA, County Library Service: New Opportunities, New Models, 29 January 2013, ACLA-WIKI; Allegheny Regional Asset District, Minutes of Board Meeting, 1 December 1998, RAD files, Contractual Assets–Folders, Library Committee ACLA 1997–2006, ARAD-OFFICE; Allegheny Regional Asset District, Final Report, 2000 Library Formula Committee of the Board, , 8 November 1999, RAD files, Contractual Assets–Folders, Library Committee ACLA 1997–2006, ARAD-OFFICE; ACLA, New Directors Handbook, July 2006, ACLA-WIKI; David Brown, "Cultural Funding Down," *Pittsburgh Tribune-Review*, 18 November 2001; Michael A. Fuoco, "Asset District Adopts Slimmer Budget," *Pittsburgh Post-Gazette*, 20 November 2001.
12 ACLA, County Wide Library Services Committee, 31 January 2000; Anders C. Dahlgren, Debra W. Johnson, and Robert Bocher, *A New Plan for a New Millennium: A Review of the Long-Range Plan for the Allegheny County Library Association* (Madison: Library Planning Associates, 2001); Treshea N. Wade, "Libraries Get Marketing Advice," *Pittsburgh Tribune-Review*, 3 June 2002; ACLA, Timeline: Review of "a New Plan for the New Millennium", 2001, ACLA historical file (Folder: Countywide Development), ACLA-OFFICE; Marilyn A. Jenkins, Letter from Marilyn Jenkins to Ad Hoc Committee Members, 4 May 2001, ACLA historical file (Folder: Countywide Development), ACLA-OFFICE; ACLA,

A Plan for the New Millennium: Ad Hoc Committees, 2001, ACLA historical file (Folder: Countywide Development), ACLA-OFFICE; Connie Galbraith, Letter from Connie Galbraith to Ann Koempel, 8 May 2001, ACLA historical file (Folder: Countywide Development), ACLA-OFFICE; ACLA, Allegheny County Library Association: A New Plan for the New Millennium, Municipal Meetings Handouts, 2001, ACLA historical file (Folder: Countywide Development), ACLA-OFFICE; ACLA east region, Notes from East Region Meeting: Combined Regional Responses, 15 November 2001, ACLA historical file (Folder: Countywide Services), ACLA-OFFICE; ACLA, County Wide Library Services Committee, 27 February 2001, ACLA historical file (Folder: Countywide Services), ACLA-OFFICE; ACLA, County Wide Library Services Committee, 22 August 2001; ACLA, County Wide Library Development Issues Committee, 10 July 2001, ACLA historical file (Folder: Countywide Development). ACLA-OFFICE; ACLA, Countywide Development Brainstorming Session, 16 November 2001, ACLA historical file (Folder: Countywide Services). ACLA-OFFICE; ACLA, Committee Retreat, Meeting Minutes, 16 November 2001, ACLA historical file (Folder: Countywide Development). ACLA-OFFICE; Thomas Mott, Letter from Thomas Mott to Marilyn Jenkins: Council of Governments, 7 May 2001, ACLA historical file (Folder: Countywide Development). ACLA-OFFICE.

13 *Dahlgren, Johnson, and Bocher, A New Plan for a New Millennium.*

14 Jenkins, Commonwealth Libraries LSTA Fund 1998: NetShare Application, 1998; EIN, Electronic Information Network: Strategic Plan 1999 – 2002, 1999, ACLA-WIKI; ACLA, acLIBS 3(1), April 1998; ACLA, New Directors Handbook, July 2006; ACLA, The ACLA Report 2(1), 20 March 1998, ACLA historical file (Folder: ACLA History), ACLA-OFFICE; Allegheny County, Review of Capital Revenues and Expenditures for the Electronic Information Network for the Period January 1, 1995 through March 31, 1998, 1 July 1998, ACLA historical file (Folder: eiNetwork Audit), ACLA-OFFICE; Terrence P. O'Horo, Letter from Terrence O'Horo to William Mckain, 9 September 1999, ACLA historical file (Folder: eiNetwork Audit), ACLA-OFFICE; EIN, EIN Promotional Video for Buhl Foundation, 28 August 2000,

15 eiNetwork, Frequently Asked Questions About eiNetwork, July 2008, ACLA-WIKI.

16 Cindi Lash, "County Restores Library Internet Funds," *Pittsburgh Post-Gazette*, 11 December 1999; Jones, "Overdue."

17 EIN, Electronic Information Network Grant Agreement, 1 January 1998, ACLA historical file (Folder: eiNetwork Agreements), ACLA-OFFICE.

18 Dahlgren, Johnson, and Bocher, *A New Plan for a New Millennium*, 44 – 45.

19 Carolyn D. Duronio, Application for Recognition of Exemption under Section 501(C)(3) of the Internal Revenue Code: The Electronic Information Network, 5 March 2002, ACLA historical file (IRS 501(c)(3) application, loosely filed), ACLA-OFFICE; eiNetwork, Frequently Asked Questions About eiNetwork, July 2008.

20 eiNetwork, Frequently Asked Questions About eiNetwork, July 2008.

21 eiNetwork, Frequently Asked Questions About eiNetwork, July 2008; EIN, Agreement between the Electronic Information Network (EIN) and Carnegie Library of Pittsburgh, 1 January 1999, ACLA historical file (Folder: eiNetwork Agreements), ACLA-OFFICE; EIN, Bylaws of the Electronic Information Network, 5 March 2002, ACLA-WIKI.

22 eiNetwork, Frequently Asked Questions About eiNetwork, July 2008.

23 Olszak Management Consulting, EIN Explanation, and ACLA/CLP/EIN Relationships, 10 April 2008, ACLA-WIKI.

24 eiNetwork, Frequently Asked Questions About eiNetwork, July 2008; eiNetwork, eiNetwork Board Meeting Minutes, 26 September 2007, ACLA-WIKI; EIN, 2007 eiNetwork Program View, 2007, ACLA-WIKI.

25 Cynthia Richey, LAC Representative on ACLA Board of Directors – Report for Librarians Advisory Council Meeting, 11 October 2007, ACLA-WIKI.

26 eiNetwork, eiNetwork Board Meeting Minutes, 26 September 2007; eiNetwork, Report to LAC, September 2007, ACLA-WIKI; Allegheny Regional Asset District, Report of the Library Committee, 1 December 2008, RAD files, Contractual Assets–Folders, Library Committee ACLA 1997–2006, ARAD-OFFICE; Allegheny Regional Asset District, Report of the Library Committee, 27 May 2008, RAD files, Contractual Assets–Folders, Library Committee ACLA 1997–2006, ARAD-OFFICE; eiNetwork, Report to LAC, October 2007, ACLA-WIKI; eiNetwork, eiNetwork Board Meeting Minutes, 17 December 2007, ACLA-WIKI.

27 ACLA, Librarians Advisory Council Minutes, 10 April 2008, ACLA-WIKI.

28 eiNetwork, eiNetwork Board Meeting Minutes, 14 November 2007, ACLA-WIKI; ACLA, Librarians Advisory Council Minutes, 10 April 2008.

29 eiNetwork, eiNetwork Board Meeting Minutes, 17 December 2007; ACLA, Librarians Advisory Council Minutes, 10 April 2008.

30 Olszak Management Consulting, EIN Explanation, and ACLA/CLP/EIN Relationships, 10 April 2008.

31 eiNetwork, eiNetwork Board Task Force: Consideration of Scenario 1, 15 April 2008, ACLA-WIKI.

32 eiNetwork, Frequently Asked Questions About eiNetwork, July 2008.

33 eiNetwork, eiNetwork Board Meeting Minutes, 14 November 2007; eiNetwork, eiNetwork Board Meeting Minutes, 28 January 2008, ACLA-WIKI; eiNetwork, eiNetwork Board Meeting Minutes, 17 December 2007; eiNetwork, eiNetwork Scenario 3, 7 May 2008, ACLA-WIKI.

34 eiNetwork, eiNetwork Board Meeting Minutes, 17 December 2007; Cynthia Richey, LAC Representative on ACLA Board of Directors Report for Librarians Advisory Council, June 2008, ACLA-WIKI.

35 eiNetwork, eiNetwork Board Meeting Minutes, 17 December 2007; eiNetwork, eiNetwork Board Task Force: Consideration of Scenario 1, 15 April 2008.

36 ACLA, Librarians Advisory Council Minutes, 10 April 2008.

37 Cynthia Richey, LAC Representative on ACLA Board of Directors–Report for Librarians Advisory Council, May 2008, ACLA-WIKI.

38 ACLA, Librarians Advisory Council Minutes, 10 April 2008.

39 Richey, LAC Representative on ACLA Board of Directors–Report for Librarians Advisory Council, May 2008.

40 ACLA, Librarians Advisory Council Minutes, 10 April 2008; eiNetwork, eiNetwork Scenario 3, 7 May 2008; ACLA, Librarians Advisory Council, 8 May 2008, ACLA-WIKI; Richey, LAC Representative on ACLA Board of Directors Report for Librarians Advisory Council, June 2008.

41 eiNetwork, eiNetwork Board Meeting Minutes, 26 September 2007; eiNetwork, Report to LAC, October 2007; eiNetwork, eiNetwork Board Meeting Minutes, 17 December 2007; eiNetwork, eiNetwork Board Meeting Minutes, 14 November 2007; Cynthia Richey, LAC Representative on ACLA Board of Directors – Report for Librarians Advisory Council Meeting, 8 November 2007, ACLA-WIKI; ACLA, Librarians Advisory Council Minutes, 10 January 2008, ACLA-WIKI; eiNetwork, eiNetwork Board Meeting Minutes, 28 January 2008; Cynthia Richey, LAC Representative on ACLA Board of Directors – Report for Librarians Advisory Council Meeting, 14 February 2008, ACLA-WIKI; Cynthia Richey, LAC Representative on ACLA Board of Directors–Report for Librarians Advisory Council Meeting, 13 March 2008, ACLA-WIKI; Cynthia Richey, LAC Representative on ACLA Board of Directors–Report for Librarians Advisory Council Meeting, 10 April 2008, ACLA-WIKI; eiNetwork, eiNetwork Board Task Force: Consideration of Scenario 1, 15 April 2008.

42 ACLA, Librarians Advisory Council Minutes, 10 April 2008; Richey, LAC Representative on ACLA Board of Directors–Report for Librarians Advisory Council, May 2008.
43 Richey, LAC Representative on ACLA Board of Directors Report for Librarians Advisory Council, June 2008; ACLA, Librarians Advisory Council Meeting Minutes, 3 June 2008, ACLA-WIKI; ACLA, eiNetwork Recommendations, June 2008, ACLA-WIKI; ACLA, eiNetwork Recommendations from ACLA Board Task Force Amended Per Member Input, 1 July 2008, ACLA-WIKI.
44 ACLA, eiNetwork Task Force Recommendations, June 2008, ACLA-WIKI; ACLA, Attachment F: eiNetwork Task Force Recommendations, June 2008, ACLA-WIKI; ACLA, eiNetwork Current Services to Libraries, 16 June 2008, ACLA-WIKI; Olszak Management Consulting, eiN Current State, April 2008, ACLA-WIKI; ACLA, eiNetwork Recommendations from ACLA Board Task Force Amended Per Member Input, 1 July 2008.
45 ACLA, *ACLA 2009 Annual Report to the Allegheny Regional Asset District* (www.aclalibraries.org/reports, 1 June 2010).
46 ACLA, Minutes of the Librarians Advisory Council, 16 October 2008, ACLA-WIKI.
47 ACLA, *Highlights from 2009* (www.aclalibraries.org/reports, 2009); ACLA, Minutes of the Librarians Advisory Council, 15 January 2009, ACLA-WIKI.
48 eiNetwork, Frequently Asked Questions About eiNetwork, July 2008.
49 ACLA, *Highlights from 2009* (2009); Cynthia Richey, LAC Representative on ACLA Board of Directors–Report for Librarians Advisory Council, November 2008, ACLA-WIKI.
50 ACLA, Librarians Advisory Council Minutes, 13 November 2008, ACLA-WIKI.
51 ACLA, *Highlights from 2009* (2009); ACLA, *ACLA 2009 Annual Report to the Allegheny Regional Asset District* (1 June 2010); eiNetwork, eiNetwork Board Meeting Minutes, 28 January 2008.
52 ACLA, Minutes of the Librarians Advisory Council, 10 June 2010, ACLA-WIKI; eiNetwork, Open Source Update, 25 January 2011, ACLA-WIKI; ACLA, Librarians Advisory Council Minutes, 13 January 2011, ACLA-WIKI; Marilyn A. Jenkins, Executive Director's Summary of Major Action and Discussion Items, 22 February 2010, ACLA-WIKI; ACLA, General Membership Meeting Minutes, 26 October 2011, ACLA-WIKI.
53 Dahlgren, Johnson, and Bocher, *A New Plan for a New Millennium*, 32.
54 ACLA, acLIBS 3(1), April 1998.
55 Dahlgren, Johnson, and Bocher, *A New Plan for a New Millennium*, 32.
56 Dahlgren, Johnson, and Bocher, *A New Plan for a New Millennium*, 2001; ACLA, County Wide Library Services Committee, Goals for 1998, 28 January 1998; ACLA, County Wide Library Services Committee, 12 December 2000; ACLA, General Expectations of ACLA Board Members, 5 January 2001, ACLA-WIKI.
57 ACLA, A Brief History of the Allegheny County Library Association, 1999; Gilbert M. Gaul, "Survey: Most Want Change in Library Funding in Pa.," *Philadelphia Inquirer*, 20 July 1998.
58 Librarian 1, Interview Transcript, 7 January 2016, interviewed by Michael M. Widdersheim.
59 Pennsylvania Library Association, *The Platform for 21st Century Libraries*, Pennsylvania Library Association (Harrisburg, 1998).
60 Jones, "Overdue."
61 Librarian 1, Interview Transcript, 7 January 2016.
62 Patricia Lowry, "Libraries Face 50 % Cut in State Funding," *Pittsburgh Post-Gazette*, 14 March 2003; Elizabeth Barczak, "Budgetary Cliffhanger," *Pittsburgh Tribune-Review*, 17 April 2003.
63 Library Administrator 1, Interview Transcript, 28 July 2015.
64 Politician 2, Interview Transcript, 23 June 2021, interviewed by Michael M. Widdersheim.
65 Barbara Cole, Letter from Barbara Cole to District Library Center Administrators, System Administrators, and Public Library Directors, 22 July 2005, ACLA historical file (Folder: Standards),

ACLA-OFFICE; Politician 2, Interview Transcript, 23 June 2021; Governor's Advisory Council on Library Development, Minutes, 11 July 2006, State Library of Pennsylvania, Archived Minutes of the Council. https://www.statelibrary.pa.gov/Libraries/LawsRegulations/GovernmentAdvisory/Pages/Archived-Minutes-of-the-Council.aspx.

66 Governor's Advisory Council on Library Development, Minutes, 11 March 2008, State Library of Pennsylvania, Archived Minutes of the Council. https://www.statelibrary.pa.gov/Libraries/LawsRegulations/GovernmentAdvisory/Pages/Archived-Minutes-of-the-Council.aspx; Governor's Advisory Council on Library Development/Federal Advisory Council, Report from Committee on Guidelines on the Distribution of State Aid by Federated Public Library Systems, 11 March 2008, Department of Education, Commonwealth Libraries, Minutes of the Governor's Advisory Council on Library Development 1962–2009 collection, Carton 2, Folder 3, PASA; Governor's Advisory Council on Library Development, Minutes, 12 January 2010, State Library of Pennsylvania, Archived Minutes of the Council. https://www.statelibrary.pa.gov/Libraries/LawsRegulations/GovernmentAdvisory/Pages/Archived-Minutes-of-the-Council.aspx.

67 Barczak, "Budgetary Cliffhanger."; Elizabeth Barczak, "Libraries Prepare to Cut Staff, Hours," *Pittsburgh Tribune-Review*, 24 April 2003; Mike Wereschagin, "RAD to Dole out More Money in 2004," *Pittsburgh Tribune-Review*, 2 December 2003; Mark Belko, "Tough Choices for Assets Board," *Pittsburgh Post-Gazette*, 30 August 2004; Pittsburgh Post-Gazette, "Allegheny Regional Asset District," *Pittsburgh Post-Gazette*, 2 October 2007; Bill Zlatos, "Parks, Carnegie Library Get Big Boost," *Pittsburgh Tribune-Review*, 2 October 2007; ACLA, Librarians Advisory Council Minutes, 14 February 2008, ACLA-WIKI; Richey, LAC Representative on ACLA Board of Directors – Report for Librarians Advisory Council Meeting, 14 February 2008; Bill Zlatos, "Shelved: Libraries Forced to Slash Hours, Staff, Popular Programs as Funding Fades," *Pittsburgh Tribune-Review*, 7 January 2010.

68 Bob Hoover, "Bad Year for Pittsburgh's Libraries with Cuts and Closings," *Pittsburgh Post-Gazette*, 13 October 2009.

69 Meela Dudley, "City Mourns Local Library Closings," *The Tartan*, 2 November 2009.

70 Marilyn A. Jenkins, Executive Director's Summary of Major Action and Summary Items: ACLA General Membership Meeting, 27 January 2009, ACLA-WIKI; Marilyn A. Jenkins, Executive Director's Summary of Major Action and Discussion Items: ACLA General Membership Meeting, 26 January 2011, ACLA-WIKI.

71 ACLA, Librarians Advisory Council Minutes, 10 February 2011, ACLA-WIKI; Kaitlynn Riely, "Allegheny County Library Systems Look to Cooperate," *Pittsburgh Post-Gazette*, 2 February 2014.

72 ACLA, Librarians Advisory Council Minutes, 12 May 2011, ACLA-WIKI; ACLA, Librarians Advisory Council Minutes, 9 June 2011, ACLA-WIKI; Joe Smydo, "Library Supporters Collect 10,000 Names for Tax Referendum," *Pittsburgh Post-Gazette*, 26 July 2011; Rich Lord, "Carnegie Library Funding Approved by Big Margin," *Pittsburgh Post-Gazette*, 9 November 2011; Riely, "Allegheny County Library Systems Look to Cooperate."

73 Cynthia Richey, Letter from Cynthia Richey to ACLA Countywide Library Development Committee, 2001, ACLA historical file (Folder: Countywide Development), ACLA-OFFICE; ACLA, ACLA Service Standards for Local Libraries, 14 May 2001, ACLA historical file (Folder: Countywide Development), ACLA-OFFICE.

74 Marilyn A. Jenkins, Letter from Marilyn Jenkins to ACLA Member Libraries, 5 November 2001, ACLA historical file (Folder: Countywide Development), ACLA-OFFICE.

75 ACLA, Standards Recommended for ACLA Libraries, December 2002, ACLA-WIKI; ACLA, How to Configure Points, 2003, ACLA historical file (Folder: Standards), ACLA-OFFICE; ACLA Standards Taskforce, Recommended Standards for ACLA Libraries, January 2003, ACLA historical file (Folder: Standards), ACLA-OFFICE; ACLA, Preliminary Report to the General Membership, March 2003,

ACLA historical file (Folder: Standards), ACLA-OFFICE; ACLA, Summary of Discussion, 23 July 2003, ACLA historical file (Folder: Standards), ACLA-OFFICE; Marilyn A. Jenkins, Letter from Marilyn Jenkins to Marcia Taylor, 19 August 2003, ACLA historical file (Folder: Standards), ACLA-OFFICE; ACLA, Summary of Discussion, 20 August 2003, ACLA historical file (Folder: Standards), ACLA-OFFICE; ACLA, West, East, North, South Region Standards Worksheets, 17 September 2003, ACLA historical file (Folder: Standards), ACLA-OFFICE; ACLA, Summary of Discussion, 17 September 2003, ACLA historical file (Folder: Standards), ACLA-OFFICE; Trustee 2, Interview Transcript, 2 July 2015, interviewed by Michael M. Widdersheim; Elliot Greenman, Letter from Elliot Greenman to Jennifer Fox Rabold, Marcia Taylor, and Marilyn Jenkins, July 2004, ACLA historical file (Folder: Standards). ACLA-OFFICE; Jennifer Fox Rabold, Letter from Jennifer Fox Rabold to Elliot Greenman, Marilyn Jenkins, and Marcia Taylor, 14 July 2004, ACLA historical file (Folder: Standards). ACLA-OFFICE; Marilyn A. Jenkins, Letter from Marilyn Jenkins to Board Presidents of ACLA Member Libraries, 27 July 2005, ACLA historical file (Folder: Standards). ACLA-OFFICE; ACLA, ACLA Service Standards for Local Libraries, 30 November 2005, ACLA-WIKI; Pennsylvania Department of Education, *Pennsylvania Library Laws* (Harrisburg, 28 August 2007).

76 K. J. Rafalko, Letter from K. J. Rafalko to ACLA Board, 20 May 2002, ACLA historical file (Folder: Countywide Services), ACLA-OFFICE.

77 ACLA, How to Configure Points, 2003; ACLA, Giant Eagle Model of Libraries, 13 October 2003, ACLA historical file (Folder: Standards), ACLA-OFFICE; Trustee 2, Interview Transcript, 2 July 2015.

78 ACLA, Standards for ACLA Member Libraries, Standards for ACLA Affiliates and Full Service Branches of Member Libraries, 31 May 2006, ACLA-WIKI; ACLA, *ACLA Strategic Plan Narrative Update–Year-End 2006* (www.aclalibraries.org/reports, December 2006); ACLA, New Directors Handbook, July 2006, 15; ACLA, Working Draft–Proposed Recommendations for Standards & Compliance, 12 September 2006, ACLA-WIKI; Olszak Management Consulting, ACLA Allocation Formula & Formula Development Process Audit Report & Recommendations, 20 August 2007, ACLA-WIKI.

79 ACLA, Formula 2006, 26 April 2006, ACLA-WIKI; Olszak Management Consulting, ACLA Allocation Formula & Formula Development Process, 20 August 2007.

80 Bill Zlatos, "Library Funding Plan Creates Winners, Losers," *Pittsburgh Tribune-Review*, 29 August 2006.

81 ACLA, ACLA Standards Compliance & Exemptions Procedures, 16 October 2006, ACLA-WIKI; ACLA, *ACLA Strategic Plan Narrative Update–Year-End 2006* (December 2006).

82 ACLA, *ACLA Strategic Plan Narrative Update–Year-End 2006* (December 2006).

83 Library Administrator 5, Interview Transcript, 16 July 2015, interviewed by Michael M. Widdersheim; Library Administrator 1, Interview Transcript, 28 July 2015.

84 Olszak Management Consulting, ACLA Allocation Formula & Formula Development Process, 20 August 2007; Library Administrator 1, Interview Transcript, 28 July 2015; Allegheny Regional Asset District, Report of the Library Committee, 3 October 2006, RAD files, Contractual Assets–Folders, Library Committee ACLA 1997–2006, ARAD-OFFICE; Allegheny Regional Asset District, Report of the Library Committee, 27 November 2006, RAD files, Contractual Assets–Folders, Library Committee ACLA 1997–2006, ARAD-OFFICE.

85 ACLA, North Region Minutes, 7 September 2007, ACLA-WIKI.

86 ACLA, *ACLA Strategic Plan Narrative Update–Year-End 2006* (December 2006); Olszak Management Consulting, ACLA Allocation Formula & Formula Development Process, 20 August 2007; Allegheny Regional Asset District, Report of the Library Committee, 27 November 2006; David Donahoe, Letter from David Donahoe to Marilyn Jenkins, 28 November 2006, RAD files, Contractual Assets–Folders, Library Committee ACLA 1997–2006, ARAD-OFFICE.

87 Library Administrator 1, Interview Transcript, 2016, January 27,

88 ACLA, *ACLA Strategic Plan Narrative Update–Year-End 2006* (December 2006); Olszak Management Consulting, ACLA Allocation Formula & Formula Development Process, 20 August 2007; ACLA, *2006 Annual Report to the Allegheny Regional Asset District* (www.aclalibraries.org/reports, 1 June 2007); ACLA, *2007 Annual Report to the Allegheny Regional Asset District* (www.aclalibraries.org/reports, 1 June 2008); ACLA, Librarians Advisory Council Minutes, 8 March 2007, ACLA-WIKI.

89 ACLA, *2007 Annual Report to the Allegheny Regional Asset District* (1 June 2008); ACLA, New Directors Handbook, July 2006; ACLA, Librarians Advisory Council Bylaws, March 2006, ACLA-WIKI; ACLA, Recommended Salary Structures, 8 March 2006, ACLA-WIKI; ACLA, Highlights from 2007, 2007, www.aclalibraries.org/reports; ACLA, Customer Services Standards and Practices, April 2007, ACLA-WIKI; Bill Zlatos, "Groups Seek $98.6 m in RAD Funds," *Pittsburgh Tribune-Review*, 21 July 2007; Bill Zlatos, "Library, Aviary, Museums Pitch for Money," *Pittsburgh Tribune-Review*, 28 August 2007; Cynthia Richey, LAC Representative on ACLA Board of Directors–Report for Librarians Advisory Council, June 2009, ACLA-WIKI; ACLA, *ACLA 2010 Annual Report* (1 June 2011), aclalibraries.org/reports.

90 ACLA, Librarians Advisory Council Minutes, 8 March 2007; Olszak Management Consulting, ACLA Allocation Formula & Formula Development Process, 20 August 2007; ACLA, North Region Minutes, 7 September 2007; ACLA, Summary of Major Action and Discussion Items: Special Meeting of the ACLA General Membership, 20 September 2007, ACLA-WIKI; Marilyn A. Jenkins, Special Membership Meeting, 20 September 2007, ACLA-WIKI; ACLA, Request for Consulting Services, 21 September 2007, ACLA-WIKI; ACLA, Librarians Advisory Council Minutes, 11 October 2007, ACLA-WIKI; Marilyn A. Jenkins, Report to the Librarians Advisory Council, 20 September 2007, ACLA-WIKI; Richey, LAC Representative on ACLA Board of Directors – Report for Librarians Advisory Council Meeting, 11 October 2007; ACLA, RAD Formula 2008, 19 December 2007, ACLA-WIKI; ACLA, *2007 Annual Report to the Allegheny Regional Asset District* (1 June 2008); Olszak Management Consulting, Formula Committee Meeting Minutes, 21 February 2008, ACLA-WIKI; Richey, LAC Representative on ACLA Board of Directors – Report for Librarians Advisory Council Meeting, 8 November 2007; Richey, LAC Representative on ACLA Board of Directors – Report for Librarians Advisory Council Meeting, 14 February 2008; Olszak Management Consulting, ACLA Formula Development Process Round 1 – Progress Summary, 27 May 2008, ACLA-WIKI; Olszak Management Consulting, ACLA Formula Development Process Round 3 – Progress Summary, 9 October 2008, ACLA-WIKI; Olszak Management Consulting, Formula Committee Meeting Minutes, 8 April 2008, ACLA-WIKI.

91 ACLA, Highlights from 2007, 2007; Olszak Management Consulting, ACLA Allocation Formula & Formula Development Process, 20 August 2007, 34.

92 Olszak Management Consulting, ACLA Allocation Formula & Formula Development Process, 20 August 2007, 35.

93 Olszak Management Consulting, ACLA Allocation Formula & Formula Development Process, 20 August 2007, 48; Olszak Management Consulting, Formula Committee Meeting Minutes, 27 August 2008, ACLA-WIKI.

94 Olszak Management Consulting, Formula Committee Meeting Minutes, 27 August 2008; Olszak Management Consulting, ACLA Formula Development Process Round 3 – Progress Summary, 9 October 2008.

95 ACLA, RAD Distribution to Member Libraries 1995 – 2016, 7 March 2016, ACLA-WIKI.

96 ACLA, FDP Discussion Chain from Wiki, 1 May 2009, ACLA-WIKI.

97 Olszak Management Consulting, Formula Committee Meeting Minutes, 29 October 2008, ACLA-WIKI; Olszak Management Consulting, ACLA Formula Development Process Round 5 – Progress Summary, 25 March 2009, ACLA-WIKI; Olszak Management Consulting, Formula Committee Meeting Minutes, 11 March 2009, ACLA-WIKI; ACLA, Report on South Region Meeting, 17 Februrary

2009, ACLA-WIKI; Olszak Management Consulting, Formula Committee Meeting Minutes, 8 April 2009, ACLA-WIKI.

98 Olszak Management Consulting, Final Recommendations for Funding Formula: Factor Calculations and Formula Selection Guide, 21 May 2009, ACLA-WIKI; Olszak Management Consulting, Formula Committee Meeting Minutes, 14 April 2009, ACLA-WIKI; Allegheny Regional Asset District, Report of the Library Committee, 30 March 2009, RAD files, Contractual Assets–Folders, Library Committee ACLA 1997–2006, ARAD-OFFICE; Mary Niederberger, "New RAD Formula May Hurt Libraries," *Pittsburgh Post-Gazette*, 14 May 2009; Mary Niederberger, "Change in Formula May Push Libraries Funds Down, Up," *Pittsburgh Post-Gazette*, 21 May 2009; Bob Hoover, "Libraries Taken Aback by RAD Reslicing," *Pittsburgh Post-Gazette*, 25 June 2009; Laura Van Wert and Stephanie Hacke, "Library Directors Brace for Deep Funding Cuts," *Pittsburgh Tribune-Review*, 23 July 2009; Bill Zlatos, "Libraries Ask RAD for 1-Time Grant to Soften Expected Cuts," *Pittsburgh Tribune-Review*, 26 July 2009; Olszak Management Consulting, FDP Meeting Minutes by Rounds, 5 October 2009, ACLA-WIKI.

99 Hoover, "Libraries Taken Aback by RAD Reslicing."; Zlatos, "Libraries Ask RAD for 1-Time Grant to Soften Expected Cuts."

100 Bob Hoover, "RAD Board Rejects Revised Funding Formula for Libraries," *Pittsburgh Post-Gazette*, 30 September 2009; RAD, Report of the Library Committee, 29 September 2009, ACLA-WIKI; Allegheny Regional Asset District, Report of the Library Committee, 29 September 2009, RAD files, Contractual Assets–Folders, Library Committee ACLA 1997–2006, ARAD-OFFICE; Pittsburgh Post-Gazette, "Open Book: County Libraries Need a Diverse Funding Plan," *Pittsburgh Post-Gazette*, 5 October 2009.

101 ACLA, ACLA Special General Membership Meeting, 6 October 2009, ACLA-WIKI; ACLA, Librarians Advisory Council Meeting Minutes, 7 October 2009, ACLA-WIKI; ACLA, ACLA RRT First Session Notes, 12 October 2009, ACLA-WIKI; Helen Palascak, Local Income Factor, 12 October 2009, ACLA-WIKI; Helen Palascak, Local Income Factor Rationale, 12 October 2009, ACLA-WIKI; ACLA, ACLA RRT Second Session Notes, 14 October 2009, ACLA-WIKI; ACLA, Letter from ACLA to RAD Board: Response to RAD Library Committee Report, 23 October 2009, ACLA-WIKI; ACLA, General Membership Meeting Formula Votes, 29 October 2009, ACLA-WIKI; Marilyn A. Jenkins, Letter from Marilyn Jenkins to David Donahoe: Response to Concerns, 2 November 2009, ACLA-WIKI; ACLA, *Highlights from 2009* (2009).

102 Bill Zlatos, "RAD Looking between the Lines," *Pittsburgh Tribune-Review*, 5 November 2009; ACLA, Minutes of the Librarians Advisory Council, 12 November 2009, ACLA-WIKI.

103 Bob Hoover, "Uncertainty About Library Funding Ends with Reslicing of Pie," *Pittsburgh Post-Gazette*, 2 December 2009.

104 Allegheny Regional Asset District, Report of the Library Committee, 1 December 2009, RAD files, Contractual Assets–Folders, Library Committee ACLA 1997–2006, ARAD-OFFICE; RAD, Library Commitee Report, 1 December 2009, ACLA-WIKI; Hoover, "Uncertainty About Library Funding Ends with Reslicing of Pie."

105 ACLA, Minutes of the Librarians Advisory Council, 14 January 2010, ACLA-WIKI.

106 Marilyn A. Jenkins, Executive Director's Summary of Major Action and Discussion Items: ACLA Board of Directors Meeting, 18 January 2010, ACLA-WIKI; Jenkins, Executive Director's Summary of Major Action and Discussion Items, 22 February 2010; David Donahoe, Letter from David Donahoe to Marilyn Jenkins: Questions About ACLA, 10 March 2010, ACLA-WIKI; ACLA, Charge to the ACLA Members, 7 April 2010, ACLA-WIKI; Marilyn A. Jenkins, Letter from Marilyn Jenkins to David Donahoe: ACLA Responses to Library Committee Questions, 30 March 2010, ACLA-WIKI; ACLA, Librarians Advisory Council: Discussion of RAD Question, 7 April 2010, ACLA-WIKI; ACLA, General Membership Meeting, 27 April 2010, ACLA-WIKI; ACLA, 2011 RAD Formula Task Force, 28 April 2010, ACLA-WIKI; Marilyn A. Jenkins, Letter from Marilyn Jenkins to Rob Jones, 6 May 2010, ACLA-WIKI; ACLA, Questions and Answers on 2011 Formula Recommendations, May 2010, ACLA-WIKI.

107 Marilyn A. Jenkins, Executive Director's Summary of Major Action and Discussion Items, 17 May 2010, ACLA-WIKI; ACLA, 2011 Proposed Formula (Based on 2008 Data), 3 May 2010, ACLA-WIKI; ACLA, 2011 Proposed Formula (Based on 2009 Data), 14 June 2010, ACLA-WIKI; Bill Zlatos, "Library Group Revises Funding Formula," *Pittsburgh Tribune-Review*, 11 June 2010; Marilyn A. Jenkins, Executive Director's Summary of Major Action and Discussion Items: ACLA General Membership Special Meeting, 21 June 2010, ACLA-WIKI; Bill Zlatos, "Library Funding Shakeup Urged," *Pittsburgh Tribune-Review*, 22 June 2010; Allegheny Regional Asset District, Report of the Library Committee, 30 September 2010, RAD files, Contractual Assets–Folders, Library Committee ACLA 1997–2006, ARAD-OFFICE; RAD, Allegheny Regional Asset District Report of the Library Committee, 30 September 2010, ACLA-WIKI.

108 RAD, RAD Approves $8.1 Million Budget for 2011, 30 November 2010, ACLA-WIKI.

109 Library Administrator 6, Interview Transcript, 6 March 2015, interviewed by Michael M. Widdersheim.

110 ACLA, Minutes of the Librarians Advisory Council, 16 October 2008; Cynthia Richey, Formula Process–Phase II Follow-Up, 2 September 2008, ACLA-WIKI; ACLA, Critical Issues Launch and Oversight Committee Meeting Summary, 24 November 2008, ACLA-WIKI; ACLA, Critical Issues Launch and Oversight Committee Meeting Summary, 7 April 2009, ACLA-WIKI.

111 Olszak Management Consulting, Formula Committee Meeting Minutes, 10 June 2008, ACLA-WIKI; Olszak Management Consulting, ACLA and CLP and EIN, 10 April 2008, ACLA-WIKI; Olszak Management Consulting, ACLA Restructure: Board of Directors, LAC, Members, Regions, July 2008, ACLA-WIKI; Olszak Management Consulting, ACLA New Structure, 17 July 2008, ACLA-WIKI; Olszak Management Consulting, Timeline to Restructure ACLA, 11 August 2008, ACLA-WIKI; ACLA, Critical Issues Process, 27 January 2010, ACLA-WIKI; ACLA, Minutes of the Librarians Advisory Council, 14 January 2010; Marilyn A. Jenkins, Executive Director's Summary of Major Action and Discussion Items, 15 March 2010, ACLA-WIKI; ACLA, ACLA 7 Regions Map with Libraries, 5 August 2008, ACLA-WIKI; ACLA, General Membership Meeting, 27 April 2010.

112 ACLA, *2007 Annual Report to the Allegheny Regional Asset District* (1 June 2008); ACLA, North Region Meeting, 9 October 2007, ACLA-WIKI; Act 57: The Library Code, Amendments to State Aid to Local Libraries, P. L. 404 (22 June 2000); Pennsylvania Department of Education, Standards for Local Libraries and Library Systems, 5 August 2008, ACLA-WIKI; Olszak Management Consulting, Formula Committee Meeting Minutes, 18 June 2008, ACLA-WIKI; ACLA, Minutes of the Librarians Advisory Council, 11 March 2010, ACLA-WIKI; General Assembly of the Commonwealth of Pennsylvania Joint State Government Commission, *The Pennsylvania Public Library Code: Findings and Recommendations* (December 2010); Marilyn A. Jenkins, Executive Director's Summary of Major Action and Discussion Items, 18 April 2011, ACLA-WIKI.

113 ACLA, Librarians Advisory Council Minutes, 11 September 2008, ACLA-WIKI.

114 Richey, LAC Representative on ACLA Board of Directors – Report for Librarians Advisory Council Meeting, 11 October 2007.

115 ACLA, 2007 Compliance (Uncertified), 17 September 2008, ACLA-WIKI.

116 ACLA, *2007 Annual Report to the Allegheny Regional Asset District* (1 June 2008).

117 Jenkins, Executive Director's Summary of Major Action and Discussion Items, 18 April 2011.

118 ACLA, ACLA Service Standards for Local Libraries, November 2009, ACLA-WIKI.

119 ACLA, Minutes of the Librarians Advisory Council, 11 March 2010.

120 ACLA, ACLA Service Standards for Local Libraries, November 2009.

121 ACLA, Minutes of the Librarians Advisory Council, 11 March 2010.
122 ACLA, System Member Standards, 1 April 2010, ACLA-WIKI.
123 ACLA, Compliance Questions, 22 June 2010, ACLA-WIKI.
124 Marilyn A. Jenkins, Executive Director's Summary of Major Action and Discussion Items: ACLA Board of Directors Meeting, 21 June 2010, ACLA-WIKI.
125 Jenkins, Executive Director's Summary of Major Action and Discussion Items: ACLA Board of Directors Meeting, 21 June 2010.
126 Marilyn A. Jenkins, Executive Director's Summary of Major Action and Discussion Items: ACLA General Membership Meeting, 20 December 2010, ACLA-WIKI; Joint State Government Commission, *The Pennsylvania Public Library Code: Findings and Recommendations* (December 2010).
127 ACLA, ACLA Strategic Plan Committee, 12 May 2010, ACLA-WIKI; ACLA, Creating the Future of Our Libraries: A Strategic Plan for 2011–2013, 1 September 2010, ACLA-WIKI; ACLA, ACLA 2011 Proposal to the Allegheny Regional Asset District, 14 July 2010, ACLA-WIKI; Marilyn A. Jenkins, Executive Director's Summary of Major Action and Discussion Items: ACLA Board of Directors Meeting, 20 September 2010, ACLA-WIKI; Hill Group, Our Libraries of Tomorrow: Final Project Report and Process; Overview with Appendices, 30 October 2012, ACLA-WIKI.

Period 9: Compliance Culture, 2011–2016

During Period 8, there were attempts by ACLA HQ to substantively modify the public library infrastructure in greater Pittsburgh, both by creating a new eiNetwork governance structure and by creating a new RAD distribution formula, one that combined state and RAD funding. These attempts were unsuccessful: eiNetwork's governance problems remained unresolved, and though usage metrics were incorporated into the formula, local income was still emphasized as it had been. The proposal to incorporate state aid was rejected. Despite the underlying contentiousness during Period 8, the infrastructure remained well-funded and was highly utilized on its surface. ACLA's 45 suburban public libraries continued to provide services from 70 locations across the region.[1] Circulation, cardholders, computer usage, and downloads of electronic items were up. Though communication problems continued to plague eiNetwork, more and more digital resources were provided through the infrastructure.[2] EiNetwork continued to manage the computers, local networks, a wide-area network, and the shared ILS for the public library infrastructure.[3] A new fiber ring connecting ACLA libraries was constructed in 2012–2013.[4] Under the free access provisions, residents continued to use libraries outside of their own municipalities. The bookmobile continued to visit preschools, senior facilities, and the general community.[5] ACLA continued to use the same funding formula for 2012, one with 7 factors: base support, population, local income, in-house circulation, non-resident circulation, computer use, and distress.[6] This formula was backed by the RAD board. Funding for the ACLA libraries came from RAD, state aid, table gaming income, local government income, and donations and gifts.[7]

Despite stable funding and use, however, ACLA member libraries nevertheless struggled to maintain uniform service quality across the region. This was due in part to underlying disparities in local funding. Of ACLA's member libraries, 11 libraries had local income that accounted for at least two-thirds of their total income, another 22 libraries had local income account for between 50 percent and 66 percent of their total income, and 11 libraries had local income account for less than half of their total income.[8] In a way, the RAD funding formula perpetuated this disparity because it continued to reward libraries that already had strong local funding. It was not apparent that RAD funding expanded services to unserved areas. In 2012, Allegheny County had 130 municipalities, but 24 of these reported that they had no local library, and of those, nearly all provided no financial support to other libraries in the county.[9] While there was some improved service in selected locations, service gaps persisted. It was puzzling how new library buildings emerged in some parts of the county while other libraries elsewhere struggled

https://doi.org/10.1515/9783111013404-011

or were consolidated. For instance, Robinson Township Library was newly established in an affluent area in 2007, yet nearby F.O.R. Sto-Rox Library, located in an economically distressed area, floundered.[10] Meanwhile, Millvale Community Library became the newest ACLA-eligible library in 2015, even though several other libraries already existed short distances away.[11] There seemed to be no effective mechanism found by ACLA to allocate resources where they were most needed.

It was at this time that there was a resurgence of interest in regional initiatives in greater Pittsburgh. There was some talk of consolidation of local governmental functions and services into larger entities.[12] The economic downturn of 2008 and 2009 in part prompted local leaders to view the economy regionally.[13] But the defining theme of Period 9 was accountability of funding. RAD fostered a culture of compliance within ACLA, and ACLA HQ, in turn, inculcated compliance in its member libraries.

Illustrative of RAD's and ACLA HQ's compliance culture was ACLA HQ's monitoring of member libraries' fiscal accounting of both state and RAD funds. As part of its responsibilities as a condition for receiving state funding, ACLA HQ collected annual reports for its member libraries and submitted them to the state. ACLA HQ also submitted financial audits for each member library. It was the responsibility of the member libraries to conduct the audits and submit them to ACLA HQ. If those audits were not submitted, ACLA HQ had the authority by the state to withhold that library's state funding. This same situation held true for RAD funding. As it did for the state, ACLA HQ collected annual audits of ACLA member libraries to submit to RAD.[14] The audits were to demonstrate that RAD funding had been used for appropriate purposes. Libraries that did not submit the audit on time or submitted one incompletely had RAD funds withheld "until compliance was established."[15] ACLA HQ enforced compliance by revising its Cooperation and Support Agreement (CASA), bylaws, and Member Library Agreement (MLA). The CASA was the contract between each of the member libraries and ACLA HQ.[16] The CASA was required by RAD as a condition for receiving RAD funding, while the MLA was required under the state library code. Bylaws were required for ACLA by the state because ACLA was registered in Pennsylvania as a 501(c)3 non-profit corporation.[17]

As part of a larger trend in compliance culture, ACLA was audited by RAD for its efficiency as an organization.[18] The audit found that ACLA's "cost reporting, funds, reporting, cost recovery, administrative costs and salary, and internal controls were adequate," and that "RAD and state fund distributions show that the distributions were in compliance with appropriate contractual requirements." At the same time, the report also found duplication of services between ACLA HQ and CLP's district services. Duplicated services included children's and young adult programming, continuing education, managing data for state reports, purchasing, and management of eiNetwork.[19] It was also found in the report that communication within ACLA HQ was not effective, transparent, or oriented to consensus-building. The audit distilled the feelings of many member libraries from Period 8, which was characterized by divisive and laborious formula revisions and secretive meetings about eiNetwork restructuring. The audit report stated that "ACLA's leadership and organizational practices have played a large role in creating and sustaining this cultural and organizational dysfunction."[20]

ACLA HQ responded to the audit by promising to increase and improve dialogue between ACLA and CLP management.[21] The RAD Library Committee encouraged ACLA and CLP to examine the issues raised in the audit during their strategic planning processes with a view to reducing overhead costs.[22] To some extent, ACLA HQ and CLP carried out this directive by better delineating their services, though their finances remained entangled.[23] In 2012 and onward, CLP was an ACLA member, but it made its own budget proposal to RAD and it received its RAD funding directly from RAD. At the same time, CLP was still designated by the state as the district library center, and, for this reason, district library aid and county coordination aid from the state went to CLP instead of ACLA. ACLA received no district services aid and no county coordination aid from the state.[24] Confusingly, however, CLP received its state-aid distribution funding through ACLA HQ. CLP used this money for electronic resources and mobile services. EiNetwork was not treated as a separate RAD grant. Instead, both ACLA and CLP funded eiNetwork, though CLP continued to be the fiscal agent of eiNetwork. On the face of things, ACLA operated as a pass-through agency for state and RAD funds, channeling funding to CLP, eiNetwork, and its member libraries.

Beneath the surface, however, ACLA's financial accounting was more complicated. It was more than just a pass-through organization. Since its creation in 1994, ACLA HQ had significantly expanded. Some of its functions included "Compliance and consulting services," "Collaborations and strategic alliances," and "Continuing education and professional development," "Adult programming," "Youth services," and "Mobile services."[25] ACLA HQ employed an Executive Director, Assistant Director, and multiple coordinators.[26] Many district library center functions had gradually transitioned from CLP to ACLA.[27] At first, the budgetary expansion exhibited by ACLA was due in part to a change in how ACLA HQ handled the state funding it received. ACLA HQ began to decide on its own how to use the portion of state funding that was not distributed to its member libraries. This practice differed from how ACLA handled state money in 1998 when the association first began receiving state aid. At that time, and until 2002, it was the LAC that recommended how to use that funding to the ACLA board. The ACLA board then made the decision. That practice ceased, and instead ACLA HQ bypassed LAC and the

ACLA board and put the state money toward what it called "system services."[28] This accounting practice was done because ACLA's original agreement with RAD was to use no more than 1 percent of its budget for funding administrative costs. Since 1995, RAD has held ACLA's administrative costs close to the 1 percent maximum. Since 1995, it identified several other revenue streams, including state aid and table gaming money. With an influx of multiple revenue streams, ACLA's administrative overhead also grew. By 2015, administrative costs represented 1.6 percent of ACLA's total budget, including distributions for its member libraries.[29] Administration used 2.6 percent of ACLA's total RAD funding, which was more than the 1 percent originally agreed upon by RAD.[30]

In the wake of the audit of ACLA by RAD, it was recommended that ACLA HQ reorganize to create two separate entities, one for managing library operations and the other for managing fundraising, data management, and fiscal accountability.[31] This recommendation was adopted by ACLA, resulting a new bifurcated budget structure, which in turn enabled ACLA HQ to grow even further. ACLA HQ split its budget into "Administration" and "System Services." This way, the costs for the "System Services" could continue to grow while the "Administrative" side remained low. The administration budget included two salaries, it was funded almost entirely by RAD money, and it remained close to RAD's 1 percent maximum.[32] System services included functions such as advocacy, capacity building, communications, professional development, and resource development, and it included several salaries. Most of the system services budget was funded through state aid (52 percent). The rest was funded through grants and contributions (43 percent) and earned revenue (5 percent). No RAD money was allocated to system services. ACLA HQ kept system services separate from administration to make its overall administrative costs appear smaller. ACLA HQ received state aid in excess of the amount distributed to member libraries because ACLA HQ produced "local financial effort."[33] This is not to mention a third section of ACLA HQ's budget dedicated to mobile services. [34] Needless to say, ACLA HQ's budget reveals that it expanded beyond the role of pass-through organization that was initially envisioned in 1994.[35]

In light of the audit, ACLA HQ delineated its services from CLP's, and it redesigned its budget to distinguish internal operations between administration and system services. These changes complied with RAD's directives. RAD also pushed ACLA to develop service standards, an initiative that failed in Period 8.[36] There was little appetite at this time within ACLA for another protracted formula debate. By 2012, the ACLA member libraries only recently emerged from a contentious formula development process that concluded in 2010. There seemed to be a consensus at that time that the formula should be left untouched for 3 years so that other issues could be addressed. There was some discussion and debate about the formula, especially concerning the allocation for e-resources, but the formula remained unchanged in Period 9.[37]

Rather than try to resuscitate formula discussions, ACLA HQ instead focused on developing service standards. These standards could be used not to distribute RAD money, but to direct how RAD money was utilized by libraries. The discussion about performance standards was revived in a roundabout way, using ACLA's audit by RAD as cover. Ostensibly, in the wake of the audit, ACLA and CLP entered into a strategic planning process to decide how to restructure to make the public library infrastructure more efficient. This might include combining the three organizations—ACLA, CLP, and eiNetwork—into a single one. It was during this strategic planning process that the issue of performance standards emerged, albeit in an oblique and seemingly unrelated way. The emergence of the issue of standards was painted as organic and member driven. In fact, it was manufactured from the start by ACLA HQ.

In the wake of the RAD audit, ACLA and CLP entered into discussions to creatively reimagine how they might move forward together, possibly as a single organization.[38] The process came to known as Library Service in the 21st Century, or LS21.[39] A task force was charged to develop recommendations for how to structure and fund comprehensive twenty-first century library services throughout the county.[40] ACLA board members hoped and expected that LS21 would result in a new overall vision for public library service in Allegheny County.[41] Discussions lasted from 2013 through 2014. Two processes ran concurrently: community discussions, which were in principle open to anyone and included representatives from the ACLA membership, were supposed to inform discussions held by a separate task force, which was charged with developing final recommendations.[42] The task force evolved into what was called the County-City Library Service Panel (CCLS).[43] The charge to the panel was to "determine if there is a better organizational model for delivering library service in Allegheny County and if so, how it should be structured, governed and funded." The purpose of the panel was, first, to "develop an overall vision for all library services in Allegheny County and the City of Pittsburgh that will best meet the needs of the public," and second, "to determine how comprehensive 21st Century library services within the county should be structured and funded." The panel focused on issues such as equitable access for all residents, standardizing the service delivery system in whole or in part, leveraging existing and new resources, collectively advocating more effectively for funding, ensuring data-based design decisions, and "maximizing collective resources of libraries while responding to unique neighborhood needs."[44] A $2.6 million deficit was projected for ACLA, CLP, and eiNetwork by 2018.[45] The CCLS panel investigated local funding patterns and alternative county-based library systems.[46] Several different organization structures were considered.[47]

One of the sub-committees in the panel remained focused on restructuring possibilities. It closely considered consolidation and analyzed from a legal standpoint what would be necessary to completely reorganize the county structure into a single system. [48] According to their analysis, it would be possible for CLP to convert from a public trust to a non-profit corporation, then merge with eiNetwork and ACLA.[49] The sub-committee also considered how ACLA could be dissolved in line with the state library code and in such a way that state funding would continue to be distributed to current ACLA member libraries.[50] Another subcommittee reviewed ACLA member library compliance with state standards. It was reported that "27 libraries failed to completely meet those standards in 2012."[51]

From the beginning of LS21, there were criticisms of the process. It was said that service problems throughout the county were the result of unequal local support, and that redistributing wealth and resources was not the solution.[52] Other commentators noted that the LS21 process was being conducted without adequate rationale.[53] The concern about the lack of municipal representation was echoed throughout the process.[54] Participants in other community meetings were also confused. There were concerns that the direction of the conversations drifted away from the actual concerns raised in the audit report that related to management and duplication between ACLA HQ, CLP, and eiNetwork. Instead, the community discussions focused on individual libraries. There were concerns that individual libraries in the county were being portrayed in an unduly negative light.[55] Participants were also concerned that the CCLS panel discussed consolidation.[56] The concern about consolidation was part of a larger problem many saw with the process, namely that the work of the CCLS panel was disconnected from the community meetings. There were several reasons for this perception, including how, at least initially, the meetings of the CCLS panel were closed and how the content of what CCLS panelists shared about their work in public did not match with what they discussed in their working sessions. There was concern that there was a foregone conclusion to the LS21 process.[57] Participants elaborated how the LS21 process had an inherently negative slant.[58] Observers were concerned that the presentation by CLP, eiNetwork, and ACLA HQ to the CCLS panel were misrepresentative and were designed to direct the attention of the panel away from those 3 entities and toward individual libraries.[59] In presentations to the CCLS panel, ACLA HQ was accused of using inaccurate and outdated information on standards compliance to shift focus to member library standards.[60]

In the end, the CCLS panel recommended centralizing some library functions, including shipping, technology, e-resources, and maybe cataloging, marketing, and fundraising. It was not clear, however, what entity would carry out these functions, whether it would be a new single entity or separate existing ones. Panelists agreed that the libraries faced a funding crisis and that a new countywide funding source should be pursued, though it was unclear what that would be.[61] The final report of the CCLS panel was released in July 2014. The report made 5 recommendations to the CLP and ACLA boards for more effective public library service: 1) establish mandatory standards for libraries, 2) combine CLP district services, ACLA HQ, and eiNetwork into a single entity responsible for centralized services, 3) enable individual libraries to opt in to centralized services, 4) apply funding to centralized services, and 5) identify new funding sources.[62] Performance standards of the independent libraries were a central focus of the report. The report proclaimed that neither ACLA, CLP, nor eiNetwork had the "authority to mandate performance expectations of the independent libraries in the current federated system."[63] The report also proclaimed that only 17 out of the 45 ACLA member libraries met the state standards for federated systems. The report claimed that the public library system in Allegheny County was underperforming.

The claims made by the report regarding standards were misleading. In fact, the state library code did not stipulate that all libraries in a federated system such as ACLA had to meet incentive-level standards, only that the system as a whole had to meet those standards. Under ACLA's CASA, libraries only had to meet minimum state standards to qualify for membership and RAD funding. The report also failed to mention that the state aid system had been broken since 2002–2003. Libraries still received some state aid on a pro-rated basis according to what they received in 2002–2003. There was not adequate state aid to distribute incentive-level funding to libraries across the state. Moreover, the report omitted how libraries routinely received waivers and were granted exceptions when standards were not met. The report made it appear that ACLA was in danger of losing state funding, but this was not true. Finally, the report made it appear that library service quality was wildly inconsistent, which was false. ACLA member libraries developed several internal policies and best practices related to circulation, cataloging, and delivery. All the libraries shared the same ILS, catalog, and technology provider, eiNetwork. There were joint decisions regarding the purchase of e-resources. In other words, a large degree of consistency was required to make ACLA's system work. This point was not emphasized in the CCLS panel report. Instead, the report focused on perceived underperformance.

As LS21 participants noted earlier, the CCLS panel portrayed ACLA libraries in an unduly negative light. The misleading and negative claims related to library performance, whether intentionally misleading or not, were used to support the report's first recommendation to establish mandatory standards for library service among ACLA member libraries. The report recommended that ACLA adopt the standards that were proposed in 2009. These standards were debated within ACLA in 2010, but they were tabled at that time pending revisions of the state library code that were underway. The ACLA standards included categories for independent libraries, administrative units, and branches. The standards for independent libraries were similar though not identical to the state's incentive-level standards. The proposed ACLA standards also included some basic state standards, and there were also additional standards specific to ACLA that related to board structure, administration, and assets.[64]

Beyond the recommendation of mandatory ACLA standards, the CCLS panel report also suggested that ACLA, district services, and eiNetwork merge to form a single entity that would provide a number of centralized services. From the report, it is unclear if the recommendation was to merge CLP entirely or just CLP's district services. The centralized services proposed in the report included human resources, finance, collections, and cataloging. Centralized human resources included uniform position descriptions, pay equity, healthcare benefits, pensions, portability, and professional development. Finance services included accounting, audits, and reports.[65] Separate subcommittees of the panel investigated these various aspects of this proposal. Finally, the panel suggested that new funding sources be pursued to support the newly structured system.

The panel called their set of recommendations an action plan, but this description is misleading. In fact, the report only offered vague and high-level suggestions for making the county library service more efficient. There were no detailed steps in the report for implementing these ideas. For instance, in the report, there was no specific action recommended to obtain additional funding. Even though a dedicated county library tax had been discussed earlier by the panel, this suggestion was not present in the report. And while there was a detailed legal analysis conducted by two panel members regarding what would have to happen legally to merge ACLA, CLP, and eiNetwork, the report seemed to suggest that only district services, ACLA, and eiNetwork should merge. It did not recommend that CLP merge in its entirety. It remained unclear how such a merger would happen.[66]

The CCLS panel was not a decision-making body, only an advisory one. After July 2014, in the months that followed the release of the LS21 report, the ACLA board, ACLA HQ, the ACLA regions, and LAC considered the report's recommendations. ACLA HQ outlined what types of services it would begin to offer on an opt-in basis. These included HR, financial support, risk management, and consulting and resources.[67] ACLA began to offer more centralized services.[68] A county library tax was proposed by CCLS, but this proposal was not pursued by ACLA.[69] In light of the CCLS panel report, ACLA HQ emphasized service standards, using those that had been tabled in 2010 as a starting point.[70] Overall, it was said that the LS21 process was "stunted," that it "wavered off of path," that it "never completely blossomed," and that it "grew into something no one planned."[71] Any momentum for a consolidated system or for a countywide tax fizzled out by the fall of 2014. This was attributed to resistance from municipalities, resistance from individual libraries, and withdrawal by CLP from discussions.[72]

Later in 2014, following the release of the CCLS panel report, discussions shifted to developing standards for ACLA member libraries.[73] ACLA's new proposed standards deviated from the state's existing incentive-level standards. Compliance with the incentive-level standards was recommended by the CCLS panel report, but ACLA HQ noted that state standards were dated and were under review by the State Librarian.[74] The new proposed ACLA standards would instead account for community needs.[75] One dilemma facing ACLA HQ and the ACLA board in their attempts to justify the new ACLA standards was how libraries were deficient in standards that were no longer perceived to be current anyway. ACLA HQ, in "talking out of both sides of their mouth," caused consternation among the membership.[76] Another difficulty for ACLA HQ and the ACLA board, in trying to develop their own standards, was that the state had not yet finalized its own. All that was known was that the state library had moved in the "direction of emphasizing community planning as they key element."[77] A draft of the new state standards was set to be released in the summer of 2015, and some preliminary discussions were held with library systems and districts prior to that.[78] Yet, it was contradictory for ACLA HQ and the ACLA board to insist that ACLA member libraries comply with state standards when those new standards were unknown. What complicated the situation even further was that the state librarian resigned in the middle of redoing the state standards.[79] The state standards revision project was thus left unfinished in early 2015.

These challenges notwithstanding, a draft of ACLA standards was created by late 2014. ACLA HQ and the ACLA board did not wait for the state to complete its own standards revision process.[80] Given the emphasis by the state library on community-responsive services, ACLA HQ projected that moving forward it would assist libraries in the community assessment process, which would form the basis for service plans, collection development plans, staffing, and technology plans.[81] Yet, this was only conjecture. Community responsiveness was only one category within the proposed ACLA standards. The draft of ACLA system standards contained 6 sections on General Compliance, Community Responsiveness, Consortium Participation, Governance, Funding, and New Libraries. The first section stated general membership conditions.[82] The section on community responsiveness outlined a process of community assessment planning. Consortium participation listed several system operations where member libraries would be expected to contribute according to specific standards. The governance section spoke to the required structure and policies for local library boards. The sections on funding and new libraries had more specific criteria than the other sections. The funding and new libraries sections in the proposed ACLA standards explicitly referenced the state's incentive-level standards.

The standards draft was circulated among ACLA member libraries in January 2015, and they would become part of ACLA's strategic plan for 2016 – 2018.[83] It was discussed not only what the standards would be but also what resources would be needed to achieve those standards.[84] It was unclear what the consequences would be for failing to meet the standards.[85] Some members felt the standards should be referred to as "best practices" instead of standards.[86] Notably, finding increased funding for libraries was not included as a goal in the draft, nor was re-assessing ACLA, eiNetwork, and CLP for cost efficiencies.[87] These points were later added in light of criticisms from the membership, but they reveal how singularly focused ACLA HQ was on adopting system standards.[88]

Resistance to the new standards grew. At their regional meeting in February 2015, the South Region unanimously voted to table the proposed ACLA standards until all ACLA member libraries achieve state standards. There were several reasons for this, including how the proposed ACLA standards might conflict with new state standards, the proposed ACLA standards were vague and not measurable, and the proposed ACLA standards were more accurately described as a checklist of best practices.[89] It was recommended that ACLA wait until the new state standards were finalized, then determine if any libraries were non-compliant.[90] Moreover, it was unclear who would vote on the standards, the ACLA membership or the ACLA board.[91] There continued to be push back over the standards—what they would be and what role ACLA HQ and the ACLA board would play in them.[92] By this time, a number of criticisms of the standards resounded throughout the membership.[93]

Nevertheless, despite extensive resistance, and despite not having updated state standards, ACLA HQ and the ACLA board continued to push through the new proposed ACLA standards.[94] The ACLA board solicited feedback, but it was clear that the board intended to pass standards in June 2015 with or without support from the membership.[95] Criticism, confusion, and worry over the standards continued to mount. The March 30 ACLA General Membership meeting was particularly contentious. There, it became clear that the decision about standards was in effect already made, and that it would be taken out of the hands of the membership. Member libraries were especially concerned that ACLA HQ and the ACLA board would withhold money from members who did not comply with standards.[96] The ACLA board acknowledged the repeated criticisms, but it committed to moving forward with the new standards anyway.[97] One trustee who was involved in both the CCLS panel and the ACLA board revealed the following mindset of the ACLA board:

> [Y]ou got a lot of disparate interest out there and there are people who are resistant to so-called central control... this issue with the standards was a very interesting issue. This organization [ACLA] has been trying to adopt standards for 15 years. And every time was to the point where they thought they had standards that made sense, roadblocks were put up, and, have you seen the 21st century library panel study? The first recommendation there is that there should be standards. I was committed that we were going to have standards. And I wanted it to bubble up organically from the library community, and we got tremendous input from a lot of libraries. But then at the end we really ran into some obstacles from the libraries we knew we would get obstacles from, but my position was, we're going to make it happen. And they were going to learn to live with it or they could go away. And I didn't tell people that, knowing full well nobody's going away.[98]

By June 15, 2015, when a final draft of system standards was released prior to the vote by the ACLA board, the membership was divided on the issue of the proposed standards, but the contentiousness from June had dissipated because member libraries were resigned to the reality that the standards would pass.[99] The ACLA board approved the standards 7–1.[100] Compliance and implementation of the standards would figure prominently in ACLA HQ's strategic plan and its communications with RAD.[101] Librarians within the membership observed that the LS21 process was engineered, and that during the subsequent discussions about standards, the ACLA board only pretended to gather feedback.[102] The ACLA board didn't let members vote on the standards because they wanted to push them through.[103] When pressed, ACLA trustees stated that one rationale for pushing the standards through was to identify shortcomings in order that funding could then be identified for improvements.[104] But this ignored the reality of the disparate funding situation in the county where some municipalities whose libraries did not meet state standards did not have additional resources to contribute. A county library tax would have been one way to ensure more equitable funding and to force freeloading municipalities to fund the county public library system.[105] But the CCLS panel report did not explicitly recommend this option, and the ACLA board did not pursue it.

Following the passage of the vote on ACLA standards, ACLA HQ moved toward monitoring and implementation of them. Compliance monitoring continued into 2016.[106] The RAD Library Committee supported moving forward with the standards.[107] In 2016, ACLA's MLA was revised again in light of the newly adopted standards.[108]

In Period 9, the passage and implementation of performance standards for ACLA member libraries solidified the decision-making core of the ACLA organization. It was shown how the membership votes, though part of the ACLA core, could be overruled by the ACLA board. The passage of standards also further solidified ACLA HQ as deferential to the RAD board, which supported and approved ACLA's standards. The passage of standards illustrated a trend of ACLA toward a consolidated governance structure and away from existing federated structure. Decision-making was exhibited to be more centralized than previously thought. The standards debate encountered high resistance and low legitimacy. It might have been thought that these conditions would lead to an outcome of continuity for this decision cycle. But as predicted by Releaser Theory, high resistance and low legitimacy can be overcome through high civil support. The ACLA core was able to implement these standards because it successfully mobilized civil support to overcome high resistance within the circuit. Though in a seemingly manufactured way, ACLA HQ did this by including community members in the LS21 process, then claiming that the issue of standards originated from civil society. Civil society was thus recruited by ACLA HQ to bolster the movement for performance standards.

1 ACLA, Library Service in the 21st Century: A County-Wide Conversation About Public Libraries, January 2014, ACLA-WIKI.

2 ACLA, ACLA South Region Meeting, 22 January 2013, ACLA-WIKI; ACLA, West Region Minutes, 23 January 2013, ACLA-WIKI; Marilyn A. Jenkins, Executive Director's Summary of Major Action and Discussion Items, 20 May 2013, ACLA-WIKI; ACLA, Librarians Advisory Council Minutes, 12 January 2012, ACLA-WIKI; Librarian 1, Interview Transcript, 7 January 2016, interviewed by Michael M. Widdersheim; eiNetwork, eiNetwork Advisory Council (EiNAC), 2013, May 24, ACLA-WIKI; ACLA, Librarians Advisory Council Minutes, 13 June 2013, ACLA-WIKI; ACLA, LAC South Region Meeting Notes, 23 July 2013, ACLA-WIKI.

3 ACLA, ACLA East Meeting Minutes, 6 January 2012, ACLA-WIKI; ACLA, Librarians Advisory Council Minutes, 12 January 2012; ACLA, DRC Member List, 2014, ACLA-WIKI; ACLA, Digital Resources Committee, 26 February 2014, ACLA-WIKI; ACLA, E-Strategy Task Force Report: Librarians Advisory Council Meeting, 14 November 2013, ACLA-WIKI; ACLA, ACLA 2011 Annual Report to the Regional Asset District, 15 February 2012, ACLA-WIKI; ACLA, Librarians Advisory Council Minutes, 13 June 2013.

4 eiNetwork, Update for ACLA General Membership, March 2013, ACLA-WIKI; Library Administrator 8, Interview Transcript, 18 April 2016, interviewed by Michael M. Widdersheim; ACLA, Librarians Advisory Council Minutes, 12 April 2012, ACLA-WIKI; ACLA, General Membership Meeting Minutes, 26 June 2012, ACLA-WIKI.

5 ACLA, ACLA 2011 Annual Report to the Regional Asset District, 15 February 2012; ACLA, ACLA CLP, 2013, ACLA-WIKI.

6 Allegheny Regional Asset District, Meeting Minutes, Board of Director's Meeting, 27 September 2011, ARAD-OFFICE.

7 ACLA, Library Incomes Sources, 2012, ACLA-WIKI.

8 ACLA, Library Incomes Sources, 2012.

9 ACLA, Municipal Support of Libraries, 2012, ACLA-WIKI.

10 Library Administrator 3, Interview Transcript, 21 July 2015, interviewed by Michael M. Widdersheim; Library Administrator 5, Interview Transcript, 16 July 2015, interviewed by Michael M. Widdersheim; Marilyn A. Jenkins, Executive Director's Summary of Major Action and Discussion Items,

26 September 2012, ACLA-WIKI; Marilyn A. Jenkins, Executive Director's Summary of Major Action and Discussion Items: ACLA Board of Directors Meeting, 29 October 2015, ACLA-WIKI; ACLA, Highlights from 2007, 2007, www.aclalibraries.org/reports.

11 Jenkins, Executive Director's Summary of Major Action and Discussion Items, 20 May 2013; Jenkins, Executive Director's Summary of Major Action and Discussion Items: ACLA Board of Directors Meeting, 29 October 2015; ACLA, 2016 RAD Proposal, August 2015, ACLA-WIKI; Library Administrator 5, Interview Transcript, 16 July 2015; Marilyn A. Jenkins, Executive Director's Summary of Major Action and Discussion Items: ACLA Board of Directors Meeting, 11 January 2016, ACLA-WIKI; ACLA, Librarians Advisory Council Minutes, 12 May 2016, ACLA-WIKI.

12 Power of 32, *Power of 32: 32 Counties, 4 States, 1 Vision*, Power of 32 (2011).

13 Allegheny Conference on Community Development, Allegheny County Economic Indicators, 2011, ACLA-WIKI; General Assembly of the Commonwealth of Pennsylvania, *The Pennsylvania Public Library Code: Findings and Recommendations* (Harrisburg, December 2010); ACLA, Librarians Advisory Council Minutes, 9 February 2012, ACLA-WIKI.

14 Library Administrator 6, Interview Transcript, 15 December 2015, interviewed by Michael M. Widdersheim.

15 Marilyn A. Jenkins, Executive Director's Summary of Major Action and Discussion Items, 29 March 2012, ACLA-WIKI; ACLA, Librarians Advisory Council Minutes, 14 June 2012, ACLA-WIKI; Marilyn A. Jenkins, Executive Director's Summary of Major Action and Discussion Items, 16 April 2012, ACLA-WIKI; Marilyn A. Jenkins, Executive Director's Summary of Major Action and Discussion Items, 10 December 2012, ACLA-WIKI.

16 ACLA, Cooperation and Support Agreement between the Allegheny County Library Association and Member Library, 1 January 2013, ACLA-WIKI; ACLA, Librarians Advisory Council Minutes, 14 February 2013, ACLA-WIKI; ACLA, ACLA South Region Meeting, 26 February 2013, ACLA-WIKI; ACLA, ACLA East Meeting Minutes, 7 June 2013, ACLA-WIKI.

17 ACLA, Member Library Agreement, 2012, ACLA-WIKI; ACLA, Service Standards for Local Libraries 2014, 2014, ACLA-WIKI; ACLA, Compliance: Independent Library Incentive Level Standards, May 2014, ACLA-WIKI; ACLA, Library Service in the 21st Century, 21 April 2014, ACLA-WIKI; ACLA, ACLA East Meeting Minutes, 1 February 2013, ACLA-WIKI; ACLA, Librarians Advisory Council Minutes, 12 January 2012; ACLA, Allegheny County Library Association Countywide Collection Management Statement, June 2012, ACLA-WIKI; ACLA, ACLA Service Standards for Local Libraries, 25 July 2012, ACLA-WIKI; ACLA, Member Library Agreement, 2013, ACLA-WIKI; Marilyn A. Jenkins, Executive Director's Summary of Major Action and Discussion Items, 21 January 2013, ACLA-WIKI; Marilyn A. Jenkins, Executive Director's Summary of Major Action and Discussion Items: ACLA General Membership Meeting, 18 March 2013, ACLA-WIKI; Jenkins, Executive Director's Summary of Major Action and Discussion Items, 26 September 2012; ACLA, Bylaws of Allegheny County Library Association, a Nonprofit Corporation (a Federated Library System), 18 June 2013, ACLA-WIKI; Marilyn A. Jenkins, Executive Director's Summary of Major Action and Discussion Items: ACLA General Membership Meeting, 18 June 2013, ACLA-WIKI; ACLA, General Membership Meeting Minutes, 18 June 2013, ACLA-WIKI; ACLA, ACLA East Meeting Minutes, 3 February 2012, ACLA-WIKI; Marilyn A. Jenkins, Executive Director's Summary of Major Action and Discussion Items: ACLA General Membershp Meeting, 30 November 2013, ACLA-WIKI; ACLA, Key Questions Regarding Member Library Agreement for Regional Discussion, April 2013, ACLA-WIKI; ACLA, 2013 Review of Member Library Agreement, September 2013, ACLA-WIKI; ACLA, ACLA South Region Meeting, 23 April 2013, ACLA-WIKI; ACLA, Librarians Advisory Council Minutes, 12 September 2013, ACLA-WIKI; ACLA, LAC South Region, 19 November 2013, ACLA-WIKI; ACLA, Librarians Advisory Council Minutes, 12 May 2016; Marilyn A. Jenkins, Executive Director's Summary of Major Action and Discussion Items: ACLA Board of Directors Meeting, 16 May 2016, ACLA-WIKI.

18 ParenteBeard, Report on Consulting Services Regarding the Allegheny County Library Association, 8 February 2012, ACLA-WIKI; Allegheny Regional Asset District, Meeting Minutes, Board of Director's Meeting & Public Hearing, 27 November 2012, ARAD-OFFICE; Allegheny Regional Asset District, Meeting Minutes, Board of Director's Meeting and Public Hearing, 26 November 2013, ARAD-OFFICE.

19 ParenteBeard, Report on Consulting Services, 8 February 2012.

20 ParenteBeard, Report on Consulting Services, 8 February 2012.

21 Marilyn A. Jenkins, Response to Parente Beard Report, 27 February 2012, ACLA-WIKI.

22 RAD, Report of the Library Committee, 27 March 2012, ACLA-WIKI; David Donahoe, Letter from David Donahoe to Mary Frances Cooper and Marilyn Jenkins: Parentebeard Report and Possible Duplication of Services, 28 March 2012, ACLA-WIKI.

23 ACLA, ACLA and the District Library Center: Delineation of Responsibilities, 2013, ACLA-WIKI; ACLA, RAD Outcomes for Regional Discussion, 14 February 2013, ACLA-WIKI; ACLA, Summary Input from Meetings with Library Directors, 8 March 2013, ACLA-WIKI; Jenkins, Executive Director's Summary of Major Action and Discussion Items: ACLA General Membership Meeting, 18 March 2013; ACLA, Librarians Advisory Council Minutes, 12 January 2012; ACLA, Librarians Advisory Council Minutes, 9 February 2012; Marilyn A. Jenkins, Executive Director's Summary of Major Action and Discussion Items, 2012, February 20, www.acla.pbworks.com; Hill Group, Our Libraries of Tomorrow Cluster Session Feedback, 12 June 2012, ACLA-WIKI; Hill Group, Our Libraries of Tomorrow: Final Project Report and Process; Overview with Appendices, 30 October 2012, 1, ACLA-WIKI; Hill Group, Our Libraries of Tomorrow Cluster Session Feedback, 12 June 2012; ACLA, ACLA East Meeting Minutes, 1 June 2012, ACLA-WIKI; ACLA, ACLA East Meeting Minutes, 4 May 2012, ACLA-WIKI; ACLA, West Region Libraries: Monthly Meeting, 27 June 2012, ACLA-WIKI; ACLA, Librarians Advisory Council Minutes, 8 March 2012, ACLA-WIKI; ACLA, ACLA South Region Meeting, 27 March 2012, ACLA-WIKI; ACLA, Librarians Advisory Council Minutes, 12 April 2012; Jenkins, Executive Director's Summary of Major Action and Discussion Items, 16 April 2012; Jenkins, Executive Director's Summary of Major Action and Discussion Items: ACLA General Membership Meeting, 18 June 2013; ACLA, Librarians Advisory Council Minutes, 14 June 2012.

24 ACLA, Librarians Advisory Council Minutes, 13 September 2012, ACLA-WIKI; ACLA, Librarians Advisory Council Minutes, 11 October 2012, ACLA-WIKI; Marilyn A. Jenkins, Executive Director's Summary of Major Action and Discussion Items: ACLA Board of Directors Meeting, 8 December 2014, ACLA-WIKI; ACLA, 2013 Budget / Comparative View, 23 October 2012, ACLA-WIKI; ACLA, General Membership Meeting Minutes, 18 June 2013; ACLA, Librarians Advisory Council Minutes, 10 October 2013, ACLA-WIKI; Jenkins, Executive Director's Summary of Major Action and Discussion Items: ACLA General Membershp Meeting, 30 November 2013.

25 ACLA, 2013 Highlights of Activity, February 2014, ACLA-WIKI.

26 Marilyn A. Jenkins, Executive Director's Summary of Major Action and Discussion Items: ACLA Board of Directors Meeting, 3 April 2014, ACLA-WIKI.

27 Librarian 1, Interview Transcript, 7 January 2016.

28 Library Administrator 1, Interview Transcript, 28 July 2015, interviewed by Michael M. Widdersheim.

29 ACLA, 2016 Budget Overview, August 2015, ACLA-WIKI.

30 ACLA, 2016 RAD Proposal, August 2015.

31 ParenteBeard, Report on Consulting Services, 8 February 2012.

32 ACLA, 2016 RAD Proposal, August 2015. ACLA, 2016 Budget Overview, August 2015.

33 ACLA, 2016 Budget Overview, August 2015; ACLA, ACLA East Meeting Minutes, 5 April 2013, ACLA-WIKI.

34 ACLA, 2016 Budget Overview, August 2015.

35 ACLA, South Region Meeting Minutes, 28 September 2015, ACLA-WIKI; ACLA, Draft 2015 Budget, 9 July 2014, ACLA-WIKI; ACLA, 2015 Budget Overview, 15 July 2014, ACLA-WIKI; Marilyn A. Jenkins, Executive Director's Summary of Major Action and Discussion Items: ACLA Board of Directors Meeting, 15 September 2014, ACLA-WIKI; Julie Goetz and David Donahoe, RAD Approves $93.7 Million Budget for 2015, 1 December 2014, www.radworkshere.org; Jenkins, Executive Director's Summary of Major Action and Discussion Items: ACLA Board of Directors Meeting, 8 December 2014; ACLA, 2015 Annual Report, February 2016, ACLA-WIKI.

36 Marilyn A. Jenkins, Executive Director's Summary of Major Action and Discussion Items, 21 May 2012, ACLA-WIKI; ACLA, 2013 RAD Distribution Formula, 13 June 2012, ACLA-WIKI; ACLA, General Membership Meeting Minutes, 29 March 2012, ACLA-WIKI; ACLA, 2013 Proposal to RAD, 21 August 2012, ACLA-WIKI; ACLA, 2013 ACLA Budget, 10 December 2012, ACLA-WIKI; ACLA, 2013 Budget, 10 December 2012, ACLA-WIKI.

37 ACLA, West Region Libraries: Monthly Meeting, 25 July 2012, ACLA-WIKI; ACLA, ACLA East Meeting Minutes, 6 November 2013, ACLA-WIKI; ACLA, Librarians Advisory Council Minutes, 8 November 2012, ACLA-WIKI; ACLA, ACLA East Meeting Minutes, 4 January 2013, ACLA-WIKI; ACLA, North Region, 8 January 2013, ACLA-WIKI; ACLA, Librarians Advisory Council Minutes, 10 January 2013, ACLA-WIKI; Jenkins, Executive Director's Summary of Major Action and Discussion Items, 21 January 2013; ACLA, Librarians Advisory Council Minutes, 14 February 2013; ACLA, ACLA South Region Meeting, 26 February 2013; Jenkins, Executive Director's Summary of Major Action and Discussion Items: ACLA General Membership Meeting, 18 March 2013; ACLA, ACLA East Meeting Minutes, 5 April 2013; ACLA, ACLA South Region Meeting, 23 April 2013; Jenkins, Executive Director's Summary of Major Action and Discussion Items: ACLA General Membership Meeting, 18 June 2013; ACLA, General Membership Meeting Minutes, 18 June 2013; ACLA, 2014 Final Formula, December 2013, ACLA-WIKI; ACLA, Librarians Advisory Council Minutes, 12 June 2014, ACLA-WIKI; Mt. Lebanon Public Library, ACLA's New Model for Using RAD Funding Re E-Resources, 24 June 2014, ; Mt. Lebanon Public Library, Summary of Issues Regarding the "Library Service in the 21st Century" Process, 20 May 2014, ; Cynthia Richey, Letter from Cynthia Richey to ACLA Library Directors, 28 May 2014, ; ACLA, Librarians Advisory Council Minutes, 9 October 2014, ACLA-WIKI; ACLA, 2016 RAD Formula, 26 October 2015, ACLA-WIKI; ACLA, 2016 RAD Proposal, August 2015.

38 ACLA, Librarians Advisory Council Minutes, 12 September 2013; ACLA, Board Resolution, 24 July 2013, ACLA-WIKI.

39 ACLA, Librarians Advisory Council Minutes, 12 September 2013.

40 ACLA, Librarians Advisory Council Minutes, 12 September 2013; Marilyn A. Jenkins, Executive Director's Summary of Major Action and Discussion Items, 16 September 2013, ACLA-WIKI.

41 ACLA, Librarians Advisory Council Minutes, 9 October 2014; Swerhun, LAC Update #2: LS21, 14 November 2013, ACLA-WIKI.

42 ACLA, Librarians Advisory Council Minutes, 9 October 2014; Swerhun, LAC Update: LS21, 10 October 2013, ACLA-WIKI; Swerhun, LS21: October Engagement Training Follow-Up, 31 October 2013, ACLA-WIKI; ACLA, Librarians Advisory Council Minutes, 14 November 2013, ACLA-WIKI; "A County-Wide Conversation About Public Libraries," January 2014, accessed February 11, 2015, https://web.archive.org/web/20140328014125/http://www.countycitylibraries.org/; Marilyn A. Jenkins, Executive Director's Summary of Major Action and Discussion Items: ACLA Board of Directors Meeting, 2 December 2013, ACLA-WIKI.

43 Swerhun, ACLA Board Update: LS21, 2 December 2013, ACLA-WIKI; ACLA, County-City Library Services Panel, 2014, ACLA-WIKI; ACLA, County-City Library Service Panel, 13 January 2014, ACLA-WIKI; Swerhun, Library Service in the 21st Century: Discussion Guide 1, January 2014, ACLA-WIKI; Swerhun, January Engagement Training Follow-Up, January 2014, ACLA-WIKI; Swerhun, Engagement Options for Part One, January 2014, ACLA-WIKI; Swerhun, Integrated Summary of Community Feedback, February 2014, ACLA-WIKI; Swerhun, LS21: Part 1 Consultation with ACLA General Membership, 13 February 2014, ACLA-WIKI; ACLA, ACLA East Meeting, 2014, March 28, ACLA-WIKI; Swerhun, Library Service in the 21st Century: Discussion Guide 2, April 2014, ACLA-WIKI; ACLA, Librarians Advisory Council Minutes, 6 March 2014, ACLA-WIKI; Swerhun, Integrated Summary of Community Feedback, May 2014, ACLA-WIKI; Swerhun, LS21: Draft Discussion Summary from Meetings with Library Directors, 8 & 9 May 2014, ACLA-WIKI; Swerhun, LS21: Working Session #3 with ACLA General Membership, 5 June 2014, ACLA-WIKI; CCLS panel, Summary of Meeting, 24 January 2014, ACLA-WIKI.

44 ACLA, County-City Library Service Panel, 13 January 2014; Marilyn A. Jenkins, Executive Director's Summary of Major Action and Discussion Items, 20 January 2014, ACLA-WIKI; CCLS panel, Summary of Meeting, 24 January 2014; CCLS panel, Summary of Meeting, 21 February 2014, ACLA-WIKI; ACLA, Library Service in the 21st Century: A County-Wide Conversation About Public Libraries March/April 2014 (Part 2 of 3), March & April 2014, ACLA-WIKI; Sisterson, Report on Review of Projections Prepared by Carnegie Library of Pittsburgh, Electronic Information Network and Allegheny County Library Association, 10 June 2014, ACLA-WIKI.

45 CCLS panel, Summary of Meeting, 11 June 2014, ACLA-WIKI; Kaitlynn Riely, "Allegheny County Library Systems Look to Cooperate," *Pittsburgh Post-Gazette*, 2 February 2014.

46 CCLS panel, Summary of Meeting, 21 February 2014.

47 CCLS panel, Summary of Meeting, 21 February 2014; ACLA, Registered Cardholders among Allegheny County Municipalities, March 2014, ACLA-WIKI; ACLA, Library Structures, 19 February 2014, ACLA-WIKI; ACLA, County Library Populations, 19 February 2014, ACLA-WIKI; Jenkins, Executive Director's Summary of Major Action and Discussion Items: ACLA Board of Directors Meeting, 3 April 2014; ACLA, Library Service in the 21st Century: A County-Wide Conversation About Public Libraries March/April 2014 (Part 2 of 3), March & April 2014; ACLA, State Requirements of ACLA & CLP, July 2014, ACLA-WIKI; ACLA, Library Service in the 21st Century, 21 April 2014.

48 ACLA, Librarians Advisory Council Minutes, 6 March 2014.

49 CCLS panel, Summary of Meeting, 19 May 2014, ACLA-WIKI.

50 CCLS panel, Summary of Meeting, 19 May 2014.

51 CCLS panel, Summary of Meeting, 19 May 2014.

52 Upper St. Clair Township Library, *Community Meetings Summary Report*, (10 February 2014).

53 Karen F. Fosbaugh, Public Statement – South Park Township – Panel Meeting of 6/11/14, 11 June 2014, ACLA-WIKI.

54 ACLA, South Region Meeting Minutes, 30 April 2014, ACLA-WIKI.

55 ACLA, South Region Meeting Minutes, 17 March 2014, ACLA-WIKI.

56 ACLA, South Region Meeting Minutes, 17 March 2014.

57 Swerhun, Library Service in the 21st Century: Part Two Consultation with ACLA General Membership, 3 April 2014, ACLA-WIKI; ACLA, South Region Meeting Minutes, 30 April 2014.

58 ACLA, South Region Meeting Minutes, 30 April 2014.

59 Library Administrator 1, Interview Transcript, 28 July 2015.

60 Library Administrator 1, Interview Transcript, 28 July 2015; Swerhun, January Engagement Training Follow-Up, January 2014; Swerhun, Library Service in the 21st Century: Part Two Consultation with ACLA General Membership, 3 April 2014; Swerhun, Frequently Asked Questions Regard-

ing the County-City Library Service Panel, 7 April 2014, ACLA-WIKI; CCLS panel, Summary of Meeting, 21 March 2014, ACLA wiki, acla.pbworks.com.
61 CCLS panel, Summary of Meeting, 11 June 2014.
62 County-City Library Service Panel, 21st Century Library Service in Allegheny County, July 2014, ACLA-WIKI.
63 County-City Library Service Panel, 21st Century Library Service in Allegheny County, July 2014, 6.
64 County-City Library Service Panel, 21st Century Library Service in Allegheny County, July 2014.
65 County-City Library Service Panel, 21st Century Library Service in Allegheny County, July 2014.
66 County-City Library Service Panel, 21st Century Library Service in Allegheny County, July 2014.
67 ACLA, ACLA Strategic Priorities: Supporting Libraries and Library Service, 19 August 2014, ACLA-WIKI; ACLA, RAD Board Presentation, 25 August 2014, ACLA-WIKI; ACLA, West Region Meeting Minutes, 20 August 2014, ACLA-WIKI.
68 Library Administrator 2, Interview Transcript, 8 July 2015, interviewed by Michael M. Widdersheim; Trustee 3, Interview Transcript, 1 July 2015, interviewed by Michael M. Widdersheim; Library Administrator 6, Interview Transcript, 6 March 2015, interviewed by Michael M. Widdersheim; Librarian 1, Interview Transcript, 7 January 2016; Jenkins, Executive Director's Summary of Major Action and Discussion Items: ACLA Board of Directors Meeting, 11 January 2016; ACLA, ACLA: An Overview of Activity and Proposed 2017 Budget, 15 June 2016, ACLA-WIKI.
69 Michael DiVittorio, "Mid-Mon Valley Libraries Weigh in on Allegheny County Recommendations," *TribLive*, 9 September 2014.
70 ACLA, South Region Meeting Minutes, 7 September 2014, ACLA-WIKI; ACLA, Librarians Advisory Council Minutes, 11 September 2014, ACLA-WIKI.
71 Library Administrator 7, Interview Transcript, 17 February 2015, interviewed by Michael M. Widdersheim.
72 Library Administrator 6, Interview Transcript, 6 March 2015; Trustee 2, Interview Transcript, 26 January 2016, interviewed by Michael M. Widdersheim; Library Administrator 6, Interview Transcript, 15 December 2015.
73 Library Administrator 7, Interview Transcript, 22 December 2014, interviewed by Michael M. Widdersheim.
74 ACLA, ACLA Service Standards: Process, Timeline, Resources, January 2015, ACLA-WIKI.
75 ACLA, Librarians Advisory Council Minutes, 11 September 2014; Jenkins, Executive Director's Summary of Major Action and Discussion Items: ACLA Board of Directors Meeting, 15 September 2014.
76 ACLA, Librarians Advisory Council Minutes, 9 April 2015, ACLA-WIKI.
77 ACLA, Minutes, General Membership Meeting, 29 October 2014, ACLA-WIKI.
78 ACLA, Librarians Advisory Council Minutes, 15 January 2015, ACLA-WIKI.
79 Library Administrator 2, Interview Transcript, 8 July 2015.
80 ACLA, Minutes, General Membership Meeting, 29 October 2014.
81 Marilyn A. Jenkins, Executive Director's Summary of Major Action and Discussion Items: ACLA General Membership Meeting, 29 October 2014, ACLA-WIKI; ACLA, Librarians Advisory Council Minutes, 13 November 2014, ACLA-WIKI.
82 ACLA, Draft System Standards, December 2014, ACLA-WIKI.
83 ACLA, Strategic Plan 2016 – 2018, 2015, ACLA-WIKI; Jenkins, Executive Director's Summary of Major Action and Discussion Items: ACLA Board of Directors Meeting, 8 December 2014; ACLA, ACLA Service Standards: Process, Timeline, Resources, January 2015; ACLA, ACLA Strategic Plan, 2015, ACLA-WIKI.

84 ACLA, Librarians Advisory Council Minutes, 15 January 2015; Marilyn A. Jenkins, Executive Director's Summary of Major Action and Discussion Items: ACLA Board of Directors Meeting, 19 January 2015, ACLA-WIKI.
85 ACLA, Regional Meeting Minutes: East, 2 January 2015, ACLA-WIKI; ACLA, Regional Meeting Minutes: East, 30 January 2015, ACLA-WIKI.
86 ACLA, Regional Meeting Minutes: South Region, 3 February 2015, ACLA-WIKI; ACLA, Draft System Standards (with Board and LAC Edits), February 2015, ACLA-WIKI; ACLA, Librarians Advisory Council Minutes, 12 February 2015, ACLA-WIKI.
87 ACLA, South Region Meeting Minutes, 26 February 2015, ACLA-WIKI.
88 ACLA, Librarians Advisory Council Minutes, 9 April 2015.
89 Library Administrator 1, Interview Transcript, 28 July 2015; ACLA, Draft Standards-Regional Feedback-North-Priorities, 4 March 2015, ACLA-WIKI.
90 ACLA, South Region Meeting Minutes, 26 February 2015; Library Administrator 2, Interview Transcript, 8 July 2015; ACLA, Draft Standards-Regional Feedback-North-Priorities, 4 March 2015.
91 ACLA, South Region Meeting Minutes, 26 February 2015.
92 Library Administrator 6, Interview Transcript, 6 March 2015; ACLA, Standards Feedback, 2 March 2015, ACLA-WIKI; ACLA, Draft Standards-Regional Feedback-North-Priorities, 4 March 2015; ACLA, South Region Meeting Minutes, 25 March 2015, ACLA-WIKI.
93 Library Administrator 2, Interview Transcript, 8 July 2015; Trustee 3, Interview Transcript, 1 July 2015.
94 ACLA, Librarians Advisory Council Minutes, 12 March 2015, ACLA-WIKI; Marilyn A. Jenkins, ACLA Board of Directors Meeting: Executive Director's Summary of Major Action and Discussion Items, 30 March 2015, ACLA-WIKI.
95 Michael M. Widdersheim, ACLA General Membership Meeting, Fieldnotes, 30 March 2015,
96 ACLA, Librarians Advisory Council Minutes, 9 April 2015; ACLA, Regional Meeting Minutes: West, 22 April 2015, ACLA-WIKI; ACLA, ACLA East Region Meeting, 24 April 2015, ACLA-WIKI.
97 ACLA, Board Response on Standards, 12 May 2015, ACLA-WIKI; ACLA, Draft System Standards, May 2015, ACLA-WIKI; ACLA, Draft System Standards Redline Version, May 2015, ACLA-WIKI; Marilyn A. Jenkins, ACLA Board of Directors Meeting: Executive Director's Summary of Major Action and Discussion Items, 11 May 2015, ACLA-WIKI; ACLA, Librarians Advisory Council Minutes, 14 May 2015, ACLA-WIKI.
98 Trustee 3, Interview Transcript, 1 July 2015.
99 ACLA, Librarians Advisory Council Minutes, 11 June 2015, ACLA-WIKI; ACLA, ACLA System Standards, 15 June 2015, ACLA-WIKI.
100 Marilyn A. Jenkins, ACLA Board of Directors Meeting: Executive Director's Summary of Major Action and Discussion Items, 15 June 2015, ACLA-WIKI.
101 ACLA, 2016 RAD Proposal, 30 June 2015, ACLA-WIKI.
102 Library Administrator 1, Interview Transcript, 28 July 2015.
103 Library Administrator 1, Interview Transcript, 28 July 2015.
104 Trustee 3, Interview Transcript, 1 July 2015.
105 Trustee 2, Interview Transcript, 2 July 2015, interviewed by Michael M. Widdersheim.
106 ACLA, Regional Meeting Minutes: East, 28 August 2015, ACLA-WIKI; ACLA, Librarians Advisory Council Minutes, 8 October 2015, ACLA-WIKI; Marilyn A. Jenkins, Executive Director's Summary of Major Action and Discussion Items: ACLA General Membership Meeting, 31 March 2016, ACLA-WIKI; ACLA, General Membership Meeting Minutes, 31 March 2016, ACLA-WIKI; ACLA, ACLA: An Overview of Activity and Proposed 2017 Budget, 15 June 2016.

107 ACLA, Librarians Advisory Council Minutes, 9 November 2015, ACLA-WIKI.
108 Marilyn A. Jenkins, Executive Director's Summary of Major Action and Discussion Items: ACLA Board of Directors Meeting, 19 September 2016, ACLA-WIKI; ACLA, General Membership Meeting, 25 October 2016, ACLA-WIKI; ACLA, Member Library Agreement, 25 October 2016, ACLA-WIKI.

Conclusion

The foregoing chapters described in a narrative style how the public library infrastructure in greater Pittsburgh developed over time. Each chapter recounted a decision cycle in the case, and decision cycles were characterized by a major issue, a build-up of support or non-support, a formal decision, and implementation or non-implementation. From a high-level view, what do the specifics of each of these periods mean? How do the cycles compare, and what do they tell us? What general knowledge can be gained from them? In other words, what factors or what general rule might be proposed to explain why each cycle resulted as it did? Why did some periods result in a major change while others did not? And what general implications do these findings suggest for library practitioners? What strategies for change management might be extrapolated?

The case of the public library infrastructure in greater Pittsburgh exhibited nine decision cycles, or periods, over the course of its development. The temporal scope of the case extended from 1924 when the infrastructure was first imagined to 2016 when the last observed decision cycle culminated. Some of these decision cycles resulted in substantive changes to the infrastructure while others resulted in continuities. The periods of the case and their outcomes are shown in Table 5. The determination of whether a decision cycle resulted in change or continuity is based on both qualitative and quantitative data. Qualitative descriptions are those presented in the foregoing chapters, but separate quantitative data previously gathered substantiates this qualitative account as well. An analysis of a representative sample of communicative events from each period, performed using systematic coding, indicates that periods with no change exhibited low implementation rates. The implementation ratio for each period is defined as a period's total successful implementations to total communicative events. The implementation ratios for Period 1, Period 4, and Period 8 were appreciably lower than the other six periods.[1] This means that the issues raised in in those periods were not implemented as often as those issues in periods with high implementation.

Table 5: Periods and outcomes in the case of greater Pittsburgh.

Period	Year range	Period title	Outcome
1	1924 – 1945	Periphery and Center	Continuity
2	1945 – 1961	County Contract	Change
3	1961 – 1970	The Rise of District Services	Change
4	1970 – 1979	Unification Revisited	Continuity

https://doi.org/10.1515/9783111013404-012

Table 5: Periods and outcomes in the case of greater Pittsburgh. *(Continued)*

Period	Year range	Period title	Outcome
5	1980 – 1993	Fiscal Crises and Digital Visions	Change
6	1993 – 1994	The Pursuit of RAD	Change
7	1994 – 1997	System Dreams	Change
8	1998 – 2011	Formula Wars	Continuity
9	2011 – 2016	Compliance Culture	Change

What explains these different outcomes? It is here that the RLCr, or "Releaser," Theory best explains why some periods in the case resulted in change while others did not. Releaser Theory focuses on four causal conditions related to public sphere communication. These conditions are Responsiveness, Legitimacy, Civil Support, and Resistance. Periods that exhibited high responsiveness, together with either high legitimation, high civil support, or low resistance, resulted in a change outcome. Conversely, periods with either low responsiveness or low legitimation, low civil support, and high resistance resulted in continuity. Responsiveness is a necessary but insufficient condition for infrastructural change. For a causal configuration sufficient for change to occur, responsiveness must be co-present with at least one of the other three factors: either high legitimacy, high civil support, or low resistance. Again, the factors of responsiveness, legitimacy, civil support, and resistance were developed and measured using a mix of both qualitative and quantitative data. Qualitative descriptions were presented in the preceding chapters as a narrative account. The data set for the quantitative analysis was created in earlier studies by systematically coding a representative sample of communicative events for each period. A period's responsiveness ratio is defined as legitimated issues that are implemented to the total number of legitimated issues. The legitimacy ratio is defined as a period's legitimated issues to total communicative events. The civil support ratio is defined as a period's communicative events by pro-change civil society actors to total number of communicative events. And the resistance ratio is defined as a period's legitimation failures due to objections to total communicative events.[2] The conditions and outcomes for each period are shown in Table 6.

Table 6: Conditions and outcomes for each period.

Period	Responsiveness	Legitimacy	Civil support	Resistance	Implementation
1	Low	High	High	Low	No change
2	High	High	High	Low	Change
3	High	Low	Low	Low	Change
4	Low	Low	High	High	No change
5	High	High	Low	Low	Change
6	High	High	Low	Low	Change
7	High	High	Low	High	Change
8	High	Low	Low	High	No change
9	High	Low	High	High	Change

What do the causal conditions in Releaser Theory mean? How were they identified as factors? How are they defined and exemplified? Responsiveness is perhaps the most important factor to consider because it is a necessary condition for change. High responsiveness is present in a tessellation when core decision making bodies recognize and react to legitimated concerns. This sensitivity to communicative power by decision makers is an essential feature of a healthy public sphere, namely that legitimated concerns are recognized, considered, taken up, and acted upon. It is appropriate to consider responsiveness a relevant factor in a change theory related to a public sphere because decisionmakers must be able to react to legitimated concerns for there to be the possibility for change. But responsiveness is not just a feature of the audience side of communication—it is due to circumstances on both the actor side and the audience side of communicative events. Responsiveness is high when both an issue is actionable, achievable, and has been presented clearly (actor side) and when the receiver of the message is tuned in to and receptive to it (audience side). As demonstrated by the case of the public library infrastructure in greater Pittsburgh, legitimated issues are not always considered or acted upon by core decision making bodies. In other words, it has been shown that the public sphere does not always function in an ideal way, where communicative power is transformed into administrative power. The comparison of Period 1 and Period 2 illustrates well different levels of audience responsiveness. Period 1 exhibited high legitimacy and high civil support, meaning that a large proportion of communicative events were supported discursively and were backed by civil society groups who mobilized for a change to the public library infrastructure. Communicative power was created by the groups using argumentation and reasoning in an open and inclusive context oriented to common interests. These messages targeted several decision-making cores: the county commissioners, the city council, and the city mayor. Despite a proliferation of legitimated concerns in Period 1, the relevant decision-making bodies did not take up and implement the issue of extending public library service into rural areas of the county. They also did not immediately see the need to combine the two city public library systems. In other words, decision makers were unresponsive, “deaf,” or “tuned out” to issues raised and legitimated in their peripheries. By contrast, in Period 2, responsiveness was high. This was because decisionmakers at the time shifted their stance to recognize and act upon the issues they received. In Period 2, communicative power was successfully transformed to administrative power. In other instances, unlike Period 1, low responsiveness can be attributed to poor messaging, not just to audience failure. In Period 4, for example, the only other decision cycle besides Period 1 where responsiveness was low, the proposal for a consolidated public library system was not adopted by decision making cores. This was not because the issue was not deemed legitimate, but because the action plan that was proposed was not carefully thought out. Lacking a clear message, decision makers were unable to implement the issue.

Legitimacy is the second condition of Releaser Theory. Legitimacy is the measure of discursively supported issues in a tessellation for a given decision cycle. It is a measure of the medium or context out of which messages arise, in other words, how the issue is presented from the actor side. Open and unrestrained discourse, oriented to reasons and consensus, is an essential feature of a functioning public sphere. Ideally, a public sphere raises issues of common concern and justifies the proposed solution through reasoned debate in a setting that is inclusive and where anyone can freely disagree and exchange their own views. It is therefore reasonable to consider legitimacy a relevant factor in a communication-focused and public sphere theory of change. The process of reasoned debate makes an issue legitimate in a public sphere, and communicative events that arise from such a context can be coded as legitimated. When the proportion of legitimated issues to total communicative events is high in a decision cycle, the tessellation exhibits high legitimacy. Period 3, for example, exhibited low legitimacy. In this decision cycle, the issue of a new state plan was introduced to greater Pittsburgh from outside, from the state library and the State Legislature. There was little to no open discussion within Pittsburgh about the state plan, meaning that legitimacy for the issue was low. In Period 6, by contrast, when library leaders petitioned the RAD board for RAD funding, legitimacy was high because discussants made extensive efforts to bolster the case publicly for the public library infrastructure as a regional cultural asset.

Civil support is the third condition of Releaser Theory. Civil support is a measure of the degree to which the communication by civil society actors and groups supports infrastructural change. Quantitatively, the civil support ratio for a period is defined as communicative events by pro-change civil society actors to total number of communicative events. Civil society actors include those in the third sector such as non-profit corporations, volunteer associations, private foundations, charities, individuals, and citizens. Civil society is distinct from businesses and government. Actors are not considered civil society if they are associated with a for-profit business or a government agency. In the case of greater Pittsburgh, voluntary library associations and committees operated as civil society groups. These included the Citizens Study Committee from Period 4, CLASP from Period 5 and Period 6, and the LS21 discussion groups from Period 9. While these were library-related groups, the groups were not strictly librarians or trustees. The groups were voluntary, their membership was open to anyone, and they were driven by values pertaining to library funding and governance. The Civic Club from Period 1 and ACCD from Period 2 are other ideal examples of civil society groups whose communication yielded high civil support. According to the circulation of power model, issues can arise from anywhere, whether from within decision making cores or from the outer or inner peripheries of circuits. But civil society groups occupy a privileged position in the public sphere. The issues raised by them sometimes carry additional weight. Civil society groups operate "on the ground," they are "closer to the people" than private sector or public sector actors, and they are sometimes perceived as more "in touch" with issues that matter, representing the voice of the people. Because the communicative power of civil society groups is often strong, it is appropriate to include civil support as a factor in a theory of change because the presence or absence of it can make a difference in outcome.

Resistance is the fourth and final condition of Releaser Theory. Resistance is a counterpart to legitimacy. It is a measure of objections to an issue. Quantitatively, the resistance ratio for a period is defined as legitimation failures due to objections to total communicative events. As demonstrated in Period 7, where ACLA HQ pushed the ACLA membership to become a state-recognized federated library system, it is possible for both legitimacy and resistance to be high. In the case of Period 7, there were both proponents of and opponents to the new proposed system. Proponents expressed a legitimated issue, the system change, while opponents such as the Library Militia criticized the plan. In a public sphere, where communication is exchanged in an open way among a variety of actors, resistance is to be expected. Resistance was included as a factor in the theory because whether resistance is high in a decision cycle affects decisionmakers' perceptions of an issue.

These four conditions—responsiveness, legitimacy, civil support, and resistance—explain the outcomes for each of the nine periods in the case of the public library infrastructure in greater Pittsburgh. By the end of each decision cycle, the tessellation underwent substantive change when high responsiveness was present together with either high legitimacy, high civil support, or low resistance. Details from the preceding narrative accounts from each of the periods show this to be so. In Period 1, the Civic Club and other civil society groups petitioned for extension of public library services into rural parts of the county, and they also petitioned for consolidation of the two city public library systems, CLP and North Side. There was high legitimacy for the issue, with all proponents offering clear rationales for why change to the system was needed. The issue originated from civil society, and this manifested as high civil support. Resistance to the issue was low because there were very few opponents to the new proposed infrastructure. Yet, Period 1 did not result in substantive change, which seems puzzling at first. The failed issue in Period 1 was due to low responsiveness. The decision-making bodies, including the county commissioners and city council, which were targeted by the civic groups, did not take up and act on the issue. This was not the fault of the civil society actors or Ralph Munn, who developed actionable plans for changing the system. Instead, the issue failed to carry due to the prevailing ideology and reigning philosophy of the decisionmakers during that period. Period 1 occurred during Pittsburgh's pre-Renaissance era. It might be said that decisionmakers were not yet receptive to regional initiatives or cultural investment. But this ideology, together with the composition of the decision-making bodies, and the dominant civil society groups, changed in Period 2. Like in Period 1, legitimacy and civil support were high, and resistance was low. The issue was the same. Period 2 resulted in substantive change because decision makers were receptive to the issue. The city systems combined, and CLP became the county library, thus providing bookmobile service to rural parts of the county. Corporate leaders in ACCD, who sponsored the issue, convinced the Allegheny County Commissioners and Pittsburgh City Council to invest in cultural and educational infrastructure, which was part of a larger movement to reshape the environment of greater Pittsburgh. In Period 2, responsiveness was high because the audience was receptive to issues that spoke to the improvement of social, educational, and cultural life.

In Period 3, the issue of a state library plan was imported from outside greater Pittsburgh from the State Library of Pennsylvania and the Pennsylvania legislature. Unlike Periods 1 and 2, the issue in Period 3 was not grassroots led, it was not supported by civil society actors, and there was little discussion about the issue in and around Pittsburgh. In Period 3, legitimacy was low and civil support was low. But resistance was also low—there were few if any detractors to the state plan. Under the state plan, CLP became the district library center and regional resource center. Together with CLP, the other libraries in the county began to receive state aid in the form of the public library subsidy. There was little room to oppose this plan. The few legitimated claims about the issue of the state plan were taken up and heard. Administrators at CLP were responsive to the issue of the state plan, and they agreed to the terms, as did the other libraries in the county. Without high resistance, there was nothing to stop the issue from carrying.

Period 4 occurred during a tumultuous time in U.S. history. There was large-scale political, economic, cultural, and technological upheaval underway. Yet, Period 4 did not result in a change outcome. Multiple studies during this time proposed a unified countywide public library system. Later in Period 4, a citizen-composed committee again proposed such a plan. Civil support was high due to the activity of this study group. But legitimacy was low, and resistance was high. The plan that was proposed was not workable. Responsiveness was also low. When the plan was presented to CLP and the county commissioners, since the plan was not actionable, it was rejected by the core decision making bodies in these circuits.

Period 5 and Period 6 had similar trajectories and the same conditions: high responsiveness, high legitimacy, low civil support, and low resistance. Both periods resulted in change outcomes. In Period 5, in the wake of *Quiet Crisis*, CLASP was formed, together with CFLAC. The county libraries began to work with CLP on Project Link-Up, which was the precursor to a shared countywide network with a union catalog. The county libraries came together to form the first regional public library organization in the area. Similarly, in Period 6, CLASP, later ACLA, petitioned the RAD board to become recognized as a regional cultural asset. The campaign had thoughtful reasons, and it met with no detractors. The RAD board was primed to allocate RAD funding to the county libraries. In both periods, civil support was low. This was because the movements were led by the libraries themselves.

In Period 7, ACLA HQ sought to transform the organization into a state-recognized federated library system. This was to expand ACLA's potential to receive state funding. Unlike in Period 5 and Period 6, however, resistance was high in Period 7. There was a concerted effort to derail the initiative to form a federated public library system. But legitimacy was also high in Period 7, and the ACLA membership and the ACLA board were receptive to acting on the legitimated claims to form the new system. As in Period 5 and Period 6, civil support was low in Period 7 because the issue was raised and legitimated internal to the library organization.

Period 8 and Period 9 were both contentious decision cycles where the issue of funding was disputed within the public library infrastructure. In Period 8, a new RAD distribution formula was proposed for ACLA member libraries, one that tied funding to performance metrics and combined state and RAD funding. In Period 9, ACLA HQ shifted attention away from itself to instead focus on implementing performance standards for member libraries. In both periods, legitimacy was low and resistance was high. Period 8 resulted in continuity, but Period 9 resulted in change. The difference that made a difference in Period 9 was civil support. Because ACLA HQ carried out the LS21 process, which included community members in the discussions, findings from the LS21 process, which included the recommendation of performance standards, built up high civil support. ACLA HQ could then point to this civil support when deciding to implement the performance standards they desired. For this reason, Period 9 resulted in a change outcome.

It has been argued that Releaser Theory is the best explanation for the changes and continuities in the case of greater Pittsburgh. This theory, which is based on the public sphere and the circulation of power model in particular, focuses on the power inherent in communication to explain change. But it is worth considering alternative possible theories that might explain why the public library infrastructure in greater Pittsburgh developed as it did. One alternative theory is that it was not communication factors, but fluctuations in wealth, that caused the infrastructure to change or remain the same at certain times. It might be argued that changes in the wealth within Allegheny County prompted changes in the infrastructure. Allegheny County was a primary funder of the infrastructure, at least until 1995 onward when RAD began to fund public libraries instead. It might be imagined that greater wealth within the county meant that there were more resources available to commit to the infrastructure, causing change. Likewise, reductions in wealth meant fewer resources were available, causing change as well. Wealth can be surmised using a variety of indicators, such as property valuation within the county, investments, or taxes. But a reliable indicator of wealth is available assets, which includes any property or funds the county government could immediately liquidate. Figure 9 shows the available assets of Allegheny County, from 1929 to 1977. Both nominal values and inflation-adjusted values are shown. If, contrary to Releaser Theory, changes in wealth were the true cause of infrastructural change, changes in the infrastructure would correspond to changes in wealth. One milestone in the case of greater Pittsburgh was when CLP became the county library and the two city systems combined. This occurred in 1956. In the year 1956, there was only a very slight increase in available assets of the county. Since a more noticeable jump occurred the following year, in 1957, it would have been thought that infrastructural change occurred then, but it did not. Likewise, when the available assets plummeted in 1965, one would expect to have observed substantive change in the public library infrastructure, but none occurred that year, either. Overall, the changes observed in wealth within Allegheny County do not correspond to the storyline or the data of the regional public library infrastructure. For these reasons, wealth is not a reliable predictor of infrastructural change.

Another alternative theory of change might posit that changes in population caused changes in the public library infrastructure. It might be postulated that as more people lived and worked in greater Pittsburgh, there was greater demand for services, and there was a greater tax base to support them. As seen in Figure 10,

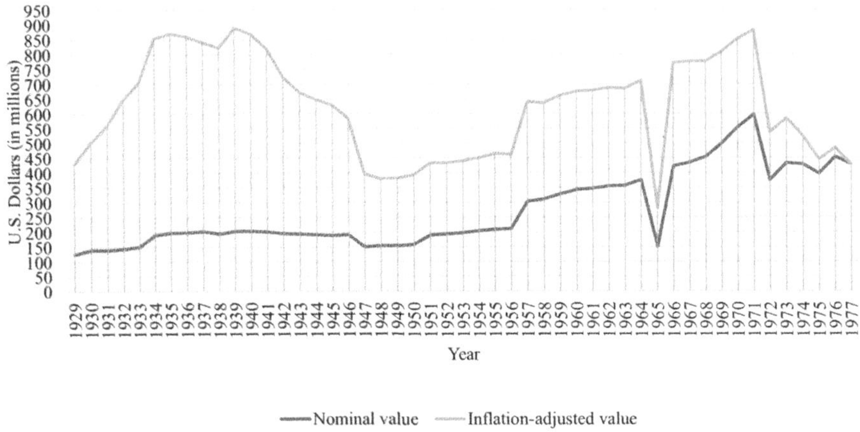

Figure 9: Available assets of Allegheny County, 1929–1977.

the populations of the City of Pittsburgh and Allegheny County rose and fell throughout the twentieth century. If infrastructural change corresponded to this wave, one would expect the infrastructure to rise and fall in a steady and gradual way corresponding to the population curve. But the infrastructure did not behave in this way. Rather, it developed in fits and starts. Unlike population, there was no discernable advancement that culminated in the infrastructure around 1960, followed by a gradual decline. To be sure, new library buildings were built around 1960 in the suburban ring outside the city where the population grew. But library buildings continued to be built beyond that time, even as the population decreased overall. The public library infrastructure continued to grow, become more complex, and obtain greater funding beyond 1960, a trend that does not parallel the population decrease. Population is therefore not an attractive explanation for infrastructural change.

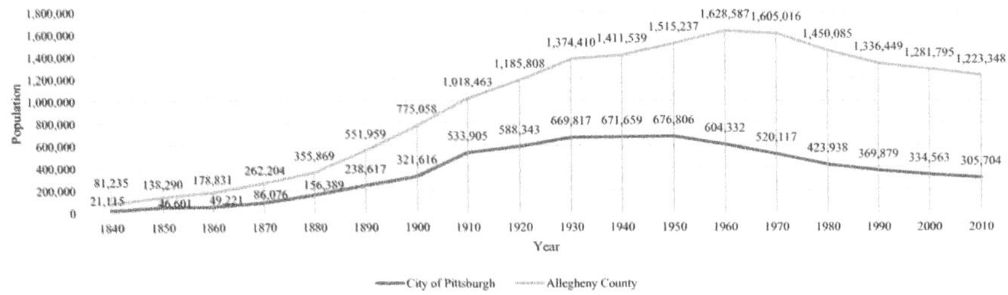

Figure 10: Population of the City of Pittsburgh and Allegheny County.

A third alternative possible theory that might be imagined is that the changes in the public library infrastructure in greater Pittsburgh were due to charismatic leadership. In other words, influential leaders with vision and innovation guided library organizations in new directions. It might even be said that this book fails to acknowledge and highlight the key contributions made by a select group of individuals. Certainly, the case of greater Pittsburgh had its share of visionary leaders. These included Lars Grondahl, member of the Civic Club of Allegheny County and proponent of extended library services; Ralph Munn, director of CLP, who imagined a completely new federated library system for Allegheny County; Lowell Martin, architect of the state plan; Dorothy Strang, independent citizen and library advocate who promoted library service locally, regionally, and at the state level; Bob Croneberger, director of CLP, who initiated the countywide network; and Frank Lucchino, local Pittsburgh politician and public library hero. This is a non-exhaustive list. Certainly, these were all influential figures. But attributing substantive change to them is problematic. For instance, in Period 1, Ralph Munn partnered with the Civic Club to petition the county commissioners for extended public library service. This initiative failed. It was not until Period 2 that the issue carried. Munn was a world-renown library leader, active in all sectors of the profession. If anyone could have been responsible for substantive change in the infrastructure, it would have been him, but his efforts in Period 1 were unsuccessful. In Period 5, Frank Lucchino authored *Looking Beyond our Past* and *Quiet Crisis*, stirring library leaders to form CLASP. CFLAC was also formed, as was Project Link-Up, which was motivated by Bob Croneberger. But in Period 6, when ACLA petitioned RAD for regional asset funding, this movement involved a wide array of individuals, not just two luminaries. Without the largescale mobilization by libraries across the county, the RAD movement would not have been successful. In fact, it might be said that both Period 5 and Period 6 resulted in change outcomes because they were grassroots-led. Though Lucchino and Croneberger were heavily involved in infrastructural change during this time, proposing catalyzing ideas for libraries to rally around, focusing only on the individuals obscures the widespread efforts that ultimately determined ACLA's success. Similarly, in Period 7, 8, and 9, there were not stand-out individuals who determined the course of events. Rather, issues were raised, discussed, criticized, and voted upon en masse. Focusing on the larger communicative activity and communicative structure, rather than individual people, offers a more informative account for why things happened as they did. These observations suggest that infrastructural change occurs due to structural, not personal, reasons.

Supposing that Releaser Theory holds true in other contexts, it offers several change management strategies that can be illustrated using examples from the case of the public library infrastructure in greater Pittsburgh. These can be seen by comparing the case's most-similar periods.[3] A change management strategy is one that can be applied by a public library administrator or other library leader to guide a decision cycle to a desired change outcome. Change management is a subfield within the larger body of scholarship on organizational management.

The first change management strategy is to clarify the issue. This means to clearly articulate the existing problem, then clearly state the proposed solution in enough detail that there is an actionable and achievable plan. A comparison between Period 4 and Period 9 illustrates this strategy. Both Period 4 and Period 9 exhibited low legitimacy, high civil support, and high resistance. The outcome of Period 4 was continuity while the outcome of Period 9 was change. The difference in Period 9 was high responsiveness. In Period 9, ACLA HQ strove to implement performance standards. They had a draft set of standards and a way to work with member libraries to adopt them. Despite high resistance and low legitimacy, the standards had been worked out through a discursive process that involved civil society actors, so civil support was high. The workable plan is what distinguishes Period 9 from Period 4. The Citizens Study Committee from Period 4 did not articulate a feasible action plan. Decision makers such as the county commissioners and CLP could not implement the plan that was not feasible, and, for this reason, the issue failed in Period 4. Clarifying the issue, therefore, and developing a workable plan to address, is an essential change management strategy.

The second strategy is to map the decision cycle. This means for change advocates to plan ahead and to orchestrate a legitimation process that builds support and momentum toward a single formal decision point. This strategy is best illustrated by comparing Period 6 with Period 8. Period 6 had high responsiveness, high legitimacy, low civil support, and low resistance. Period 8 had high responsiveness, low legitimacy, low civil support, and high resistance. In Period 6 there was a change outcome, but Period 8 resulted in continuity. What was the difference in Period 6? In Period 6, ACLA rallied for RAD funding. It was a short decision cycle —less than a year—and there was great uncertainty about the outcome, but campaigners carefully crafted their presentations, gathered support from municipal leaders, politicians, and citizens, and oriented their full efforts toward the RAD hearings in late 1994. The outcome was successful because change advocates mapped out the process in advance and backward-designed it. This created high legitimacy. By contrast, Period 8 witnessed protracted debates about ACLA's distribution formula for RAD funding. There was no identified terminus to these discussions, and, because of this, the discussions continued unfruitfully. There was little to no coordination with the RAD board until it was too late and, without a clearly defined goal, the discussions in Period 8 just led to in-fighting. In Period 6, it was beneficial for change advocates to plan out their campaign in advance and orient their efforts toward a set decision point. This allows change advocates to build legitimation efforts in a way that does not stall momentum.

The third strategy is to identify circuits and tessellations. To identify circuits and tessellations means to determine who and where the core decision makers are within the political landscape. These decision-making bodies can then be targeted so communicative power can be transformed into administrative power. This strategy is best observed by comparing Period 8 and Period 9. The issue in both periods related to the adoption of performance standards by the ACLA membership. In both periods, responsiveness was high, legitimacy was low, and resistance was high. In Period 9, civil support was high. ACLA HQ took pains to establish themselves as the ultimate decisionmakers, not the ACLA membership or LAC. In Period 8, the debate about the funding formula and performance standards did not draw civil support, and there was confusion about who actually called the shots, ACLA HQ or RAD. RAD overturned decisions by the ACLA membership. ACLA members did not have a clear understanding of how to build communicative power and where the decision-making cores were located. In Period 9, ACLA HQ orchestrated the decision cycle to build civil support and make the final decision.

The fourth strategy is to target receptive audiences. This means to transmit communicative power to those decision-making bodies who are primed and prepared to consider implementing plans to address an issue, but to avoid efforts that target cores that are unreceptive. This strategy can be seen in the comparison between Period 1 and Period 2. In Period 1, the county commissioners were targeted as potential sponsors for extended county library services. This effort failed, not due to lack of resources or need, but because the timing and ideology was such that the commissioners were not ready to hear the proposal. By contrast, in Period 2, the outlook of the commissioners had changed, and the timing was right to transmit communicative power to them. This change of circumstance raised the responsiveness in Period 2, resulting in a change outcome.

The fifth strategy is to mobilize civil society support. Mobilizing civil society support means to harness the support of civic groups for an issue. Building communicative power with civil society actors can mean a difference in outcome. This can be seen by comparing Period 8 with Period 9. In Period 8 and Period 9, all conditions were the same except civil support, which was high in Period 9. It is no coincidence that Period 9 resulted in a change outcome. This was because ACLA HQ successfully drew communicative power from civil society actors during the LS21 process. This communicative power was then taken up directly by the ACLA board, which formally decided to implement performance standards. Period 8 did not involve civil society actors, and, for this reason, Period 8 resulted in continuity.

The sixth strategy is to minimize resistance with win-win issues. A win-win issue is one that reduces resistance, and low resistance increases the chances of a successful change outcome, all other things being equal. This strategy is best illustrated by comparing Period 3 with Period 8. Both periods had high responsiveness, low legitimacy, and low civil support. But Period 3 exhibited low resistance while in Period 8 resistance was high. Period 3 saw the implementation of the state plan. This plan was uncontroversial, and there were no detractors. The plan meant the introduction of state aid to all county libraries. CLP became the district library center and the regional resource center, and, with these new roles, the other public libraries in the county only stood to benefit. It was a win-win issue. Period 8, by contrast, saw infighting between ACLA member libraries over proposed revisions to the RAD funding formula. Whatever the outcome of the debates in Period 8, there would be winners and losers: some libraries would receive more funding and others less. This was a no-win issue because not everyone could be satisfied. The nature of the issue was such that it was met with high resistance.

The seventh and final strategy is to maintain legitimizing processes. This means to facilitate open and inclusive discussions oriented to reasons when raising issues of common concern. Legitimation processes in the public sphere are important for change management because high legitimacy is able to overcome low civil support and high resistance. This strategy is illustrated by comparing Period 7 and Period 8. Both periods exhibited high responsiveness, low civil support, and high resistance, but Period 7 also had high legitimacy. The issue in Period 7 was the formation of a federated public library system. There were a number of detractors of this issue. But the issue was thoroughly discussed in open and inclusive contexts, and there was support for it, building legitimacy for the issue and resulting in a change outcome. Period 8, by contrast, had low legitimacy because some discussions were held behind closed doors. There was not enough legitimacy for the issue of a revised formula in Period 8 to result in a change outcome.

The story of the regional public library infrastructure in greater Pittsburgh is a story of shifting circuitry. It is the story of debates about where the infrastructure should reside, which circuit it should occupy, and how it should be funded, governed, and supported over time. The circulation of power model, especially the concepts of circuits and tessellations, makes it possible to see this shifting circuitry more clearly. The infrastructure was first implemented in 1956 when CLP contracted with Allegheny County to become the county library. As county library, CLP resided in the position of inner periphery of the county government. Through its contract with the county government, CLP acted on behalf of county government, serving social welfare needs, but it also had a degree of autonomy about how best to achieve these ends. When the Pennsylvania state plan was implemented in 1961, the infrastructure then resided within the inner periphery of the state circuit as well, since CLP participated in the state plan as a county library, district library center, and regional resource center. Other libraries within the district also carried out state initiatives by adhering to state rules as a condition for receiving state aid. The other libraries were monitored in part by CLP. Subsequently, in Period 4, CLP's status as the agency charged with overseeing the infrastructure was challenged. It was considered to move the infrastructural governance more squarely within the inner periphery of the county government and with more input from other independent libraries in the county. But, in Period 5, even with the creation of CFLAC, the proposed transition of the infrastructure to squarely within county government did not happen. Instead, In Period 6, with the creation of RAD, the county library membership organization, ACLA, moved into the inner periphery of the RAD circuit. The county divested itself of funding the public library infrastructure. CLP, which for so long functioned as a catalyst for a regional infrastructure, operated within the RAD circuit alongside ACLA. Together, CLP and ACLA oversaw the infrastructure, but only under RAD's oversight. But by this time, the independent libraries in the county were already involved in multiple other circuits: local municipalities and school districts, the state plan, and ACLA. Now they became nested in the inner periphery of RAD, a special authority tax district. The simultaneous existence of the infrastructure in multiple circuits caused confusion and tension, especially after the formation of ACLA. Within ACLA alone, it was unclear what the core of ACLA of the organization really was, whether it was membership, LAC, or the board. In Period 9, the board prevailed as the primary decisionmaker within ACLA. Yet, it served effectively as a functionary of the RAD board. As demonstrated in Period 8, the ACLA board did not have the final say —the RAD board did. Meanwhile, due to defunding of the public library subsidy at the state level since 2003, the state circuit became in effect defunct in Period 8 and Period 9. In those periods, therefore, RAD stood as the primary decisionmaker for the infrastructure. While it is true that each local library retained its own board and thus retained its own decision-making power for local policies, the public library infrastructure at a regional scale moved and developed according to RAD's dictates, not the policies of each individual library. With the acceptance of RAD money, the federated system status of ACLA is federated in name only. In reality, it has come to operate more as a single unified system. With the adoption of eiNetwork across the county, and with the gradual implementation of some centralized services, individual libraries have become increasingly intertwined in such a way that it is difficult if not impossible to reverse the trend toward consolidation. ACLA's membership may have a say in some policy decisions, but membership votes amount to relatively less decision-making power compared to the ACLA board, and the ACLA board defers to the RAD board, since ACLA's funding is tied directly to RAD, not the budgets of individual member libraries. The difficulty with the infrastructure's nested occupation within RAD's inner periphery is that RAD is a relatively insulated governing body. The representatives on its board are appointed by the city and county. This means that while administrative power travels outward from the RAD board, it is difficult for communicative power to travel inward to RAD's decision-making core. Communicative power from RAD's outer periphery reaches the RAD core only once it passes through CLP HQ and ACLA HQ. Communicative power must either be funneled to RAD through these "mediating cores." This presents the opportunities for distorted communication. The quickest and sometimes most effective influence channel between the RAD board and its outer periphery is not through communicative power, but social power. Social power includes transmissions through backroom deals and backchannel communication. These are forms of non-legitimate communication because they are not open to objection, discussion, and accountability. The problem with the adoption of social power is that it can potentially override communicative power. RAD's decisions, therefore, may be seen as questionable because it is difficult to see how opportunities for open and undistorted communication, open to anyone, are possible within the RAD circuit. RAD enforces a culture of compliance, but there are few if any opportunities for public sphere oversight of the RAD board itself. Regional assets such as CLP and ACLA may be in touch with their constituents, through their outer peripheries, but RAD is not. The RAD board "governs at a distance," but without the necessary circuitry to act in accordance with public sphere dictates. This situation creates the potential for illegitimate decision-making.

The current circuitry of the infrastructure is shown in Figure 11. There, it can be seen that the ACLA board and CLP board occupy the inner periphery of the RAD circuit. The RAD, CLP, and ACLA circuits share the same outer periphery, composed of the constituency of the public library infrastructure, including patrons and taxpayers. CLP and ACLA circuits each have their own inner peripheries. The channels of communicative power within the CLP and ACLA circuits are strong. Communicative power flows to the inner peripheries and cores unproblematically. But the line of communicative power to the RAD board from its outer periphery is weak. The public library constituency communicates to the RAD board through the buffers of the ACLA board and CLP board. The same holds true for the inner peripheries of the CLP and ACLA circuits: the RAD core does not, for the most part, communicate directly with inner peripheries of the ACLA and CLP circuits. The inner peripheries of the CLP circuit and ACLA circuit are not, in principle, a functional outer periphery within the RAD circuit. This situation heightens the potential for bypassing the CLP core and ACLA core by CLP's and ACLA's inner periphery actors. Rather than transmit communicative power through the CLP and ACLA cores, these inner periphery actors potentially bypass them to transmit social power, lowering the legitimacy of the tessellation during a decision cycle. The existing circuitry of the public library infrastructure poses challenges for governance and legitimation.

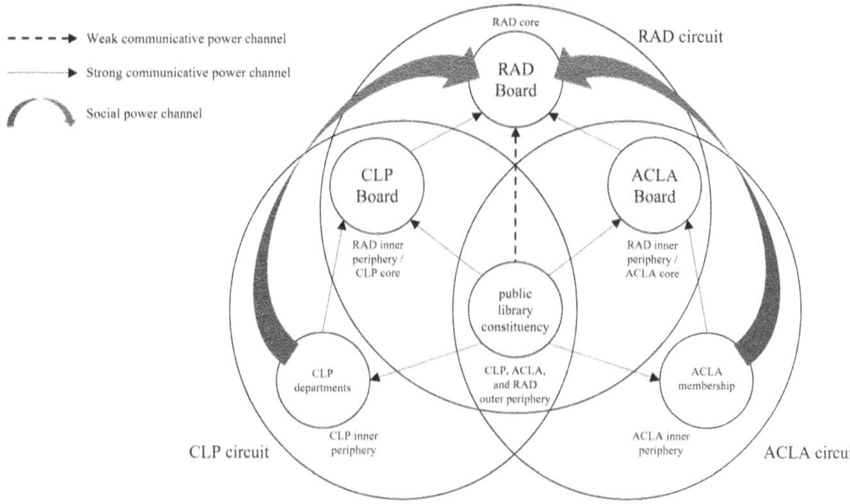

Figure 11: Current circuitry of the public library infrastructure.

Early in the infrastructure's development, library leaders and regional proponents advocated for one of two options for the circuitry of the infrastructure: either position the infrastructure within the inner periphery of the county circuit or the inner periphery of CLP's circuit. In the end, neither option obtained. Instead, the infrastructure came to occupy the inner periphery of RAD, once removed from direct governance because the ACLA board and CLP board serve as mediators between the decisions of the RAD board and concerns of constituents and libraries. Two questions now face policymakers: 1) to what extent should the infrastructure continue toward centralization, and 2) to what extent should the infrastructure continue to be administered by the RAD board? Four general scenarios result from the answers to these two choices. These four scenarios are shown in Figure 12. The central ideological divide in each of these scenarios remains the existing disparity in resources of the libraries across the county. For each of these scenarios, what is the obligation of the larger and more affluent libraries to the smaller and less affluent ones? Should there be a greater redistribution and equalization of wealth and resources? Or should local property and resources stay local?

In Scenario 1, CLP, ACLA, and eiNetwork remain much as they are now within the RAD circuit. They form federation of entities, each governed directly by their own cores and outer peripheries. They operate independently and share some centralized services but are ultimately governed by the RAD board. The problem with the tessellation in Scenario 1, as noted above, is governance. The cores within the CLP, ACLA, and eiNetwork circuits are in effect "false" cores because they do not

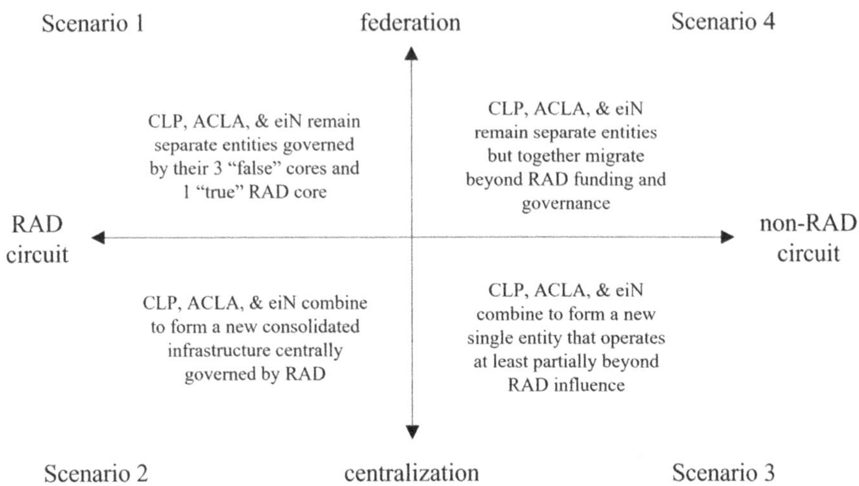

Figure 12: Potential future scenarios for the public library infrastructure.

have ultimate decision-making power over the infrastructure. The RAD board stands as the one "true" core because it has the last word on how the infrastructure is governed.

In Scenario 2, CLP, ACLA, and eiNetwork combine to form a new, single, consolidated infrastructure without its own core. Instead, RAD directly governs the infrastructure. The advantage of this arrangement is efficiency: there is no longer the middlemen of the CLP board, ACLA board, or eiNetwork board. The RAD board can move the infrastructure in new directions quickly, and it can direct resources as it sees fit. The challenge for this scenario to occur is to solve the problem of local library governance and funding. It is not clear what would happen to local funding and local governance of the independent libraries within this scenario.

In Scenarios 1 and 2, the infrastructure remains within the RAD circuit with more or less centralized services. The trend seems to be toward Scenario 2, since regional services have more and more come under centralized control, and RAD has already directly intervened in the infrastructural decision making. But Scenarios 3 and 4 represent a departure, at least in part, from the RAD circuit. In Scenario 3, CLP, ACLA, and eiNetwork combine to form a single and centralized entity, but one that operates wholly or in part outside of RAD influence. To some extent, the entities already do this. Most infrastructural funding comes through local municipal support, and RAD cannot control how this money is spent. There is thus a mix of control over the distribution of resources. In Scenario 3, the difference would be that some new entity is formed from a consolidation of CLP, ACLA, and eiNetwork. One possibility is a new special governmental district with taxation power. This new library authority could dictate infrastructural developments regionally. It might receive partial funding from RAD, and the RAD board might therefore control how those resources were used, but supposing a new library authority were formed, dedicated taxes for that district could gradually come to replace RAD funding, thus reducing the infrastructure's dependence on RAD funding. As with Scenario 2, the challenge in Scenario 3 would be for local municipal libraries to cede their own local control to the new library authority. The new authority would have the power to mandate taxes across the county, thus solving the free rider dilemma, but local libraries in affluent areas, which have invested heavily in local library infrastructure, would hesitate to see those resources potentially taken away and distributed elsewhere. A new funding formula would have to be developed to address the distribution problem.

In Scenario 4, CLP, ACLA, and eiNetwork remain a federation of entities, forming a single regional infrastructure, but the federation identifies alternative funding sources beyond RAD that allow it to operate outside of RAD's constraints. To some extent, this exists, for instance, with ACLA's status as a federated library system in the eyes of the state. ACLA HQ seeks out its own grant opportunities and manages its own state aid. ACLA's use of this funding is beyond RAD's control. A dedicated county library tax, if adopted, would fulfill the conditions necessary for the infrastructure to move closer toward Scenario 4.

It is not certain which if any of the imagined scenarios will obtain in the future. But in order for there to be infrastructural change, the tessellation must exhibit high responsiveness, together with either high legitimacy, high civil support, and low resistance. The direction of change will be dictated by which issues are taken up and supported in future decision cycles. Debates about which of the scenarios obtains will revolve around local funding disparities, which is the central ideological fault line within the infrastructure.

This book examined the historical development of the case of the regional public library infrastructural in greater Pittsburgh, Pennsylvania. A new public sphere framework, the circulation of power model, was introduced to make sense of the infrastructure's trajectory over time. Sequentially, it was shown that the infrastructure moved in accordance with decision cycles whereby major issues were raised, discussed, legitimated, and either implemented or not. Topographically, the infrastructure was shown to operate in a tessellation structure, composed of circuits. Within these circuits, actors exchange different types of power with one another. The types of power transmissions and their prevalence determined the outcome of each decision cycle. A new theory, Releaser Theory, was proposed to explain why some decision cycles resulted in change while others resulted in continuity. The four conditions of note in the theory are responsiveness, legitimacy, civil support, and resistance. As illustrated by the circulation of power model, the story of the public library infrastructure in greater Pittsburgh is one of circuitry: in which circuit the infrastructure is located, how it is governed and funded over time. The circuitry of the infrastructure will remain a central concern in future decision cycles. It remains to be seen where the infrastructure will be positioned and how the tessellation will evolve.

Several research avenues remained outside the scope of this book, but they offer future potential research directions. Limitations of this research include the generalizability of the circulation of power model (decision cycles, circuits, tessellations) and Releaser Theory. Future work can apply the circulation of power model to new contexts, both those related to public libraries and to other types of infrastructures. International comparisons of infrastructures from various political landscapes can be made. The circulation of power model presented in this book was developed to account for an American political context, which is federated, but does the model fare well in nations with more centralized political systems? The Releaser Theory can be used to make predictions, and hypotheses deduced from Releaser Theory can be tested by studying new cases, thereby potentially modifying and improving the theory's generalizability. Additionally, future work can revise and refine the concepts of circuits and tessellations. One potentially productive research direction is to identify new ways to measure the various types of power within tessellations. This may borrow tools from network analysis. New concepts, new measurements, and findings may result from pursuing this angle. The invention of new ways to measure public sphere communication would yield further insight into how infrastructures develop over time.

1 Michael M. Widdersheim, "A Political Theory of Public Library Development," *Libri* 68, no. 4 (2018); Michael M. Widdersheim, "Libraries and the Circulation of Power: A Historical Case Study of Pittsburgh, 1924–2016" (PhD thesis, University of Pittsburgh, 2017), http://d-scholarship.pitt.edu/32925/.

2 Widdersheim, "A Political Theory of Public Library Development."

3 Michael M. Widdersheim, Brady D. Lund, and Betty J. Kemboi, "Change Management in Public Libraries: Research-Based Political Strategies," *Journal of Library Administration* 59, no. 7 (2019).

Bibliography

"'Access Pennsylvania' Proposal Endorsed by Governor Thornburgh." *Pennsylvania Library Association Bulletin* (October 1984): 1.

ACLA. *2006 Annual Report to the Allegheny Regional Asset District.* www.aclalibraries.org/reports. 1 June 2007.

ACLA. *2007 Annual Report to the Allegheny Regional Asset District.* www.aclalibraries.org/reports. 1 June 2008.

ACLA. *ACLA 2009 Annual Report to the Allegheny Regional Asset District.* www.aclalibraries.org/reports. 1 June 2010.

ACLA. *ACLA 2010 Annual Report.* 1 June 2011. aclalibraries.org/reports.

ACLA. *ACLA Strategic Plan Narrative Update–Year-End 2006.* www.aclalibraries.org/reports. December 2006.

ACLA. *Highlights from 2009.* www.aclalibraries.org/reports. 2009.

Advisory Council on Library Development. *Rules and Regulations Governing Payment of State-Aid to Public Libraries and Approval of Plans for Its Use.* Harrisburg: Commonwealth of Pennsylvania, April 1962.

Allegheny County. *Home Rule Charter of Allegheny County.* 19 May 1998.

"Allegheny County Bookmobile Schedule." *Carnegie Magazine*, 1957.

Allegheny County Industrial Development Authority. *Carnegie Library of Pittsburgh, Allegheny County, Pennsylvania, Regional Asset District Revenue Bonds, Series of 2002.* 2002.

Allegheny Regional Asset District. *RAD Fax.* August 1998.

American Association of State Libraries. *Standards for Library Functions at the State Level.* Chicago: American Library Association, 1970.

American Library Association. "Council." *ALA Bulletin* 36, no. 10 (1942): P-39-P-48.

American Library Association. "The Library's Bill of Rights: Minutes of the Council of the American Library Association, June 19, 1939." *ALA Bulletin* 33, no. 11 (1939): P-55-P-63.

American Library Association. "Midwinter Council Minutes." *ALA Bulletin* 36, no. 2 (1942): 145–52.

American Library Association. "Midwinter News." *ALA Bulletin* 36, no. 1 (1942): 16–17.

American Library Association. "A National Plan for Libraries." *Bulletin of the American Library Association* 29, no. 2 (1935): 91–98.

American Library Association. "A National Plan for Libraries." *ALA Bulletin* 33, no. 2 (1939): 136–50.

American Library Association. *Public Library Service: A Guide to Evaluation, with Minimum Standards.* Chicago: ALA, 1956.

American Library Association. *Standards for Library Functions at the State Level.* Chicago: American Library Association, 1963.

American Library Association. *Standards for Library Functions at the State Level.* Chicago: American Library Association, 1985.

American Library Association. "Standards for Public Libraries." *Bulletin of the American Library Association* 27, no. 11 (1933): 513–14.

Anderson, Edward Park. "The Intellectual Life of Pittsburgh, 1786–1836: V: Literature." *Western Pennsylvania Historical Magazine* 14, no. 2 (1931): 100–14.

Angotti, Rose. "Ross Township Library Questions RAD Mandates." *Pittsburgh Tribune-Review*, 26 March 1995, D5.

Arthur D. Little. *A New Governance Structure for OCLC: Principles and Recommendations.* Metuchen: Scarecrow, 1977.

https://doi.org/10.1515/9783111013404-013

Associated Press. “Libraries across the State Struggling with Funding.” *Citizen Voice* (Wilkes Barre), 12 June 1992.
Association of College and Research Libraries. “People.” *College & Research Libraries News* 47, no. 2 (1986).
Avram, Henriette D., and David C. Hartmann. “Objectives and Accomplishments of the Network Technical Architecture Group.” *Program: electronic library and information systems* 13, no. 1 (1979): 1–13.
Avram, Henriette D., and Lenore S. Maruyama, eds. *Toward a National Library and Information Service Network: The Bibliographic Component.* Washington, DC: Library of Congress, 1977.
Avram, Henriette D., and Beacher Wiggins. “The Linked Systems Project: Introduction and Background.” In *The Linked Systems Project: A Networking Tool for Libraries*, edited by Judith G. Fenly and Beacher Wiggins, 1–6. Dublin: OCLC, 1988.
Barczak, Elizabeth. “Budgetary Cliffhanger.” *Pittsburgh Tribune-Review*, 17 April 2003.
Barczak, Elizabeth. “Libraries’ High-Tech Offerings Draw in Patrons.” *Pittsburgh Tribune-Review*, 18 August 2002.
Barczak, Elizabeth. “Libraries Prepare to Cut Staff, Hours.” *Pittsburgh Tribune-Review*, 24 April 2003.
Barnes, Tom. “Asset Tax Board Gathers Data.” *Pittsburgh Post-Gazette*, 31 August 1994, B4.
Barnes, Tom. “County Sales Tax May Be 7% by July.” *Pittsburgh Post-Gazette*, 23 February 1994, A1.
Barnes, Tom. “Ex-Revenue Secretary Now Has Job of Doling out Funds.” *Pittsburgh Post-Gazette*, 16 December 1994.
Barnes, Tom. “First Asset District Check: $2.1 Million.” *Pittsburgh Post-Gazette*, 13 September 1994.
Barnes, Tom. “Lots of Hands out, Not Enough Funds in Assets District.” *Pittsburgh Post-Gazette*, 17 September 1994.
Barnes, Tom. “Officials Fear Asset Tax Will Not Meet Their Needs.” *Pittsburgh Post-Gazette*, 11 September 1994.
Barnes, Tom. “RAD Board to Hear Money Pleas.” *Pittsburgh Post-Gazette*, 8 September 1998.
Barnes, Tom. “RAD Members Say They’re Open-Minded on Plan B.” *Pittsburgh Post-Gazette*, 30 April 1998.
Barnes, Tom. “Stadium’s the Winner in Sales Tax Handout.” *Pittsburgh Post-Gazette*, 4 October 1994.
Barnes, Tom, and John M. R. Bull. “6 Named to Board to Divide Regional Assets Tax.” *Pittsburgh Post-Gazette* (ACLA historical file (Folder: Black Binder). Allegheny County Library Association Office, Pittsburgh, PA), 22 April 1994, B1.
Beasley, Kenneth E., and Carl E. Robinson. *A Study and Recommendations of Library Districts for Pennsylvania.* Institute of Public Administration, The Pennsylvania State University, 1962.
Bebout, John E., and David Popenoe. “America as an Urban Society: A Review of Some Trends, Problems and Responses.” In *Research on Library Service in Metropolitan Areas: Report of a Rutgers Seminar, 1964/1965*, edited by Ralph Upshaw Blasingame, 1–19. New Bruswick: Graduate School of Library Service, Rutgers, the State University, 1967.
Becker, Geof. “Allegheny County to Cut 5 Mills.” *North Hills News Record*, 15 December 1994, A1.
Belko, Mark. “Lucchino Quitting Politics.” *Pittsburgh Post-Gazette*, 26 November 1998.
Belko, Mark. “Tough Choices for Assets Board.” *Pittsburgh Post-Gazette*, 30 August 2004.
Bell, Daniel. *The Coming of Post-Industrial Society: A Venture in Social Forecasting.* New York: Basic Books, 1973.
Bell, Daniel. “The Post-Industrial Society.” In *Technology and Social Change*, edited by Eli Ginzberg. New York: Columbia University Press, 1964.

Berelson, Bernard. *The Library's Public: A Report of the Public Library Inquiry.* New York: Columbia University Press, 1949.

Blasingame, Ralph Upshaw. "Chapter II: Libraries in an Urbanized Society." In *Research on Library Service in Metropolitan Areas: Report of a Rutgers Seminar, 1964/65*, edited by Ralph Upshaw Blasingame, 20 – 24. New Bruswick: Graduate School of Library Service, Rutgers, the State University, 1967.

Blasingame, Ralph Upshaw. "Equalization of Opportunity." *Library Journal* 90, no. 9 (1965): 2071 – 75.

Blasingame, Ralph Upshaw. "The Future of the Urban Main Library: I." *Library Trends* 20, no. 4 (1972): 769 – 73.

Blasingame, Ralph Upshaw. "Libraries in a Changing Society." *Library Journal* 97, no. 9 (1972): 1667 – 71.

Blasingame, Ralph Upshaw. "Pennsylvania Library Trends." *Pennsylvania Library Association Bulletin* 13, no. 2 (Fall 1957): 14 – 16.

Blasingame, Ralph Upshaw. "The Public Library as an Urban Phenomenon." Doctor of Library Service, Columbia University, 1973.

Blasingame, Ralph Upshaw, and Ernest R. DeProspo. "Effectiveness in Cooperation and Consolidation in Public Libraries." In *Advances in Librarianship*, 189 – 206. 1970.

Board of Education for Librarianship. "First Annual Report of the Board of Education for Librarianship." *Bulletin of the American Library Association* 19, no. 4 (1925): 226 – 63.

Board of Education for Librarianship. "The Librarian." *ALA Bulletin* 42, no. 11 (1948): 445 – 47.

Board of Education for Librarianship. "Minimum Requirements for Library Schools." *Bulletin of the American Library Association* 27, no. 13 (1933): 610 – 13.

Bobinski, George S. *Carnegie Libraries: Their History and Impact on American Public Library Development.* Chicago: American Library Association, 1969.

"Book-Mobile Service May Be Extended." *Pittsburgh Press*, 7 October 1955.

"Bookmobile for County Service Ok'd." *Pittsburgh Post-Gazette* (Pittsburgh, PA), 20 June 1956, 28.

Bourne, Charles P., and Trudi Bellardo Hahn. *A History of Online Information Services, 1963 – 1976.* Cambridge: MIT Press, 2003.

Bradshaw, R. V. "Pittsburg's First Library." *Pittsburg Dispatch*, 20 February 1898, 17.

Brahm, Walter T., and Mildred W. Sandoe. *Allegheny County Library Survey.* 1950.

Breeding, Marshall. "Telecommunications Options Connect OCLC and Libraries to the Future: The Co-Evolution of OCLC Connectivity Options and the Library Computing Environment." In *OCLC 1967 – 1997: Thirty Years of Furthering Access to the World's Information*, edited by K. Wayne Smith, 111 – 28. New York: Haworth, 1998.

Brown, Carole Gilbert. "Libraries Making Wish Lists." *Pittsburgh Post-Gazette*, 2 March 1995.

Brown, Charles. "Turning a Page for Funding." *Valley News Dispatch*, 20 May 1997, A3.

Brown, David. "Cultural Funding Down." *Pittsburgh Tribune-Review*, 18 November 2001.

Brown, David M. "RAD Releases Early Budget." *Pittsburgh Tribune-Review*, 2 October 1998.

Brown, Kenneth. "Books on Wheels." *Carnegie Magazine*, 1958.

Brown, Kenneth. "The New Library Service Plan." *Carnegie Magazine*, 1956, 227 – 29.

Brunell, David H. "The Strategic Alliance between OCLC and Networks: Partnerships That Work." In *OCLC 1967 – 1997: Thirty Years of Furthering Access to the World's Information*, edited by K. Wayne Smith, 19 – 29. New York: Haworth, 1998.

Bryant, Jean. "County Library Chief Has Volumes of Ideas to Sort." *Pittsburgh Press*, 26 March 1992, A2.

Buckland, Lawrence F., and William L. Basinski. *The Role of the Library of Congress in the Evolving National Network: Final Report*. National Commission on Libraries and Information Science. Washington, DC: US GPO, 1978.

Bull, John M. R. "7% Tax Is Moving on Fast Track." *Pittsburgh Post-Gazette*, 25 February 1994, B1.

Bull, John M. R. "Sales Tax Boost Near." *Pittsburgh Post-Gazette*, 30 March 1994, C1.

Campbell, Henry Cummings. *Public Libraries in the Urban Metropolitan Setting*. The Management of Change: Studies in the Evolution of Library Systems. Hamden: Linnet Books, 1973.

Carl, Herbert A. "Division of Library Services and Educational Facilities, U.S. Office of Education." In *The Bowker Annual of Library and Book Trade Information*, edited by Phyllis B. Steckler, 127. New York: Bowker, 1967.

Chapman, Edward A. "Theory and Practice in the Organization and Operation of WPA Library Service Projects." Chap. 24–28 In *NASL Proceedings and Papers, 1938*. Springfield: National Association of State Libraries, 1939.

Chute, Eleanor. "Asset Tax Formula for Library Criticized." *Pittsburgh Post-Gazette*, 2 March 1995, E1.

Cole, John Y., and Jane Aikin. "Cataloging Distribution Service." In *Encyclopedia of the Library of Congress: For Congress, the Nation, & the World*, edited by John Y. Cole and Jane Aikin, 187–90. Lanham: Bernan, 2004.

Cole, John Y., and Jane Aikin. "Machine-Readable Cataloging (MARC)." In *Encyclopedia of the Library of Congress: For Congress, the Nation & the World*, edited by John Y. Cole and Jane Aikin, 321–22. Lanham: Bernan, 2004.

"Combined Library System Needed in County, Specialists Contend." *Pittsburgh Press*, 4 October 1951.

Commings, Karen. "Libraries of the Future ... Carnegie Library of Pittsburgh Expands on-Ramps to Its Information Highway." [In English]. Article. *Computers in Libraries* 15, no. 1 (1995): 14.

Committee on Post-War Planning. *A National Plan for Public Library Service: Prepared for the Committee on Postwar Planning of the American Library Association*. Chicago: American Library Association, 1948.

Committee on Post-War Planning. *Post-War Standards for Public Libraries*. Chicago: American Library Association, 1943.

"Committee on Standards and Certification." *Pennsylvania Library and Museum Notes* 16, no. 5 (1938): 14–15.

Committee on Standards and Certification for the Public Libraries and Librarians of Pennsylvania. "Standards and Certification for the Public Libraries and Librarians of Pennsylvania." *Pennsylvania Library and Museum Notes* 16, no. 1 (January 1937): 29–47.

"Commonwealth Libraries Report on Activities." *PaLA Bulletin* (Februrary 1994): 1.

Commonwealth of Pennsylvania. *Second Amendment and Restatement of the Articles of Incorporation of Allegheny County Library Association*. Pennsylvania Department of State, 26 December 1997.

Community Councils, and Group Work Division of the Federation of Social Agencies. "Resources for Leisure Time: Circulating Libraries in Pittsburgh." *The Federator* 9, no. 8 (1935): 4–6.

Compton, Charles H. "Looking toward Greater Federal Participation in Library Development." *Bulletin of the American Library Association* 29, no. 2 (1935): 57–59.

Conant, Ralph W. "Sociological and Institutional Changes in American Life: Their Implications for the Library." *ALA Bulletin* 61, no. 5 (1967): 528–36.

Cook, Dan, and Jennifer Sabol. "Jackpot for Jocks, Snake-Eyes for Books." *In Pittsburgh*, October 1994.

Cook, Desmond L. *Educational Project Management*. Columbus: Merrill, 1971.

Corrigan, Ruth, and Stephen Folts. "Programs, Resources, Librarians, Communications." *PLA Bulletin* 32, no. 6 (November 1977): 125–26.

"Council Acts for Merger of Libraries." *Pittsburgh Post-Gazette*, 5 June 1956.
Council on Library Resources. *1st Annual Report.* Washington, DC, 1957.
Council on Library Resources. *14th Annual Report.* Washington, DC, 1970.
"County-Wide Library Plan Advocated." *Pittsburgh Post-Gazette*, 17 November 1955.
"County Notes." *Pennsylvania Library Association Bulletin* 4, no. 3–4 (Autumn-Winter 1948): 9.
Courtright, Harry R., and Annette McAlister. "Pennsylvania's Interlibrary TWX Demonstration." *PLA Bulletin* 24, no. 4 (May 1969): 145–46.
Curley, Arthur. "Anatomy of an Advocate." *American Libraries*, 1994, November, 962.
Dahlgren, Anders C., Debra W. Johnson, and Robert Bocher. *A New Plan for a New Millennium: A Review of the Long-Range Plan for the Allegheny County Library Association.* Madison: Library Planning Associates, 2001.
Dahlgren, Anders C., Debra Wilcox Johnson, and Robert Bocher. *A Plan for Library Service in Allegheny County.* Streamwood: Library Planning Associates, May 1996.
Daub, Ella English. "The Allegheny County Library Survey." *Carnegie Magazine* 26, no. 1 (1952): 377–79, 81.
Davis, Marie A. *Libraries in Allegheny County Today and Tomorrow: A Summary of the Allegheny County Library Survey by Walter T. Braham and Mildred W. Sandoe.* 1951.
DiVittorio, Michael. "Mid-Mon Valley Libraries Weigh in on Allegheny County Recommendations." *TribLive*, 9 September 2014.
Doerschuk, Ernest E. *Amendments to Rules and Regulations Governing Payment of State Aid to Public Libraries and Approval of Plans for Its Use.* Harrisburg: State Library, June 24, 1965.
Doerschuk, Ernest E. "Library Services Act in Pennsylvania." *PLA Bulletin* 16, no. 3 (1961): 142–45.
Doerschuk, Ernest E. "View from the State Library." *PLA Bulletin* 28, no. 5 (September 1973): 208–09.
Doerschuk, Ernest E. "View from the State Library." *PLA Bulletin* 30, no. 4 (July 1975): 82–83.
Doms, Keith. "Pittsburgh Regional Library Center." *PLA Bulletin* 23, no. 3 (February 1968): 168–71.
Donovan, Sandra Fischione. "One for the Books." *Beaver County Times*, 25 February 1995.
Dow, Victoria. "Caucus Held on the Public Library Project." *PLA Bulletin* (August 1993): 8.
Dowlin, Charles Edwin. "A New Approach to the Evaluation of Public Libraries: A Pilot Study." PhD, University of Pittsburgh, 1980.
Dowlin, Kenneth Everett. *Noisy River: The Saga of Captain Paul Dowlin.* Bloomington: AuthorHouse, 2013.
Dragon, Andrea. "In Memoriam: Ernest R. Deprospo, 1937–1983." *Journal of Library Administration* 5, no. 1 (1984): 3–4.
Dudley, Meela. "City Mourns Local Library Closings." *The Tartan*, 2 November 2009.
Duerst, Kendra. "Pleasant Hills Approves County Sales Tax." 2 March 1994.
Dvorchak, Robert. "RAD Board Played Hand It Was Dealt." *Pittsburgh Post-Gazette*, 12 July 1998.
Earle, George H. *The Budget of the Commonwealth of Pennsylvania for the Fiscal Biennium June 1, 1935, to May 31, 1937.* February 1935.
Eberhart, W. Lyle. "Standards for State Library Agencies." *Library Trends* 27, no. 2 (1978): 209–12.
Ellingson, Margaret W., and Susan D. Morris. "Interlibrary Loan: Evolution to Revolution." In *Interlibrary Loan Practices Handbook*, edited by Cherié L. Weible and Karen L. Janke, 1–16. Chicago: American Library Association, 2011.
Epstein, Hank. "The Technology of Library and Information Networks." *Journal of the American Society for Information Science* 31, no. 6 (1980): 425–37.
"Essential Reading." *Pittsburgh Post-Gazette*, 15 May 1990.

Federal Reserve Bank of St. Louis. *Unemployment Rate: Aged 15–64: All Persons for the United States.* 2020, January 19. https://fred.stlouisfed.org/series/LRUN64TTUSA156N.

Fink, Ron. *Libraries: The Story of State Aid 1962.* Harrisburg: Pennsylvania State Library, Public Information Office, May 8, 1963.

Fink, Ron. "State Library News." *PLA Bulletin* 20, no. 1 (August 1964): 39.

Fletcher, Mildred Stahl. "Library Service in Allegheny County Outside Pittsburgh." *The Federator* 12, no. 11 (1937): 253–56.

Flynn, Matthew. "Draft Legislation for a New Library Code." *PLA Bulletin* (December 1994): 1, 10–11.

Frantz, John C. "The Changing Environment and Changing Institution: The Urban Library." *Library Trends* 20, no. 2 (1971): 367–75.

Friedmann, John, and John Miller. "The Urban Field." *Journal of the American Institute of Planners* 31, no. 4 (1965): 312–20.

Fry, James W. "LSA and LSCA, 1956–1973: A Legislative History." *Library Trends* 24, no. 1 (1975): 7–26.

Fuoco, Linda Wilson. "Commission Probing Ways to Fund Libraries." *Pittsburgh Post-Gazette*, 10 July 1991, 3.

Fuoco, Linda Wilson. "Panel Urges Eventual End to Personal Property Levy." *Pittsburgh Post-Gazette*, 24 March 1993, B4.

Fuoco, Michael A. "Asset District Adopts Slimmer Budget." *Pittsburgh Post-Gazette*, 20 November 2001.

Garceau, Oliver. *The Public Library in the Political Process.* New York: Columbia University Press, 1949.

Gaul, Gilbert M. "Survey: Most Want Change in Library Funding in Pa." *Philadelphia Inquirer*, 20 July 1998.

Gaul, Gilbert M., and Peter Tobia. "Libraries in Distress: A Reprint of a Series Published in the Philadelphia Inquirer from June 1 through June 4, 1997." *The Philadelphia Inquirer*, June 1997.

General Assembly of the Commonwealth of Pennsylvania. *The Pennsylvania Public Library Code: Findings and Recommendations.* Harrisburg: December 2010.

Goodfellow, Donald M. "Centenary of a Pittsburgh Library." *Western Pennsylvania Historical Magazine* 31, no. 1–2 (1948): 21–25.

Government Studies and Systems. *Alternatives for Financing the Public Library.* Washington, DC: US GPO, 1974.

Government Studies and Systems. *Evaluation of the Effectiveness of Federal Funding of Public Libraries.* Washington, DC: US GPO, 1976.

Government Studies and Systems. *Improving State Aid to Public Libraries.* Washington, DC: National Commission on Libraries and Information Science, 1977.

"Governor Appoints Twenty-Five Member Commission to Study Pennsylvania Libraries." *Pennsylvania Library Association Bulletin* 13, no. 4 (Spring 1958): 8.

"Governor Robert P. Casey Signs FY 1989–90 PA Budget Including $5 Million for Library Access." *PLA Bulletin* (September 1989): 1.

Haas, Warren J., Nancy E. Gwinn, and C. Lee Jones. "Managing the Information Revolution: CLR's Bibliographic Service Development Program." *Library Journal* 104 (1979): 1867–70.

Harris, Michael H. *History of Libraries in the Western World.* 4th ed. Lanham: Scarecrow, 1999.

Haynes, Monica L. "Library Group Awaits Meeting before It Budgets Asset Money." *Pittsburgh Post-Gazette*, 13 October 1994.

Hildreth, Charles R. *Library Automation in North America: A Reassessment of the Impact of New Technologies on Networking.* München: Saur, 1987.

Hirsch, Donald B. "Community Center on Wheels." *Carnegie Magazine*, 1954.

Hodiak, Bohdan. "Libraries' Lending, Use Increase." *Pittsburgh Post-Gazette*, 12 July 1990, 1, 5.

Homyak, Joan B. "About Northland Library." *Pittsburgh Post-Gazette*, 27 April 1997, E2.

Hoover, Bob. "Bad Year for Pittsburgh's Libraries with Cuts and Closings." *Pittsburgh Post-Gazette*, 13 October 2009.

Hoover, Bob. "Countywide Library System Urged." *Pittsburgh Post-Gazette*, 3 May 1990.

Hoover, Bob. "Libraries Taken Aback by RAD Reslicing." *Pittsburgh Post-Gazette*, 25 June 2009.

Hoover, Bob. "RAD Board Rejects Revised Funding Formula for Libraries." *Pittsburgh Post-Gazette*, 30 September 2009.

Hoover, Bob. "Uncertainty About Library Funding Ends with Reslicing of Pie." *Pittsburgh Post-Gazette*, 2 December 2009.

Joeckel, Carleton Bruns. *Library Extension: Problems and Solutions: Papers Presented before the Library Institute at the University of Chicago, August 21–26, 1944.* Chicago: University of Chicago Press, 1946.

Joeckel, Carleton Bruns. *Library Service.* Washington, DC: US GPO, 1938.

Joeckel, Carleton Bruns. "Postwar Planning." *ALA Bulletin* 41, no. 11 (1947): 351–52.

Joeckel, Carleton Bruns. "Postwar Planning." *ALA Bulletin* 36, no. 12 (1942): 656–57.

Joeckel, Carleton Bruns, and Helen A. Ridgway. "Postwar Planning Committee." *ALA Bulletin* 42, no. 11 (1948): 422.

Joint State Government Commission, General Assembly of the Commonwealth of Pennsylvania. *The Pennsylvania Public Library Code: Findings and Recommendations.* December 2010.

Jones, Abbie. "Overdue." *Pittsburgh*, January, 2001, 74–81.

Kamper, Albert F. "Why Debate over Funds for Library?" *Pittsburgh Post-Gazette*, 5 April 1995.

Kane, Mary. "Rumor of Closing Creates Sounding Board for Library Needs." *Pittsburgh Press*, 1990, April 19, B5.

Karlson, Marjorie. "A New National Interlibrary Loan Code." *PLA Bulletin* 23, no. 6 (November 1968): 363–71.

Kennedy, Matthew. "Study of County Libraries Criticizes Lack of Progress." *Pittsburgh Press*, 15 January 1979.

Kenneweg, Dorothy. "Here Comes Your Library on Wheels." *Carnegie Magazine*, 1952.

Kilgour, Frederick G. "Historical Note: A Personalized Prehistory of OCLC." *Journal of the American Society for Information Science* 38, no. 5 (1987): 381–84.

King, Gilbert W., Harold P. Edmundson, Merrill M. Flood, Manfred Kochen, Richard L. Libby, Don R. Swanson, and Alexander Wylly. *Automation and the Library of Congress: A Survey Sponsored by the Council on Library Resources.* Washington, DC: US GPO, 1963.

King, Peter B. "A Modem for the Common Man." *Pittsburgh Post-Gazette*, 1994, September 11, J1, J4.

King Research. *Interlibrary Loan Compensation Plan for the Commonwealth of Pennsylvania.* Rockville: King Research, 15 October 1985.

Knight, Douglas N., and E. Shepley Nourse, eds. *Libraries at Large: Tradition, Innovation, and the National Interest.* New York: Bowker, 1969.

Ladenson, Alex. "Overhauling the Library Services & Construction Act." *Library Journal* 103, no. 10 (1978): 1025–29.

Lane, Rodney P. *Basic Issues in the Governmental Financing of Public Library Services.* New York: Teachers College, Columbia University, 1973.

Lane, Rodney P. *The Role of the State in the Development of Public Library Services.* Washington, DC: Division of Library Programs, 1974.

Lash, Cindi. "County Restores Library Internet Funds." *Pittsburgh Post-Gazette*, 11 December 1999, B1.

Lawrence, David L. "Address of Governor David L. Lawrence." *PLA Bulletin* 16, no. 2 (1960): 113–16.

Lawrence, David L. "Rebirth." In *Pittsburgh: The Story of an American City*, edited by Stefan Lorant, 372–455. Garden City: Doubleday, 1964.

"Leadership on Libraries." *Pittsburgh Post-Gazette*, 19 March 1991.

Lear, Bernadette A. *Made Free and Thrown Open to the Public: Community Libraries in Pennsylvania from the Colonial Era through World War II.* Pittsburgh: University of Pittsburgh Press, 2021.

Lear, Bernadette A. "A State Library Transformed: Pennsylvania, 1878–1921." *Information & Culture: A Journal of History* 48, no. 1 (2013): 26–49.

Leeper, Dennis P. *Pennsylvania Citizens for Better Libraries Newsletter* 14, no. 2 (Spring 1991).

Leigh, Robert D. *The Public Library in the United States: The General Report of the Public Library Inquiry.* New York: Columbia University Press, 1950.

Leigh, Robert D. "The Public Library Inquiry." *Library Journal* (1947, May 1): 698, 720–24.

Leonard, Kim Timlin. "Libraries Vote on Power Structure." *Pittsburgh Tribune Review*, 27 April 1997, A1, A6.

Leonard, Kim Timlin. "Library Delegates Ok Plan for County." *Pittsburgh Tribune Review*, 18 June 1997, B1.

Leonard, Kim Timlin. "Plan Suggests Allegheny Libraries Form Central System." *Pittsburgh Tribune Review*, 23 June 1996, B1, B3.

Leonard, Kim Timlin. "Tuesday Vote to Decide Future of Library System." *Pittsburgh Tribune Review*, 16 June 1997, B1, B3.

"Library Access (Statewide Library Card Program)." *PLA Bulletin* (April 1993): 6.

"Library Expansion, Unification Planned." *Pittsburgh Post-Gazette*, 7 February 1957.

Library of Congress. *Annual Report of the Librarian of Congress for the Fiscal Year Ending September 30, 1979.* Washington, DC: US GPO, 1980.

Library Planning Associates. *A Plan for Library Service in Allegheny County.* Streamwood: Library Planning Associates, March 1996).

"Library Services Bill." *Pennsylvania Library Association Bulletin* 11, no. 3 (Winter 1956): 2–3.

Library Services Branch, US Office of Education. *Library Research in Progress.* April 1960.

"Library Shift May Be Illegal." *Pittsburgh Sun-Telegraph*, 18 February 1954.

Loeffler, William. "RAD Money Not a Deal for Library." *North Hills News Record*, 6 March 1995.

Lord, Rich. "Carnegie Library Funding Approved by Big Margin." *Pittsburgh Post-Gazette*, 9 November 2011.

Lowry, Patricia. "Libraries Face 50% Cut in State Funding." *Pittsburgh Post-Gazette*, 14 March 2003.

Lubove, Roy. *Twentieth-Century Pittsburgh: Government, Business, and Environmental Change.* Pittsburgh: Wiley, 1969.

Lubove, Roy. *Twentieth-Century Pittsburgh: The Post-Steel Era.* Vol. 2, Pittsburgh: University of Pittsburgh Press, 1996.

MacDonald, Anna A. "The County Library Movement in Pennsylvania." *Pennsylvania Library Notes* 12, no. 2 (1929): 39–41.

Maciuszko, Kathleen L. *OCLC: A Decade of Development, 1967–1977.* Littleton: Libraries Unlimited, 1984.

Mandak, Joe. *Valley News Dispatch*, 8 March 1994, A1, A6.

Mann, Lawrence D. "Studies in Community Decision-Making." In *Readings in Community Organization Practice*, edited by Ralph M. Kramer and Harry Specht. Englewood Cliffs: Prentice Hall, 1969.

Marcum, Deanna B. “Automating the Library: The Council on Library Resources.” *IEEE Annals of the History of Computing* 24, no. 3 (2002): 2–13.
Martin, Lowell A. “The Desireable Minimum Size of Public Library Units.” Ph.D., University of Chicago, 1945.
Martin, Lowell A. “The Future of the Urban Main Library: II.” *Library Trends* 20, no. 4 (1972): 774–87.
Martin, Lowell A. *Library Response to Urban Change: A Study of the Chicago Public Library.* Chicago: American Library Association, 1969.
Martin, Lowell A. “A Plan for Public Library Service in America.” *ALA Bulletin* 40, no. 8 (1946): 276–83.
Martin, Lowell A. *Progress and Problems of Pennsylvania Libraries: A Re-Survey.* Harrisburg: Pennsylvania State Library, 1967.
Martin, Lowell A. “Standards for Public Libraries.” *Library Trends* 21, no. 2 (1972): 164–77.
Martin, Lowell A., Grace W. Estes, David C. Palmer, and Frank W. Summers. *Library Service in Pennsylvania, Present and Proposed: A Survey Commissioned by the Pennsylvania State Librarian at the Request of the Honorable George M. Leader, Governor of the Commonwealth of Pennsylvania.* Harrisburg: Pennsylvania State Library, 1958.
Massey, Steve. “Tax Hike Praised by Economists.” *Pittsburgh Post-Gazette*, 16 December 1993, C8.
Matthews, Joseph R. “The Four Online Bibliographic Utilities: A Comparison.” *Library Technology Reports* 15, no. 6 (1979): 665–838.
McClure, Charles R., John Carlo Bertot, and John C. Beachboard. *Internet Costs and Cost Models for Public Libraries.* Washington, DC: US GPO, 1995.
McClure, Charles R., John Carlo Bertot, and Douglas L. Zweizig. *Public Libraries and the Internet: Study Results, Policy Issues, and Recommendations.* Washington, DC: National Commission on Libraries and Information Science, 1994.
McClure, Charles R., Amy Owen, Douglas L. Zweizig, Mary Jo Lynch, and Nancy A. Van House. *Planning and Role Setting for Public Libraries: A Manual of Options and Procedures.* Chicago: American Library Association, 1987.
McCrossan, John A. “Cooperative Library Networks–Today and Tomorrow.” *PLA Bulletin* 27, no. 5 (September 1972): 250–55.
McDonough, Roger H., and David C. Palmer. “Standards for State Libraries.” *Library Trends* 21, no. 2 (1972): 178–89.
McMahon, Jim. “New Sales Tax Gets Grudging Support.” *Pittsburgh Post-Gazette*, 2 March 1994.
McNulty, Timothy. “Rapping RAD Is Part of Election Debate Mix.” *Pittsburgh Post-Gazette*, 9 April 1998.
Metropolitan Study Commission of Allegheny County. *An Urban Home Rule Charter for Allegheny County: A Report of the Metropolitan Study Commission of Allegheny County.* Pittsburgh, 1955.
Michelmore, David L. “The 1% Solution.” *Pittsburgh Post-Gazette*, 16 January 1994, A1, A10.
Miles, Arnold, and Lowell A. Martin. *Public Administration and the Library.* Chicago: University of Chicago Press, 1941.
“Milestones in Legislative Support of Pennsylvania Public Libraries.” *PLA Bulletin* (January 1981): 5.
Miller, Glenn R. “FCC Endorses Deep Discounts for Libraries.” *PaLA Bulletin* (December 1996): 1.
Miller, Glenn R. “State PUC Adopts Telecommunications Discounts for Libraries, Schools.” *PaLA Bulletin* (June/July 1997): 4.
Miller, Glenn R. “Telecomm Discount Applications Mailed to Libraries.” *PaLA Bulletin* (January 1998): 1, 7.
Miller, Glenn R. “Two Year Saga Ends as LSTA Becomes Law.” *PaLA Bulletin* (November 1996): 1.

Molz, Redmond Kathleen. *National Planning for Library Service, 1935–1975.* Chicago: American Library Association, 1984.

Molz, Redmond Kathleen, and Phyllis Dain. *Civic Space/Cyberspace: The American Public Library in the Information Age.* Cambridge: MIT Press, 1999.

Mounce, Marvin W. "Library Development in Pennsylvania: A Review." *PLA Bulletin* 27, no. 5 (September 1972): 241–49.

Munn, Ralph. "Libraries for Allegheny County." *Carnegie Magazine*, 1958.

Munn, Ralph. *A Plan for the Federation of Libraries in Allegheny County.* Pittsburgh: Carnegie Library of Pittsburgh, 21 April 1958.

Munn, Ralph. *Special Report: 1928–1964.* 1964.

Munn, Ralph, and John Barr. *New Zealand Libraries: A Survey of Conditions and Suggestions for Their Improvement.* Christchurch: Libraries Association of New Zealand, 1934.

Munn, Ralph, and Ernest Roland Pitt. *Australian Libraries: A Survey of Conditions and Suggestions for Their Improvement.* Melbourne: Australian Council for Education Research, 1935.

Murphy, Patti. "Library Cooperation Works–for Some." *Pittsburgh Post-Gazette*, 28 April 1988.

Myers, Joseph H. "The Library Services and Construction Act: Title II, Construction, in Pennsylvania, 1966." *PLA Bulletin* 22, no. 2 (November 1966): 70–72.

National Commission on Libraries and Information Science. *Annual Report, 1971–1972.* Washington, DC: US GPO, 1973.

National Commission on Libraries and Information Science. *The National Information Infrastructure and the Recommendations of the 1991 White House Conference on Library and Information Services.* Silver Spring: Washington Information Services, 1994.

National Commission on Libraries and Information Science. *Public Sector/Private Sector Interaction in Providing Information Services: Report to the Nclis from the Public Sector/Private Sector Task Force.* Washington, DC: US GPO, 1982.

National Commission on Libraries and Information Science. *Toward a National Program for Library and Information Services: Goals for Action.* National Commission on Libraries and Information Science. Washington, DC: US GPO, 1975.

National Research Council. *Evolving the High Performance Computing and Communications Initiative to Support the Nation's Information Infrastructure.* Washington, DC: National Academy Press, 1995.

Nelson Associates. *Public Library Systems in the United States: A Survey of Multijurisdictional Systems.* Chicago: American Library Association, 1969.

Ness, Charles H. "The President's Page." *PLA Bulletin* 30, no. 2 (March 1975): 31–32.

Network Advisory Committee. *Networks for Networkers II Conference: A Synthesis of Conference Papers and a Summary of Conference Resolutions.* Washington, DC: Library of Congress, 1991.

Network Technical Architecture Group. *Message Delivery System for the National Library and Information Service Network: General Requirements.* Washington, DC: Library of Congress, 1978.

"New Library Plan." *Pittsburgh Press*, 17 November 1955.

"New Report Lists Many Objections to Charter Plan." *Pittsburgh Press*, 1 October 1955, 2.

"News: Carnegie Library Changes to LC." *PLA Bulletin* 28, no. 3 (May 1973): 144.

"News: New Union List." *PLA Bulletin* 29, no. 1 (January 1974): 58.

Niederberger, Mary. "Change in Formula May Push Libraries Funds Down, Up." *Pittsburgh Post-Gazette*, 21 May 2009.

Niederberger, Mary. "New RAD Formula May Hurt Libraries." *Pittsburgh Post-Gazette*, 14 May 2009.

Northland Public Library. *Northland Public Library Strategic Plan, 2016–2020.* 26 January 2016.

"Not So Quiet Library Crisis." *Pittsburgh Press*, April 1991.

O' Matz, Megan. "Funding Proposed for County Libraries." *Pittsburgh Press*, 13 March 1991.

O' Toole, James. "Allegheny County Has a Long History of Commissioners." *Pittsburgh Post-Gazettte*, 2 January 2000.

OCLC. *Annual Report 1977/1978.* Columbus: OCLC, 1978.

OCLC. *Annual Report 1978/1979.* Columbus: OCLC, 1979.

OCLC. *Annual Report 1980/1981.* Dublin: OCLC, 1981.

Office of the Federal Register. *United States Government Organization Manual, 1972/73.* Washington, DC: US GPO, 1972.

Opalka, William. "Assets Tax Leaves Library Funding in Limbo." *Pittsburgh Tribune Review*, 6 March 1994, A8.

Ovenshine, Gordon. "Area Libraries Protest New System." *North Hills News Record*, 30 August 1997, A3.

Palmour, Vernon E., Marcia C. Bellassai, and Nancy V. De Wath. *A Planning Process for Public Libraries.* Chicago: American Library Association, 1980.

Palmour, Vernon E., Marcia C. Bellassai, and Nancy K. Roderer. *Resources and Bibliographic Support for a Nationwide Library Program: Final Report to the National Commission for Libraries and Information Science.* Washington, DC: US GPO, 1974.

Pennsylvania Department of Education. *Pennsylvania Library Laws.* Harrisburg, 28 August 2007).

Pennsylvania Department of Labor & Industry. *Center for Workforce Information & Analysis.* 2020. https://paworkstats.geosolinc.com.

Pennsylvania Economy League. "The League Gets a Big New Job: Research for the Metropolitan Study Commission of Allegheny County." *P.E.L. Newsletter for Western Pennsylvania*, September-October 1953.

Pennsylvania Economy League. *Regional Asset Nature of the Pittsburgh Zoo, Phipps Conservatory, and the Pittsburgh Aviary.* Pittsburgh: Pennsylvania Economy League, October 1990.

Pennsylvania Library Association. *The Platform for 21st Century Libraries.* Harrisburg: Pennsylvania Library Association, 1998.

Pennsylvania Library Association Public Library Survey Committee. *How Good Are Pennsylvania's Public Libraries? (Exclusive of Pittsburgh and Philadelphia).* 1951, April.

Pennsylvania Library Master Plan Committee. *Pennsylvania Library Master Plan Committee Report.* Harrisburg: State Library of Pennsylvania, 1974.

Pennsylvania Library Master Plan Committee. *Working Report of the Pennsylvania Library Master Plan Committee.* Harrisburg: Bureau of Library Development, State Library of Pennsylvania, 1973.

Pennsylvania Library Master Plan Committee. *Working Report of the Pennsyvania Library Master Plan Committee.* Harrisburg: Bureau of Library Development, September 1973.

Petrosky, Georgianne. "Project Outreach." *PLA Bulletin* 23, no. 5 (August 1968): 308 – 09.

"Pilot Project Termed a Success." *Pennsylvania Library Association Bulletin* (May/June 1986): 1, 6.

Pittsburgh Board of Public Education. *The Quest for Racial Equality in the Pittsburgh Public Schools: Problems, Principles, Practices, Plans.* Pittsburgh: Pittsburgh Board of Public Education, 1965.

Pittsburgh Post-Gazette. "Allegheny Regional Asset District." *Pittsburgh Post-Gazette*, 2 October 2007.

Pittsburgh Post-Gazette. "Allegheny Regional Asset District Allocations." *Pittsburgh Post-Gazette*, 1997, October 2.

Pittsburgh Post-Gazette. "A County Library System?" *Pittsburgh Post-Gazette*, 10 August 1973.

Pittsburgh Post-Gazette. "Open Book: County Libraries Need a Diverse Funding Plan." *Pittsburgh Post-Gazette*, 5 October 2009.

Pittsburgh Post-Gazette. “RAD Waste: Why Northland Library Should Treat All Patrons Equally.” *Pittsburgh Post-Gazette*, 9 September 1995.

PLA Board of Directors. “Board Endorses Proposed State Aid Formula.” *PLA Bulletin* (October 1995): 5.

“Potpourri: $$ Cutback in Allegheny County.” *PLA Bulletin* 31, no. 3 (May 1976): 62.

Potter, Donald C. “Catching up with Yesterday.” *PLA Bulletin* 24, no. 5 (July 1969): 182–86.

Potter, Donald C. “New Directions for the State Library’s Bureau of Library Development.” *PLA Bulletin* 23, no. 5 (August 1968): 296–98.

Power of 32. *Power of 32: 32 Counties, 4 States, 1 Vision.* Power of 32. 2011.

Public Library Association. *Minimum Standards for Public Library Systems, 1966.* Chicago: American Library Association, 1967.

Public Library Association. *The Public Library Mission Statement and Its Imperatives for Service.* Chicago: American Library Association, 1979.

“Quiet Crisis Getting Noisy: Library Community Fights Back; Bucks National Trends.” *PLA Bulletin* (March 1992): 3.

Raber, Douglas. *Librarianship and Legitimacy: The Ideology of the Public Library Inquiry.* Westport: Greenwood, 1997.

“RAD Budget Applications Due July 21.” *Pittsburgh Courier*, 11 June 1997.

Rafter, Joseph L. “Standards and Certification for the Public Libraries and Librarians of Pennsylvania.” *Pennsylvania Library and Museum Notes* 16, no. 1 (1937): 29–47.

Ragin, Charles C. *The Comparative Method: Moving Beyond Qualitative and Quantitative Strategies.* Berkeley, CA: University of California Press, 1987.

Reed, Thomas H. “Pittsburgh Plans the World’s Most Complete Experiment in Federated City Government.” *The American City* 36 (1927): 596–98.

Reeves, Tim. “Is Plan to Raise Sales Tax an Asset?” *Pittsburgh Post-Gazette*, 12 December 1993, A1, A20.

Reeves, Tim. “Senate OKs Sales Tax.” *Pittsburgh Post-Gazette*, 14 December 1993, A1.

Reeves, Tim. “Singel Says He’ll Ok Bill for County Sales Tax.” *Pittsburgh Post-Gazette*, 9 December 1993, C1.

Reeves, Tim. “Tax Push Campaign a Lesson in Lobbying.” *Pittsburgh Post-Gazette*, 12 December 1993, A21.

Riddle, Rebecca. “SHARECO Program Expands for Residents, Businesses.” *Daily News*, 17 October 2000.

Riely, Kaitlynn. “Allegheny County Library Systems Look to Cooperate.” *Pittsburgh Post-Gazette*, 2 February 2014.

Rihoux, Benoît, and Charles C. Ragin, eds. *Configurational Comparative Methods: Qualitative Comparative Analysis (QCA) and Related Techniques.* Los Angeles, CA: SAGE, 2009.

Robbins, Louise S. *Censorship and the American Library: The American Library Association’s Response to Threats to Intellectual Freedom, 1939–1969.* Westport: Greenwood, 1996.

Robinet, Jane-Ellen. “Winning Asset District Funding Could Be Revenue Drag.” *Pittsburgh Business Times*, 17–23 October 1994.

Robinson, Barbara M. “Cooperation and Competition among Library Networks.” *Journal of the American Society for Information Science* 31, no. 6 (1980): 413–24.

Salmon, Stephen R. “LITA’s First Twenty-Five Years: A Brief History.” *Information Technology and Libraries* 12, no. 1 (1977): 15–35.

Santanam, Ramesh. “O’hara Man Appointed to Asset Board.” *North Hills News Record*, 8 May 1998.

Schmitz, Jon. "Asset District Adopts $53 Million '95 Budget." *Pittsburgh Post-Gazette*, 2 December 1994.
Schmitz, Jon. "Positions on Assets Board Are a Hot Item." *Pittsburgh Post-Gazette*, 17 April 1994, A1, A6.
Schmitz, Jon, and Tom Barnes. "David Donahoe Named to Head Asset District." *Pittsburgh Post-Gazette*, 15 December 1994.
Schmitz, Jon, and Frank Reeves. "Push Is on for Early Pa. Sales Tax Increase." *Pittsburgh Post-Gazette*, 3 December 1993.
Schreier, Margrit. *Qualitative Content Analysis in Practice.* Los Angeles, CA: SAGE, 2013.
Seikel, Michele, and Thomas Steele. "How MARC Has Changed: The History of the Format and Its Forthcoming Relationship to RDA." *Technical Services Quarterly* 28, no. 3 (2011): 322–34.
Shapiro, Harold T., Vivian B. Shapiro, and Stephen Ferguson. *The Davies Project at Princeton University.* 2015. https://daviesproject.princeton.edu/databases/index.html.
Sheldon, Brook E. "Application of an Operational Model for Library Project Evaluation." PhD, University of Pittsburgh, 1977.
Shenk, Gretchen Knief. *County and Regional Library Development.* Chicago: American Library Association, 1954.
Shorey, Katharine. "Milestones on the Path to Success with H.B. 132, 1933–1961." *PLA Bulletin* 17, no. 1 (1961): 186–91.
Silver, Jonathan D. "RAD Formula Called Unfair to Poorer Libraries." *Pittsburgh Post-Gazette*, 13 November 1998.
Smydo, Joe. "Library Supporters Collect 10,000 Names for Tax Referendum." *Pittsburgh Post-Gazette*, 26 July 2011.
Spicher, Karen M. "The Development of the MARC Format." *Cataloging & Classification Quarterly* 21, no. 3–4 (1996): 75–90.
Stalley, Marshall. "Pittsburgh and the Allegheny Conference Effort toward a Unified Community Program for the Region." *Landscape Architecture Magazine* 38, no. 4 (1948, July): 148–53.
State Library of Pennsylvania. *Access Pennsylvania: An Agenda for Knowledge and Information through Libraries.* Harrisburg: State Library of Pennsylvania, September 1984.
State Library of Pennsylvania. *A Five Year Plan: Pennsylvania Library Development, 1971–1976.* Harrisburg: State Library of Pennsylvania, 1 September 1972.
State Library of Pennsylvania. *LSCA Long Range Plan, 1982–1987.* Harrisburg: Pennsylvania State Library, 1981.
State Library of Pennsylvania. *Public Library Project: Concepts for Discussion.* Harrisburg: State Library of Pennsylvania, 11 June 1993.
"Step to Merge Libraries Due in Council." *Pittsburgh Press*, 11 March 1956.
Stevens, William F. *The Keystone State Library Association, 1901–1915.* 1916.
Stevens, William F. *The Keystone State Library Association, 1916–1922: Supplement to 1901–1915.* Altoona: Tribune Print, 1923.
Stewart, Robert C. "Cataloging with a Computer–OCLC Comes to Pennsylvania." *PLA Bulletin* 28, no. 1 (1973): 9–15.
Strasser, Alexander. *Bibliographic Access in Pennsylvania.* Pittsburgh: Pittsburgh Regional Library Center, January 1979.
Strasser, Alexander. *Bibliographic Access in Pennsylvania.* Washington, DC: U.S. Office of Education, 1979.
Sullivan, Peggy. *Carl H. Milam and the American Library Association.* New York: H. W. Wilson, 1976.

"A County-Wide Conversation About Public Libraries." January 2014, accessed February 11, 2015, https://web.archive.org/web/20140328014125/http://www.countycitylibraries.org/.

The American City. "Why the Federated Pittsburgh Plan Failed to Carry: Defeated Plan to Be Resubmitted." *The American City* 41 (1929): 150.

The Times. "Good Returns." *Beaver County Times*, 10 March 1995.

Tipping, Emily. "RAD Board: Libraries Must Open Doors or Skip Funding." *Pittsburgh Tribune-Review*, 12 March 1995, A4.

Titmus, David. "Area Libraries Share and Shareco Alike." *South Hills Record Star*, 10 August 2000, 8.

Toth, Carolyn A. "Libraries and RAD." *Pittsburgh Post-Gazette*, 13 August 1995.

"Traveling Library Stations." *Pennsylvania Library Notes* 3, no. 3 (July 1910): 7–15.

U.S. Department of Health Education and Welfare Office of Education. *State Plans under the Library Services Act, Supplement 1: A Summary of Programs for Fiscal 1958 Submitted under Public Law 597, 84th Congress*. Washington, DC: US GPO, 1959.

U.S. Department of Health Education and Welfare Office of Education. *State Plans under the Library Services Act: A Summary of Plans and Programs for Fiscal 1957 Submitted under Public Law 597, 84th Congress*. Washington, DC: US GPO, 1958.

United States Census Bureau. *The Great Migration, 1910 to 1970*. 13 September 2012. https://www.census.gov/dataviz/visualizations/020/.

"Unmet Library Needs." *Pittsburgh Post-Gazette*, 20 May 1960.

Upper St. Clair Township Library. *Community Meetings Summary Report*. 10 February 2014.

US Bureau of Labor Statistics. 2020. https://www.bls.gov/data/.

US Bureau of the Census. *Population of the United States: Trends and Prospects: 1950–1990*. Washington, DC, 1974.

Van House, Nancy A., Mary Jo Lynch, Charles R. McClure, Douglas L. Zweizig, and Eleanor Jo Rodger. *Output Measures for Public Libraries*. 2nd ed. Chicago: American Library Association, 1987.

Van Wert, Laura, and Stephanie Hacke. "Library Directors Brace for Deep Funding Cuts." *Pittsburgh Tribune-Review*, 23 July 2009.

Vann, Sarah K. *Southeastern Pennsylvania Processing Center Feasibility Study: Final Report*. Harrisburg: Pennsylvania State Library, 1967.

"Viewpoints Offered on Public Library Project." *PLA Bulletin* (July 1994): 1, 7.

Vinas, Tonya. "Four Agencies Make Pitch Fo Rextra Slice of Assets Tax." *North Hills News Record*, 11 September 1994.

Vorce, Christine. "Library Panel Catalogs Shortcomings of County System." *Pittsburgh Press*, 23 April 1992.

Wade, Treshea N. "Libraries Get Marketing Advice." *Pittsburgh Tribune-Review*, 3 June 2002.

Wall, Joseph Frazier. *Andrew Carnegie*. New York, NY: Oxford University Press, 1970.

Walsh, Lawrence. "Computer Equipment Is Missing from Library." *Pittsburgh Post-Gazette*, 24 December 1997.

Wasserman, Paul. *The New Librarianship: A Challenge for Change*. New York: R. R. Bowker, 1972.

Wereschagin, Mike. "RAD to Dole out More Money in 2004." *Pittsburgh Tribune-Review*, 2 December 2003.

Widdersheim, Michael M. "From Two-Track to Tessellation: A Revised Circulation of Power Model." *Journal of Documentation* 78, no. 6 (2022): 1389–1419.

Widdersheim, Michael M. "Governance, Legitimation, Commons: A Public Sphere Framework and Research Agenda for the Public Library Sector." *Libri* 65, no. 4 (2015): 237–45.

Widdersheim, Michael M. "Historical Case Study: A Research Strategy for Diachronic Analysis." *Library & Information Science Research* 40, no. 2 (2018): 144 – 52.

Widdersheim, Michael M. "Late, Lost, or Renewed? A Search for the Public Sphere in Public Libraries." *Proceedings of the Ninth International Conference on Conceptions of Library and Information Science.* 2017. http://informationr.net/ir/22-1/colis/colis1644.html.

Widdersheim, Michael M. "Libraries and the Circulation of Power: A Historical Case Study of Pittsburgh, 1924 – 2016." PhD thesis, University of Pittsburgh, 2017. http://d-scholarship.pitt.edu/32925/.

Widdersheim, Michael M. "A Political Theory of Public Library Development." *Libri* 68, no. 4 (2018): 269 – 89.

Widdersheim, Michael M., and Masanori Koizumi. "A Communication System Approach to the Problem of Public Library Legitimacy." *Library and Information Science Research* 39, no. 1 (2017): 23 – 33.

Widdersheim, Michael M., and Masanori Koizumi. "Conceptual Modelling of the Public Sphere in Public Libraries." *Journal of Documentation* 72, no. 3 (2016): 591 – 610.

Widdersheim, Michael M., and Masanori Koizumi. "Methodological Frameworks for Developing a Model of the Public Sphere in Public Libraries." CoLIS 9 conference, Uppsala, Sweden, June 27 – 29, 2016.

Widdersheim, Michael M., Brady D. Lund, and Betty J. Kemboi. "Change Management in Public Libraries: Research-Based Political Strategies." *Journal of Library Administration* 59, no. 7 (2019): 693 – 742.

Williams, James G., and Roger Flynn. "Network Topology: Functions of Existing Networks." In *The Structure and Governance of Library Networks: Proceedings of the 1978 Conference in Pittsburgh, Pennsylvania, Co-Sponsored by National Commission on Libraries and Information Science and University of Pittsburgh*, edited by Allen Kent and Thomas J. Galvin, 49 – 83. New York: Marcel Dekker, 1979.

Williams, Tawanda D. "Ending Nonresident Library Fees Fuels Debate." *Pittsburgh Post-Gazette*, 15 March 1995, N2.

Williams, Tawanda D. "Libraries Catalog Their Plans for Money." *Pittsburgh Post-Gazette*, 16 March 1995, N1, N2.

Wingate, Robert Bray. "The Pennsylvania State Library: 225 Years of Service." *PLA Bulletin* 25, no. 6 (1970): 340 – 42.

Wood, Stephen. "Developing Library Cooperation through IDS." *PLA Bulletin* 28, no. 2 (March 1973): 61 – 63.

Wooster, Harold A. "Planning Committee, PLA, Report." *Pennsylvania Library and Museum Notes* 17, no. 7 (July 1940): 4 – 6.

Zlatos, Bill. "Groups Seek $98.6 m in RAD Funds." *Pittsburgh Tribune-Review*, 21 July 2007.

Zlatos, Bill. "Libraries Ask RAD for 1-Time Grant to Soften Expected Cuts." *Pittsburgh Tribune-Review*, 26 July 2009.

Zlatos, Bill. "Library Funding Plan Creates Winners, Losers." *Pittsburgh Tribune-Review*, 29 August 2006.

Zlatos, Bill. "Library Funding Shakeup Urged." *Pittsburgh Tribune-Review*, 22 June 2010.

Zlatos, Bill. "Library Group Revises Funding Formula." *Pittsburgh Tribune-Review*, 11 June 2010.

Zlatos, Bill. "Library, Aviary, Museums Pitch for Money." *Pittsburgh Tribune-Review*, 28 August 2007.

Zlatos, Bill. "Parks, Carnegie Library Get Big Boost." *Pittsburgh Tribune-Review*, 2 October 2007.

Zlatos, Bill. "RAD Looking between the Lines." *Pittsburgh Tribune-Review*, 5 November 2009.

Zlatos, Bill. "Shelved: Libraries Forced to Slash Hours, Staff, Popular Programs as Funding Fades." *Pittsburgh Tribune-Review*, 7 January 2010.

Zweizig, Douglas L., and Eleanor Jo Rodger. *Output Measures for Public Libraries*. Chicago: American Library Association, 1982.

List of Figures

https://doi.org/10.1515/9783111013404-014

List of Tables

https://doi.org/10.1515/9783111013404-015

List of Archives Consulted

Acronyms appear in the notes.

Allegheny County Law Library (ACLL)
City-County Building, Pittsburgh, Pennsylvania, United States

Allegheny County Library Association offices (ACLA-OFFICE)
West End, Pittsburgh, Pennsylvania, United States

Allegheny County Library Association wiki (ACLA-WIKI)
acla.pbworks.com

Allegheny Regional Asset District office (ARAD-OFFICE)
Pittsburgh, Pennsylvania, United States

Heinz History Center, Detre Library & Archives (DETRE)
Pittsburgh, Pennsylvania, United States

Lehigh County Law Library (LCLL)
Allentown, Pennsylvania, United States

Lehigh University Libraries (LUL)
Bethlehem, Pennsylvania, United States

Carnegie Library of Pittsburgh, Pennsylvania Room (CLP-PENN)
Oakland, Pittsburgh, Pennsylvania, United States

Pennsylvania State Archives (PASA)
Harrisburg, Pennsylvania, United States

State Library of Pennsylvania, Rare and Special Collections (SLPARSC)
Harrisburg, Pennsylvania, United States

University of Pittsburgh Archives & Special Collections, Archives Service Center (PITT-ASC)
Thomas Boulevard, Pittsburgh, Pennsylvania, United States

University of Pittsburgh Special Collections, Hillman Library (PITT-HILL)
Oakland, Pittsburgh, Pennsylvania, United States

Carnegie Library of Pittsburgh, William R. Oliver Special Collections Room (CLP-OLIVER)
Oakland, Pittsburgh, Pennsylvania, United States

https://doi.org/10.1515/9783111013404-016

List of Acronyms

ACCD	Allegheny Conference on Community Development
ACLA	Allegheny County Library Association
ACSLA	Association of Cooperative and Specialized Library Associations
ALA	American Library Association
ARAD	Allegheny Regional Asset District *see* RAD
ARL	Association of Research Libraries
ARPA	Advanced Research Projects Agency
BALLOTS	Bibliographic Automation of Large Library Operations using a Time-sharing System
BRS	Bibliographic Retrieval Services
BSDP	Bibliographic Service Development Program, CLR
CAG	Circulation Advisory Group, LAC
CASA	Cooperation and Support Agreement, ACLA
CCAC	Community College of Allegheny County
CFLAC	Commission on the Future of Libraries in Allegheny County
CILOC	Critical Issues Launch and Oversight Committee, LAC
CLASP	County Library Association Serving the People
CLP	Carnegie Library of Pittsburgh
CLR	Council on Library Resources
COG	Council of Government
CRT	Cathode-ray tube
CSSP	Customer Service Standards and Practices, ACLA
CWLS	County Wide Library Services Committee, LAC
CWRT	County Wide Resources Taskforce, LAC
DLC	District Library Center
DRC	Digital Resources Committee, LAC
EIN	Electronic Information Network
EiNAC	eiNetwork Advisory Council
EREC	Electronic Resources Committee, LAC
FDP	Formula Development Process, ACLA
HEW	U.S. Department of Health, Education, and Welfare
ILL	Interlibrary loan
ILLINET	Illinois Library Information Network
IMLS	Institute for Museum and Library Services
INCOLSA	Indiana Cooperative Library Services Authority
ISAD	Information Science and Automation Division, ALA
LAC	Librarians Advisory Committee, ACLA
LAMs	Libraries, archives, and museums
LAN	Local area network
LASH	Library Association of South Hills
LC	Library of Congress
LITA	Library and Information Technology Association, ALA
LMAC	Library Militia of Allegheny County
LS21	Library Services in the 21st Century
LSA	Library Services Act

https://doi.org/10.1515/9783111013404-017

LSCA	Library Services and Construction Act
LSP	Linked Systems Project, LC
LSTA	Library Services and Technology Act
MARC	Machine-Readable Cataloging
NAC	Network Advisory Committee, LC
NBS	National Bureau of Standards
NCLIS	National Commission on Libraries and Information Science
NELINET	New England Library and Information Network
NLM	National Library of Medicine
NLN	National Library Network
NII	National Information Infrastructure
NREN	National Research and Education Network
NSF	National Science Foundation
NTAG	Network Technical Architecture Group, LC
OCLC	Ohio College Library Center; Online Computer Library Center
PaLA	Pennsylvania Library Association
PALINET	Philadelphia Area Library Network, Pennsylvania Area Library Network
PCBL	Pennsylvania Citizens for Better Libraries
PFLC	Pennsylvania Free Library Commission
PLA	Pennsylvania Library Association, Public Library Association
PRLC	Pittsburgh Regional Library Center
RAD	Regional Asset District
RLCr	Releaser Theory
RLG	Research Libraries Group
RLIN	Research Libraries Information Network
RRT	Rapid Response Team, ACLA
SDC	System Development Corporation
SHARECO	South Hills Area Resources Cooperative
SHLA	South Hills Library Association
SOLINET	Southeastern Library Information Network
SUTF	Service to Underserved Task Force, LAC
TCP/IP	Transmission control protocol/Internet protocol
TRFN	Three Rivers Free Net
TSAG	Technical Services Advisory Group, LAC
UTLAS	University of Toronto Library Automation System
WAN	Wide area network
WLN	Washington Library Network

Index

https://doi.org/10.1515/9783111013404-018

Printed and bound by CPI Group (UK) Ltd, Croydon, CR0 4YY
13/09/2026
14961091-0003